Developments in Mathematics

VOLUME 43

More information about this series at http://www.springer.com/series/5834

Developments in Mathematics

VOLUME 4

Series Editors

Krishnaswami Alladi, *University of Florida, Gainesville, FL, USA*

Hershel M. Farkas, *Hebrew University of Jerusalem, Jerusalem, Israel*

Anthony Mendes • Jeffrey Remmel

Counting with Symmetric Functions

Springer

Anthony Mendes
Mathematics Department
California Polytechnic State University
San Luis Obispo, CA, USA

Jeffrey Remmel
Department of Mathematics
University of California at San Diego
La Jolla, CA, USA

ISSN 1389-2177 ISSN 2197-795X (electronic)
Developments in Mathematics
ISBN 978-3-319-23617-9 ISBN 978-3-319-23618-6 (eBook)
DOI 10.1007/978-3-319-23618-6

Library of Congress Control Number: 2015953218

Mathematics Subject Classification (2010): 05A05, 05E05, 05A15, 05A19, 05E18

Springer Cham Heidelberg New York Dordrecht London

Printed on acid-free paper

Springer International Publishing AG Switzerland is part of Springer Science+Business Media (www.springer.com)

Preface

This book is about how symmetric functions can be used in enumeration. The development is entirely self-contained, including an extensive introduction to the ring of symmetric functions. Many of the proofs are combinatorial and involve bijections or sign-reversing involutions. There are numerous exercises with full solutions, many of which highlight interesting mathematical gems.

The intended audience is graduate students and researchers in mathematics or related subjects who are interested in counting methods, generating functions, or symmetric functions. The mathematical prerequisites are relatively low; we assume the readers possess a knowledge of elementary combinatorics and linear algebra. We use the basic ideas of group theory and ring theory sparingly in the book, using them mostly in Chapter 6.

Chapter 1 introduces fundamental combinatorial objects such as permutations and integer partitions. Statistics on permutations and rearrangements are defined and relationships between q-analogues of n, $n!$, and $\binom{n}{k}$ are given, as these are used in later chapters. We also provide an introduction to generating functions. Much of the material in this introductory chapter is classic.

Symmetric functions are introduced in Chapter 2. Our development emphasizes the combinatorics of the transition matrices between bases of symmetric functions in a way that cannot be found elsewhere. Readers may find this approach more accessible than those in other books that discuss symmetric functions. This material is essential to understanding the later chapters in the book; after all, this book is all about how to use the relationships between symmetric functions to solve counting problems.

One of the major ideas this book highlights is that ring homomorphisms applied to the ring of symmetric functions can be used to find interesting generating functions. This is first applied in Chapter 3, where we use the background material introduced in Chapters 1 and 2 to find an assortment of generating functions for permutation statistics. We are able to count and refine permutations according to restricted appearances of descents and prove a number of results about words.

In Chapter 4, the techniques introduced in Chapter 3 are extended to find generating functions for a variety of objects. The exponential formula and the generating functions derived from linear recurrence equations can be found with the methods introduced in Chapter 4.

The Robinson-Schensted-Knuth algorithm is presented in Chapter 5, an important algorithm which needs to be included in any book on symmetric functions and enumeration. Connections are made to increasing subsequences in permutations and words and the Schur symmetric functions. A q-analogue of the celebrated hook length formula is proved.

Symmetric functions are used to prove Pólya's enumeration theorem in Chapter 6, allowing us to count objects modulo symmetries. This is a standard topic in many courses on combinatorics, but too often it is not made clear that Pólya's enumeration theorem can be properly phrased using the language of symmetric functions. We also give a new combinatorial proof of the Murnaghan-Nakayama rule from the Pieri rules.

Chapters 7 and 8 are more specialized chapters than the others, and may appeal to researchers in this area. In Chapter 7 we study consecutive pattern matches in permutations, words, cycles, and alternating permutations. Chapter 8 introduces the reciprocity method, an approach which can provide a way to define ring homomorphisms with desirable properties.

Most of the results and exercises found in Chapters 3, 4, 7, and 8 are appearing in book form for the first time.

The chapter dependency chart for the text is as follows:

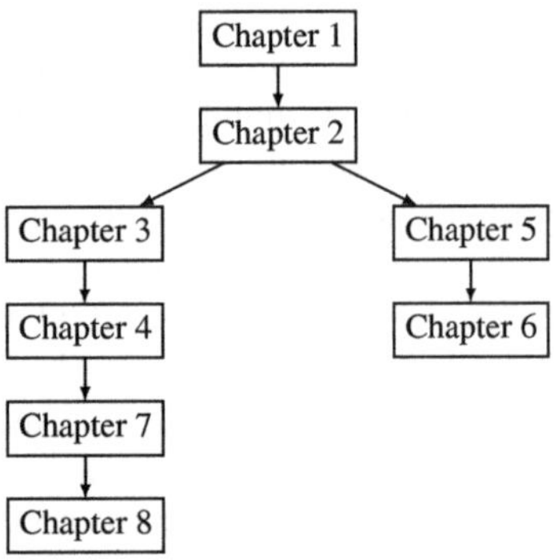

Anthony Mendes thanks Jeff Remmel for introducing him to some wonderful mathematics and for working with him over the years. He thanks all students who took Math 435 or Math 530 in the fall of 2014 at Cal Poly San Luis Obispo for carefully reading a preliminary copy of this text. Thanks also go to the following people who pointed out at least one typographical error or suggested a specific improvement to the text: Shelby Burnett, Maggie Conley, Saba Gerami, Mike LaMartina, Amanda Lombard, Thomas Stienke, and Thomas J. Taylor. Most importantly, Anthony Mendes thanks his wife Amy and daughters Ava, Tabitha, and Ruby for their support.

Jeff Remmel would like to thank Adriano Garsia, Domique Foata, and Ian Macdonald who introduced him to the theory of symmetric functions and enumerative combinatorics. He also thanks the following Ph.D. students who helped him over the years to develop parts of the theory presented in this book: Tamsen Whitehead, Diseree Beck, Tom Langley, Jennifer Wagner, Tony Mendes, Amanda Riehl, Jeff Liese, Evan Fuller, Andy Niedermaier, Andre Harmse, Miles Jones, Adrian Duane and Quang Bach. Finally, he thanks his family members, especially his wife Paula, for their continuing support which made his research career possible.

San Luis Obispo, CA, USA
San Diego, CA, USA
July 2015

Anthony Mendes
Jeffrey Remmel

Contents

Chapter 1
Permutations, Partitions, and Power Series

Permutations, integer partitions, and power series are three fundamental topics that are central to combinatorics. This chapter introduces these ideas, providing the mathematical infrastructure for our future work.

1.1 Permutations and Rearrangements

The symmetric group S_n is the set of all bijections $\sigma : \{1,\ldots,n\} \to \{1,\ldots,n\}$ under the operation of composition. Elements of S_n will be called permutations of n. Permutations have a wide variety of applications; they are essential in algebra, computer science, and statistics.

There are at least three different ways to write a permutation $\sigma \in S_n$. First, if $\sigma(i) = \sigma_i$ for $i = 1,\ldots,n$, then we can write σ in two-line notation:

$$\sigma = \begin{matrix} 1 & 2 & \cdots & n \\ \sigma_1 & \sigma_2 & \cdots & \sigma_n. \end{matrix}$$

Second, a permutation can be written in one-line notation by only writing the second row of the two rows in two-line notation: $\sigma = \sigma_1\sigma_2\cdots\sigma_n$. Third, permutations can be written in cyclic notation by letting

$$\sigma = (a_1\ a_2\ a_3\ \cdots\ a_k)(b_1\ b_2\ \cdots\ b_j)\cdots$$

represent the permutation which has a_1 in position a_k, a_2 in position a_1, a_3 in position a_2, and so on.

For example, if $\sigma \in S_5$ is defined by $\sigma(1) = 2$, $\sigma(2) = 4$, $\sigma(3) = 5$, $\sigma(4) = 1$, and $\sigma(5) = 3$, then in two-line notation we have

$$\sigma = \begin{matrix} 1\,2\,3\,4\,5 \\ 2\,4\,5\,1\,3 \end{matrix},$$

in one-line notation we have $\sigma = 2\ 4\ 5\ 1\ 3$, and in cycle notation we have $\sigma = (1\,2\,4)(3\,5)$.

A. Mendes, J. Remmel, *Counting with Symmetric Functions*, Developments in Mathematics 43, DOI 10.1007/978-3-319-23618-6_1

The main advantage of two-line notation is that the inverse function σ^{-1} can be easily found by interchanging the rows of each of the pairs in two-line notation and then sorting the pairs so that the top row reads $12\cdots n$. In the above example,

$$\sigma^{-1} = \begin{matrix} 1\,2\,3\,4\,5 \\ 4\,1\,5\,2\,3 \end{matrix}.$$

A permutation statistic is a function mapping S_n into the nonnegative integers. Four important examples of permutation statistics are the descent, excedance, inversion, and major index statistics, denoted by des, exc, inv, and maj. For any permutation $\sigma = \sigma_1 \cdots \sigma_n \in S_n$, we define these statistics by

$$\begin{aligned} \mathrm{des}(\sigma) &= (\text{the number of indices } i \text{ with } \sigma_i > \sigma_{i+1}), \\ \mathrm{exc}(\sigma) &= (\text{the number of indices } i \text{ with } \sigma_i > i), \\ \mathrm{inv}(\sigma) &= (\text{the number of indices } i \text{ and } j \text{ with } i < j \text{ and } \sigma_i > \sigma_j), \text{ and} \\ \mathrm{maj}(\sigma) &= (\text{the sum of the indices } i \text{ with } \sigma_i > \sigma_{i+1}). \end{aligned}$$

These same definitions make sense if the permutation $\sigma = \sigma_1 \cdots \sigma_n$ is replaced with any finite sequence of integers.

These four statistics on S_3 are displayed below:

σ	des(σ)	exc(σ)	inv(σ)	maj(σ)
1 2 3	0	0	0	0
1 3 2	1	1	1	2
2 1 3	1	1	1	1
2 3 1	1	2	2	2
3 1 2	1	1	2	1
3 2 1	2	1	3	3

It is no accident that the first two and the last two columns of the table are equidistributed, that is, they have the same number of 0s, 1s, 2s, and 3s.

One observation about inversions which will be used later in our work is that $\mathrm{inv}(\sigma) = \mathrm{inv}(\sigma^{-1})$ for all $\sigma \in S_n$. To see this, denote the inverse to $\sigma = \sigma_1 \cdots \sigma_n$ as $\sigma^{-1} = \sigma_1^{-1} \cdots \sigma_n^{-1}$ in one-line notation. By considering our two-line notation method of finding the inverse function, it can be seen that $i < j$ and $\sigma_i > \sigma_j$ if and only if $\sigma_j^{-1} < \sigma_i^{-1}$ and $j > i$. This says that positions i and j cause an inversion in σ if and only if the values σ_j^{-1} and σ_i^{-1} cause an inversion in σ^{-1}.

Understanding the properties of these statistics, subsequent generalizations of these statistics, and many new permutation statistics is still an active area of mathematical research. In only the past few decades, beautiful combinatorial and bijective proofs of classical and new results have been published. One of the first along these lines proves that the inversion and major index statistics are equidistributed over the symmetric group, a result of our Theorems 1.2 and 1.3.

Theorem 1.1. *Descents and excedances are equidistributed over S_n.*

Proof. We will define a bijection $\varphi : S_n \to S_n$ such that if $\sigma = \sigma_1 \cdots \sigma_n \in S_n$ has k descents, then $\operatorname{exc}(\varphi(\sigma))$ will also equal k.

Suppose $\sigma_j = 1$. Erase the first j integers in σ and begin to construct $\varphi(\sigma)$ with the cycle $(\sigma_j \cdots \sigma_2 \sigma_1)$. Continue this process iteratively with the next smallest integer in what remains in σ, building up $\varphi(\sigma)$ cycle by cycle. For example, if $\sigma = 9\ 3\ 1\ 6\ 8\ 2\ 7\ 4\ 5$, then $\varphi(\sigma)$ is the permutation $(1\ 3\ 9)(2\ 8\ 6)(4\ 7)(5)$ in cyclic notation. This construction does not break σ at a descent and writes cycles in $\varphi(\sigma)$ in such a way that if $\sigma_i > \sigma_{i+1}$ if and only if $i < \varphi(\sigma)_i$. Therefore $\operatorname{des}(\sigma) = \operatorname{exc}(\varphi(\sigma))$. □

The q-analogue of 0 is $[0]_q = 0$ and the q-analogue of the positive integer n is

$$[n]_q = 1 + q + \cdots + q^{n-1} = \frac{1-q^n}{1-q},$$

where q is an indeterminate. The q-analogue of n is a generalization of n. After all, by taking $q = 1$ (or, if we write $[n]_q = (1-q^n)/(1-q)$, by taking the limit as $q \to 1$), we recover n. However, we should refrain from thinking about q as a variable. The indeterminate q is simply a device to track the operations performed on n. This q is our bookkeeper.

The q-analogues of $n!$ and the binomial coefficient $\binom{n}{k}$ can be defined in straightforward ways; we define $[0]_q! = 1$ and for integers $0 \le k \le n$ we define

$$[n]_q! = [n]_q[n-1]_q \cdots [1]_q \qquad \text{and} \qquad \begin{bmatrix} n \\ k \end{bmatrix}_q = \frac{[n]_q!}{[k]_q![n-k]_q!}.$$

Theorem 1.2. *If n is a positive integer, then $[n]_q! = \sum_{\sigma \in S_n} q^{inv(\sigma)}$.*

Proof. The statement is true for $n = 1$. We proceed by induction on n.

There are n places to insert the integer n into a permutation $\sigma = \sigma_1 \cdots \sigma_{n-1} \in S_{n-1}$ in order to create an element of S_n. We can insert n in the position immediately preceding σ_i for each $1 \le i \le n-1$ or we can insert n after σ_{n-1}. If we insert n in the position immediately before σ_i, then we have introduced $n-i$ new inversions into the permutation. If we place the n at the end of σ, we introduce no new inversions. The exponents on q in $1 + q^1 + \cdots + q^{n-1}$ correspond to the inversions caused by the insertion of n. We now have that

$$\sum_{\sigma \in S_n} q^{\operatorname{inv}(\sigma)} = \left(1 + q + \cdots + q^{n-1}\right) \sum_{\sigma \in S_{n-1}} q^{\operatorname{inv}(\sigma)} = [n]_q[n-1]_q!.$$

Since $[n]_q! = [n]_q[n-1]_q!$, this completes the proof. □

Theorem 1.3. *If n is a positive integer, then* $[n]_q! = \sum_{\sigma \in S_n} q^{maj(\sigma)}$.

Proof. The statement is true for $n = 1$. We proceed by induction on n.

There are n places to insert the integer n into a permutation of $\sigma \in S_{n-1}$ in order to create an element of S_n. Label these places with the integers $0, \ldots, n-1$ in the following way:

1. Label the last place with 0.
2. Label the places falling between descents from right to left with $1, \ldots, \mathrm{des}(\sigma)$.
3. Label the remaining places from left to right with $\mathrm{des}(\sigma)+1, \ldots, n-1$.

For example, if $\sigma = 1\ 8\ 7\ 5\ 3\ 6\ 2\ 4$, then our labeling is

$$\underbrace{\quad}_{5}\,1\,\underbrace{\quad}_{6}\,8\,\underbrace{\quad}_{4}\,7\,\underbrace{\quad}_{3}\,5\,\underbrace{\quad}_{2}\,3\,\underbrace{\quad}_{7}\,6\,\underbrace{\quad}_{1}\,2\,\underbrace{\quad}_{8}\,4\,\underbrace{\quad}_{0}$$

Let $\sigma^{(i)}$ denote the permutation of S_n that results from σ by inserting n into the space labeled with an i. We claim that $\mathrm{maj}(\sigma^{(i)}) = i + \mathrm{maj}(\sigma)$ for all i.

Inserting the n in the last position does not change the major index, and so $\mathrm{maj}(\sigma^{(0)}) = \mathrm{maj}(\sigma)$.

Next, suppose $\sigma_j > \sigma_{j+1}$ is a descent and the space between σ_j and σ_{j+1} is labeled with i. Then there are $i-1$ descents to the right of σ_{j+1} in σ. Before inserting n, there was a contribution of j to $\mathrm{maj}(\sigma)$ arising from the descent at position j. After inserting n, there is a rise at position j in $\sigma^{(i)}$, a descent at position $j+1$, and each descent to right σ_j will have shifted its position to the right, resulting in an additional contribution of i to $\mathrm{maj}(\sigma^{(i)})$. Thus $\mathrm{maj}(\sigma^{(i)}) = i + \mathrm{maj}(\sigma)$.

Similarly, inserting n at the start of σ causes an increase to maj of 1 for the new descent at position 1 and $\mathrm{des}(\sigma)$ accounting for the additional contribution of each descent in σ to $\mathrm{maj}(\sigma^{(\mathrm{des}(\sigma)+1)})$ and therefore $\mathrm{maj}(\sigma^{(\mathrm{des}(\sigma)+1)}) = \mathrm{des}(\sigma) + 1 + \mathrm{maj}(\sigma)$.

Lastly, we consider how the major indices of $\sigma^{(i)}$ and $\sigma^{(i+1)}$ differ when $i \geq \mathrm{des}(\sigma)+1$. That is, suppose that $i \geq \mathrm{des}(\sigma)+1$ and the space before σ_j is labeled with i so that either $j = 1$ or $\sigma_{j-1} < \sigma_j$. Assume that $\sigma_j > \sigma_{j+1} > \cdots > \sigma_k < \sigma_{k+1}$ so that the space following σ_k is labeled with $i+1$. The situation is pictured here:

$$\sigma^{(i)} = \cdots n > \sigma_j > \sigma_{j+1} > \cdots > \sigma_{k-1} > \sigma_k < \sigma_{k+1} \cdots$$
$$\sigma^{(i+1)} = \cdots \sigma_j > \sigma_{j+1} > \cdots > \sigma_{k-1} > \sigma_k < n > \sigma_{k+1} \cdots$$

We claim that $\mathrm{maj}(\sigma^{(i+1)}) = 1 + \mathrm{maj}(\sigma^{(i)})$. The contribution from the sequences $\sigma_1 \cdots \sigma_{j-1}$ and $\sigma_{k+1} \cdots \sigma_{n-1}$ to the major index is the same in both $\sigma^{(i)}$ and $\sigma^{(i+1)}$. The contribution of the sequence $n > \sigma_j > \sigma_{j+1} > \cdots > \sigma_{k-1} > \sigma_k < \sigma_{k+1}$ to $\mathrm{maj}(\sigma^{(i)})$ is $j + (j+1) + (j+2) + \cdots + k$ while the contribution from the sequence $\sigma_j > \sigma_{j+1} > \cdots > \sigma_{k-1} > \sigma_k < n > \sigma_{k+1}$ to $\mathrm{maj}(\sigma^{(i+1)})$ is $j + (j+1) + \cdots + (k-1) + (k+1)$. Thus it follows that $\mathrm{maj}(\sigma^{(i+1)}) = 1 + \mathrm{maj}(\sigma^{(i)})$ and, hence, $\mathrm{maj}(\sigma^{(i+1)}) = 1 + i + \mathrm{maj}(\sigma^{(i)})$. By induction on i we have shown that $\mathrm{maj}(\sigma^{(i)}) = i + \mathrm{maj}(\sigma)$ for all i.

The exponents on q in $1+q^1+\cdots+q^{n-1}$ correspond to the increase to the major index caused by the insertion of n. Thus, just like the situation for inversions,

$$\sum_{\sigma\in S_n} q^{\text{maj}(\sigma)} = \left(1+q+\cdots+q^{n-1}\right)\sum_{\sigma\in S_{n-1}} q^{\text{maj}(\sigma)} = [n]_q[n-1]_q!.$$

Since $[n]_q! = [n]_q[n-1]_q!$, this completes the proof. □

Theorems 1.2 and 1.3 together imply that the inversion and major index permutation statistics are equidistributed. The proofs of these two theorems can be modified to arrive at a bijective proof, that is, we can create a bijection $\varphi : S_n \to S_n$ such that $\text{inv}(\sigma) = \text{maj}(\varphi(\sigma))$ for all $\sigma \in S_n$.

Create the permutation σ, starting with the permutation $1 \in S_1$, by inserting the integers $2,\ldots,n$ into the previous permutation and keeping track of inversions along the way. For example, if $\sigma = 8\ 1\ 2\ 7\ 6\ 4\ 3\ 5$, we have

σ	increase to inv caused by inserting i
1	0
1 2	0
1 2 3	0
1 2 4 3	1
1 2 4 3 5	0
1 2 6 4 3 5	3
1 2 7 6 4 3 5	4
8 1 2 7 6 4 3 5	7

To find $\varphi(\sigma)$, build a permutation in S_n by using the labeling scheme in the proof of Theorem 1.3 while forcing the major index statistic to be the same as the inversion statistic at each step. In the example of $\sigma = 8\ 1\ 2\ 7\ 6\ 4\ 3\ 5$, we have

σ	increase to maj caused by inserting i
1 1 0	0
1 2 1 2 0	0
1 2 3 1 2 3 0	0
4 1 2 3 2 1 3 4 0	1
4 1 2 3 5 2 1 3 4 5 0	0
4 1 6 2 3 5 3 2 4 1 5 6 0	3
4 1 7 6 2 3 5 4 3 5 2 1 6 7 0	4
4 1 7 6 2 3 8 5	7

This shows that $\varphi(8\ 1\ 2\ 7\ 6\ 4\ 3\ 5) = 4\ 1\ 7\ 6\ 2\ 3\ 8\ 5$.

Let $R(0^k, 1^{n-k})$ denote the set of all possible rearrangements of k 0s and $n-k$ 1s. The definitions of descent, excedance, inversions, and major index are still valid for elements of $R(0^k, 1^{n-k})$ as well as for permutations of n.

Theorem 1.4. *If* $0 \le k \le n$*, then* $\begin{bmatrix} n \\ k \end{bmatrix}_q = \sum_{r \in R(0^k, 1^{n-k})} q^{inv(r)}$.

Proof. Rewritten, the statement in this theorem is

$$[n]_q! = [k]_q![n-k]_q! \left(\sum_{r \in R(0^k, 1^{n-k})} q^{\mathrm{inv}(r)} \right).$$

Using Theorem 1.2, this is equivalent to showing

$$\left(\sum_{\sigma \in S_n} q^{\mathrm{inv}(\sigma)} \right) = \left(\sum_{\alpha \in S_k} q^{\mathrm{inv}(\alpha)} \right) \left(\sum_{\beta \in S_{n-k}} q^{\mathrm{inv}(\beta)} \right) \left(\sum_{r \in R(0^k, 1^{n-k})} q^{\mathrm{inv}(r)} \right).$$

This will be done bijectively by displaying a bijection $\varphi : S_k \times S_{n-k} \times R(0^k, 1^{n-k}) \to S_n$ such that

$$\mathrm{inv}(\alpha) + \mathrm{inv}(\beta) + \mathrm{inv}(r) = \mathrm{inv}(\varphi((\alpha, \beta, r)))$$

for all $(\alpha, \beta, r) \in S_k \times S_{n-k} \times R(0^{n-k}, 1^k)$.

Given $(\alpha, \beta, r) \in S_k \times S_{n-k} \times R(0^{n-k}, 1^k)$, write down r. Write down α underneath the 0s in r. Add k to each integer in β and write down the result underneath the 1s in r. Define $\varphi((\alpha, \beta, r))$ to be the permutation σ now written underneath r. For example, when $\alpha = 3\ 1\ 2\ 4$, $\beta = 6\ 4\ 1\ 3\ 5\ 2$, and $r = 1\ 0\ 0\ 1\ 1\ 1\ 0\ 1\ 0\ 1$, we have

$$\begin{matrix} 1 & 0 & 0 & 1 & 1 & 1 & 0 & 1 & 0 & 1 \\ 10 & 3 & 1 & 8 & 5 & 7 & 2 & 9 & 4 & 6 \end{matrix}$$

This process is reversible and therefore a bijection. The total number of inversions is the correct number since the integers in α and the integers in β keep their relative order and there are additional inversions in the resulting permutation every time a 1 appears before a 0 in r. □

Theorem 1.4 implies $\begin{bmatrix} n \\ k \end{bmatrix}_q$ must be a polynomial in q for all $n \ge k$, a fact which does not immediately follow from the definition of the q-binomial coefficient.

Theorem 1.5. *If* $0 \le k \le n$*, then*

$$\begin{bmatrix} n \\ k \end{bmatrix}_q = \sum_{r \in R(0^k, 1^{n-k})} q^{maj(r)}.$$

Proof. Theorem 1.4 allows us to prove this result by exhibiting a bijection $\varphi : R(0^k, 1^{n-k}) \to R(0^k, 1^{n-k})$ such that $\mathrm{maj}(r) = \mathrm{inv}(\varphi(r))$ for all $r \in R(0^k, 1^{n-k})$.

We first define a bijection $\Gamma : R(0^k, 1^{n-k}) \to R(0^k, 1^{n-k})$. If r ends with a 0, define $\Gamma(r)$ to be r with every consecutive substring of the form $1 \cdots 10$ changed to $01 \cdots 1$.

If r ends with a 1, define $\Gamma(r)$ to be r with every consecutive substring of the form $0\cdots01$ changed to $10\cdots0$. For example, $\Gamma(1100010100) = 0110001010$.

If r ends with a 0, then $\mathrm{inv}(\Gamma(r)) = \mathrm{inv}(r) - (n-k)$ because changing $1\cdots10$ into $01\cdots1$ for all 1s in r decreases the number of inversions in r by 1 for each of the $n-k$ 1s in r. Similarly, if r ends with a 1, then $\mathrm{inv}(\Gamma(r)) = \mathrm{inv}(r) + k$.

If r contains no 0s, then we define $\varphi(r) = r$. Otherwise, let w be r with the last 0 and all trailing 1's deleted. This way r can be written as $w01\cdots1$. For any rearrangement $r \in R(0^k, 1^{n-k})$ we define $\varphi(r)$ recursively by $\varphi(r) = \Gamma(\varphi(w))01\cdots1$. It can be checked that $\varphi(10110100011) = 00111010011$. By definition, $\varphi(r)$ ends with a 0 if and only if r ends with a 0.

The fact that φ is a bijection follows from the fact that Γ is a bijection. To complete the proof, we will show that $\mathrm{maj}(r) = \mathrm{inv}(\varphi(r))$ by induction on the length of r. Suppose we add a 0 to the end of $r \in R(0^k, 1^{n-k})$. Then we have

$$\begin{aligned}
\mathrm{inv}(\varphi(r0)) &= \mathrm{inv}\left(\Gamma(\varphi(r))0\right) \\
&= \mathrm{inv}\left(\Gamma(\varphi(r))\right) + (n-k) \\
&= \begin{cases} \mathrm{inv}\left(\varphi(r)\right) - (n-k) + (n-k) & \text{if } \varphi(r) \text{ ends in } 0, \\ \mathrm{inv}\left(\varphi(r)\right) + k + (n-k) & \text{if } \varphi(r) \text{ ends in } 1. \end{cases}
\end{aligned}$$

Using the induction hypothesis and the fact that $\varphi(r)$ ends in a 0 if and only if r does, this is equal to

$$\begin{cases} \mathrm{maj}(r) & \text{if } r \text{ ends in } 0, \\ \mathrm{maj}(r) + n & \text{if } r \text{ ends in } 1. \end{cases}$$

In both cases, this is equal to $\mathrm{maj}(r0)$. We have shown that $\mathrm{inv}(\varphi(r0)) = \mathrm{maj}(r0)$.

Now suppose we add a 1 onto the end of r. Since $\varphi(r1) = \Gamma(\varphi(w))01\cdots11 = \varphi(r)1$, we have

$$\mathrm{inv}(\varphi(r1)) = \mathrm{inv}(\varphi(r)1) = \mathrm{inv}(\varphi(r)) = \mathrm{maj}(r) = \mathrm{maj}(r1).$$

This completes the proof. □

Theorem 1.6. *Let n and k be positive integers with $k \leq n$ and let q be a prime number. Then $\begin{bmatrix} n \\ k \end{bmatrix}_q$ is equal to the number of k-dimensional vector subspaces of the vector space of dimension n over the finite field with q elements.*

Proof. We first find the number of ways of selecting k linearly independent vectors. There are $q^n - 1$ choices for the first vector since we can freely select any of the q^n vectors in the vector space except for the zero vector. There are $q^n - q$ choices for the second vector since we can select any of the $q^n - q$ vectors which are not linear combinations of our first choice of vectors. Continuing this idea, there are

$$(q^n - 1)(q^n - q)\cdots(q^n - q^{k-1})$$

possible sets of k linearly independent vectors.

This same counting argument implies that there are $(q^k-1)(q^k-q)\cdots(q^k-q^{k-1})$ possible bases for a k-dimensional subspace. Every set of k linearly independent vectors can serve as a basis for a k-dimensional subspace, so the total number of k-dimensional subspaces is

$$\begin{aligned}\frac{(q^n-1)(q^n-q)\cdots(q^n-q^{k-1})}{(q^k-1)(q^k-q)\cdots(q^k-q^{k-1})} &= \frac{(q^n-1)(q^{n-1}-1)\cdots(q^{n-k+1}-1)}{(q^k-1)(q^{k-1}-1)\cdots(q-1)} \\ &= \frac{[n]_q[n-1]_q\cdots[n-k+1]_q}{[k]_q[k-1]_q\cdots[1]_q} \\ &= \begin{bmatrix} n \\ k \end{bmatrix}_q,\end{aligned}$$

as desired. □

1.2 Integer Partitions and Tableaux

An integer partition of n, written $\lambda \vdash n$, is a finite sequence of weakly decreasing nonnegative integers. If $\lambda = (\lambda_1,\ldots,\lambda_k) \vdash n$ with $\lambda_k \neq 0$, then we write $|\lambda| = n$, $\ell(\lambda) = k$, and $\max(\lambda) = \lambda_1$. Below are all 7 integer partitions of 5:

$$(5),\ (4,1),\ (3,2),\ (3,1,1),\ (2,2,1),\ (2,1,1,1),\ (1,1,1,1,1).$$

In this order, the lengths $\ell(\lambda)$ are $1,2,2,3,3,4,5$ while the maximum parts are $5,4,3,3,2,2,1$. Occasionally it may be convenient to denote λ as $1^{m_1}2^{m_2}3^{m_3}\cdots$ if λ has m_i parts of size i. Using this notation, the integer partitions of 5 are

$$5^1,\ 1^14^1,\ 2^13^1,\ 1^23^1,\ 1^12^2,\ 1^32^1,\ 1^5.$$

Integer partitions can be identified by the corresponding Young diagram; this is a collection of left-justified rows of boxes where row i has λ_i boxes reading from bottom to top. The Young diagrams for the integer partitions of 5 are

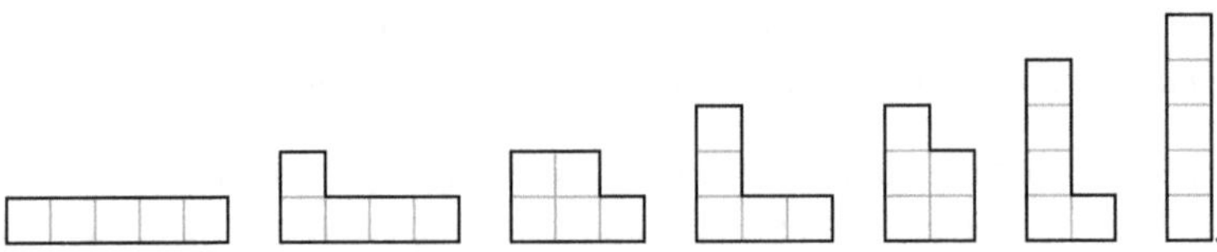

In many places Young diagrams are drawn with the largest row on top; in this way the integer partition $(5,3,2)$ would be drawn as

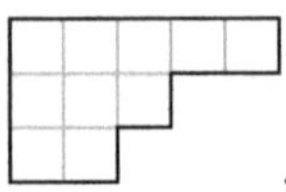

The mathematics is indifferent to the manner in which Young diagrams are drawn, and so the choice whether to draw them with maximum part on the bottom or on the top is a matter of personal preference. We prefer drawing Young diagrams with largest row on the bottom since it then appears as if the cells of the diagrams are affected by gravity.

Integer partitions can be ordered by the reverse lexicographic order. We define integer partitions $\lambda, \mu \vdash n$ to satisfy the relation $\lambda \leq \mu$ if the largest part of λ is greater than the largest part of μ. If the largest parts in λ and μ are the same, then inductively consider the second largest parts of these partitions. The integer partitions of 5 listed above are already written in increasing reverse lexicographic order.

Theorem 1.7. *The number of integer partitions $\lambda \vdash n$ with $\ell(\lambda) = k$ is equal to the number of integer partitions $\lambda \vdash n$ with* $\max(\lambda) = k$.

Proof. Take $\lambda \vdash n$ with $\ell(\lambda) = k$. Interchange rows and columns in the Young diagram of λ to create the integer partition λ'. For instance,

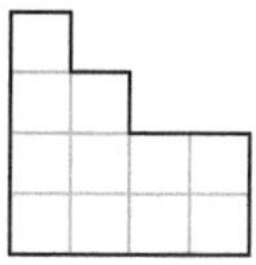

becomes

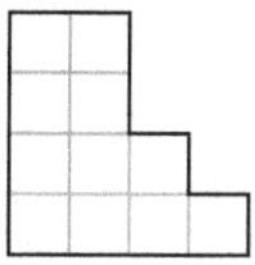

and so if $\lambda = (4,4,2,1)$, then $\lambda' = (4,3,2,2)$. Pairing λ with λ' proves the result bijectively. □

The partition λ' in the proof of Theorem 1.7 is called the conjugate of λ.

The only known direct formula for the number of integer partitions of n is

$$\sum_{k=1}^{\infty} \frac{\sqrt{k}}{\pi\sqrt{2}} \sum_{\substack{0<h<k \\ (h,k)=1}} e^{\frac{-2\pi inh}{k} + \pi i \sum_{j=1}^{k-1} \frac{j}{k}\left(\frac{hj}{k} - \left\lfloor \frac{hj}{k} \right\rfloor - \frac{1}{2}\right)} \frac{d}{dz} \frac{\sinh\left(\frac{\pi}{k}\sqrt{\frac{2}{3}\left(z - \frac{1}{24}\right)}\right)}{\sqrt{z - \frac{1}{24}}}\Bigg|_{z=n}$$

where (h,k) denotes the greatest common divisor of h and k. This formula is too complicated for everyday use, but from it comes the fact that the number of integer partitions of n is approximately equal to

$$\frac{1}{4n\sqrt{3}} e^{\pi\sqrt{2n/3}}.$$

Information about the number of integer partitions of n can be encoded as an elegant infinite product, as seen in our next theorem.

Theorem 1.8. *Let $p(n)$ be the number of integer partitions of n and z an indeterminate. Then*

$$\sum_{n=0}^{\infty} p(n) z^n = \prod_{i=1}^{\infty} \frac{1}{1 - z^i}.$$

Proof. Expanding each term in the infinite product as a geometric series, we have

$$\prod_{i=1}^{\infty}\frac{1}{1-z^i} = \left(\frac{1}{1-z}\right)\left(\frac{1}{1-z^2}\right)\left(\frac{1}{1-z^3}\right)\cdots$$
$$= \left(1+z+z^2+\cdots\right)\left(1+(z^2)+(z^2)^2+\cdots\right)\left(1+(z^3)+(z^3)^2+\cdots\right)\cdots.$$

Selecting one factor of $(z^i)^{m_i}$ between each pair of parentheses, each term in the expansion of this product is of the form $(z)^{m_1}(z^2)^{m_2}(z^3)^{m_3}\cdots$. There is a 1–1 correspondence between these terms and integer partitions written with the notation $1^{m_1}2^{m_2}3^{m_3}\cdots$. The result follows since the size of this integer partition, namely $1m_1+2m_2+3m_3+\cdots$, matches the exponent of z in $(z)^{m_1}(z^2)^{m_2}(z^3)^{m_3}\cdots$. □

There are numerous partition identities which can be found either bijectively, like our proof of Theorem 1.7, or by manipulating expressions involving indeterminates, like our proof of Theorem 1.8. Some may be found in the exercises.

Theorem 1.9. *If* $0 \le k \le n$, *then*

$$\begin{bmatrix} n \\ k \end{bmatrix}_q = \sum_{\substack{\text{partitions } \lambda \text{ with Young diagrams} \\ \text{fitting in an } k\times(n-k) \text{ rectangle}}} q^{|\lambda|}.$$

Proof. Theorem 1.4 allows us to prove this result by exhibiting a bijection φ between the partitions λ with a Young diagram fitting inside a $k\times(n-k)$ rectangle and rearrangements $r \in R(0^k, 1^{n-k})$ such that $|\lambda| = \mathrm{inv}(\varphi(\lambda))$.

Following along the northeast edge of its Young diagram, each integer partition λ can be interpreted as a path which travels south or east in unit steps, starts at $(0, n-k)$, and ends at $(k, 0)$. For example, the path created by the integer partition $(3,3,1,1)$, which fits inside a 4×5 rectangle, is the path which starts at $(0,4)$ and then moves one unit east, south, south, east, east, south, south, east, east, ending at $(5,0)$.

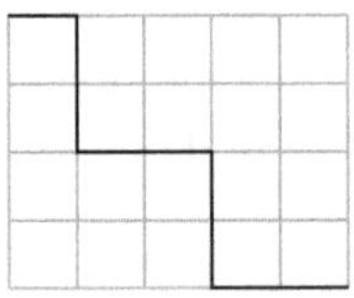

Let $\varphi(\lambda)$ by the rearrangement created by writing a 0 for each south step and a 1 for an east step in the path associated with λ. In this way, $\varphi((3,3,1,1)) =$ 100110011. Each 1 in a rearrangement r corresponds to a column in the Young diagram of λ with height equal to the number of 0s appearing after the 1 in r. This implies that $|\lambda| = \mathrm{inv}(\varphi(\lambda))$. □

A mathematical abacus of length n is a depiction of a string with either beads or empty places at positions $1,\ldots,n$ reading left to right. Below we display an abacus of length 10 with beads at positions $2,3,4,6$, and 9:

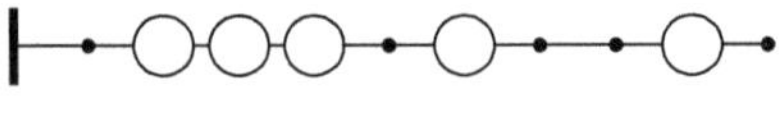

Although mathematical abaci are nothing more than a fancy way of writing down an element in $R(0^k, 1^{n-k})$, we will see in future chapters that it can be convenient to represent integer partitions using mathematical abaci.

Let $b_1, \ldots, b_k$ be the beads on a mathematical abacus a reading left to right and let $\mathrm{empty}(b_i)$ denote the number of empty places to the left of b_i. The mathematical abacus a naturally corresponds to the integer partition

$$\lambda_a = (\mathrm{empty}(b_k), \mathrm{empty}(b_{k-1}), \ldots, \mathrm{empty}(b_1))$$

where we possibly allow parts of size 0 at the end of λ_a. Then λ_a can have at most k parts of maximum size $n-k$. The bijection in Theorem 1.9, when integer partitions are interpreted as mathematical abaci, tells us that

$$\begin{bmatrix} n \\ k \end{bmatrix}_q = \sum_{\text{abaci } a \text{ of length } n \text{ with } k \text{ beads}} q^{\mathrm{empty}(b_1)+\cdots+\mathrm{empty}(b_k)}.$$

This gives us our fifth combinatorial interpretation for the q-binomial coefficient; the other four are given in Theorems 1.4, 1.5, 1.6, and 1.9.

The cycle type of a permutation $\sigma \in S_n$ is the integer partition found by writing down lengths of the cycles in σ in decreasing order. For example, the cycle type of the permutation $\sigma = (1\ 2\ 3)(4)(5\ 6\ 7\ 8)$ is the integer partition $(4,3,1)$.

Theorem 1.10. *The number of* $\sigma \in S_n$ *with cycle type* $\lambda = 1^{m_1}2^{m_2}\cdots$ *is* $n!/z_\lambda$, *where, for an integer partition* $\lambda = 1^{m_1}2^{m_2}3^{m_3}\cdots$, *we define* z_λ *to be the number* $1^{m_1}2^{m_2}3^{m_3}\cdots m_1!m_2!m_3!\cdots$.

Proof. Take any permutation in S_n written in one-line notation and place parentheses around the integers as to create a permutation of cycle type $\lambda = 1^{m_1}2^{m_2}\cdots$. There are $n!$ ways to do this. Any one of i cyclic rearrangements of a cycle of length i leaves the permutation unchanged; divide by $1^{m_1}2^{m_2}\cdots$ to account for this. Any permutation of the m_i cycles of length i will also not change the permutation; division by $m_1!m_2!\cdots$ will resolve this. Therefore the number of permutations with cycle type λ is

$$\frac{n!}{1^{m_1}2^{m_2}\cdots m_1!m_2!\cdots} = \frac{n!}{z_\lambda},$$

as desired. □

A tableau is a filling of the cells of a Young diagram with positive integers. The tableau is called column strict if these integers satisfy two restrictions:

1. The integers strictly increase when reading bottom to top within columns, and
2. the integers weakly increase when reading left to right within rows.

For example, all possible column strict tableaux of shape $(4,2,1)$ and filled with the integers $1,1,1,2,2,3,4$ are

3			
2	2		
1	1	1	4

4			
2	2		
1	1	1	3

3			
2	4		
1	1	1	2

4			
2	3		
1	1	1	2

We define CS_λ to be the set of all possible tableaux of shape λ. Let T_c denote the integer in the cell c of $T \in CS_\lambda$. The weight of T is defined to be

$$w(T) = \prod_{\text{cells } c \text{ in } T} x_{T_c}.$$

The four column strict tableaux displayed above all have weight $x_1^3 x_2^2 x_3 x_4$. The content of the tableau is the integer partition found by sorting the exponents on this weight in nonincreasing order; for instance, all four column strict tableaux shown above have content $(3,2,1,1)$.

1.3 Generating Functions

Generating functions will enable us to answer the question "How many are there?" when simple and direct formula may not exist. They provide such elegant and succinct answers to enumeration problems that, once understood, generating functions often become a preferred way of counting.

Although we did not identify them at the time, q-analogues are generating functions and we used generating functions in the proof of Theorem 1.8. Let z be an indeterminate. The generating function for the sequence $a_0, a_1, \ldots$ is

$$a_0 + a_1 z^1 + a_2 z^2 + \cdots = \sum_{n=0}^{\infty} a_n z^n.$$

For example, the generating function for the sequence $1,3,5,7,\ldots$ is

$$1 + 3z^1 + 5z^2 + 7z^3 + \cdots = \sum_{n=0}^{\infty} (2n+1) z^n$$

and the generating function for the sequence $1,1,1,1,\ldots$ is

$$1 + 1z^1 + 1z^2 + 1z^3 + \cdots = \sum_{n=0}^{\infty} z^n.$$

Generating functions have been a standard tool to the combinatorialist since the times of Euler and Laplace. They can be added, multiplied, differentiated, and integrated, so they are more than just a special way to write down sequences. Once the generating function for a sequence is known, properties such as averages, variances, and asymptotics can often be easily understood.

The generating function

$$a_0z^0 + a_1z^1 + a_2z^2 + \cdots = \sum_{n=0}^{\infty} a_n z^n$$

for the sequence $a_0, a_1, \ldots$ is also known as a formal power series. The adjective "formal" refers to the fact that we are not necessarily performing the operation of addition. We are simply presenting the sequence $a_0, a_1, \ldots$ in a specific way, using plus symbols to separate terms and using powers of z as placeholders.

The ring of formal power series, denoted $R[[z]]$, is the set of all formal power series in z representing sequences with entries in R. The ring of formal power series will help us make precise the notions of addition, multiplication, and other operations on generating functions.

For each nonnegative integer j, define a function $\cdot|_{z^j}$ from $R[[z]]$ to R such that

$$\left.\sum_{n=0}^{\infty} a_n z^n\right|_{z^j} = a_j.$$

The element in R found by an application of $\cdot|_{z^j}$ is the coefficient of z^j. Two elements in $R[[z]]$ are equal provided the coefficients of z^j in each formal power series are equal for all $j \geq 0$.

The sum of two formal power series is defined by

$$\left(\sum_{n=0}^{\infty} a_n z^n\right) + \left(\sum_{n=0}^{\infty} b_n z^n\right) = \sum_{n=0}^{\infty} (a_n + b_n) z^n,$$

where the plus symbol on the right hand side of the equation denotes the sum of two elements in R. The product of two formal power series is defined by

$$\left(\sum_{n=0}^{\infty} a_n z^n\right)\left(\sum_{n=0}^{\infty} b_n z^n\right) = \sum_{n=0}^{\infty} (a_0b_n + a_1b_{n-1} + \cdots + a_{n-1}b_1 + a_nb_0)\, z^n,$$

where the plus symbols and the adjacent elements on the right-hand side denote the sum and product of elements in R.

Let 1 represent the formal power series $1 + 0z + 0z^2 + \cdots$. If

$$\left(\sum_{n=0}^{\infty} a_n z^n\right)\left(\sum_{n=0}^{\infty} b_n z^n\right) = 1,$$

then we say that $\sum_{n=0}^{\infty} b_n z^n$ is the reciprocal of $\sum_{n=0}^{\infty} a_n z^n$ and write

$$\sum_{n=0}^{\infty} b_n z^n = \left(\sum_{n=0}^{\infty} a_n z^n\right)^{-1} = \frac{1}{\sum_{n=0}^{\infty} a_n z^n}.$$

For example, the product of the two formal power series $1-z$ and $1+z+z^2+\cdots$ is equal to 1 and so

$$1+z+z^2+\cdots=(1-z)^{-1}=\frac{1}{1-z}.$$

If this formal power series is interpreted as a complex-valued function, only certain values of z would make $1+z+z^2+\cdots=1/(1-z)$ true; namely, those values of z with $|z|<1$. However, since we are using generating functions to formally encode the values of a sequence, we can be cavalier about such issues and simply state that the generating function for $1,1,1,\dots$ is equal to $1/(1-z)$ without reference to a radius of convergence. After all, we generally do not care about evaluating generating functions at any particular value of z. On the other hand, we do reserve the right to interpret generating functions as functions of a complex variable when doing so is beneficial.

The composition of $\sum_{n=0}^{\infty} a_n z^n$ and $\sum_{m=1}^{\infty} b_m z^m$ is the formal power series

$$\sum_{n=0}^{\infty} a_n \left(\sum_{m=1}^{\infty} b_m z^m\right)^n.$$

A potential problem in this definition arises if any coefficient in the above formal power series is an infinite sum of elements in R. However, we have

$$\begin{aligned}\left.\sum_{n=0}^{\infty} a_n \left(\sum_{m=1}^{\infty} b_m z^m\right)^n\right|_{z^j} &= \left.\sum_{n=0}^{\infty} a_n \left(b_1 z^1+\cdots+b_j z^j\right)^n\right|_{z^j}\\ &= \left.\sum_{n=0}^{j} a_n \left(b_1 z^1+\cdots+b_j z^j\right)^n\right|_{z^j},\end{aligned}$$

where in the last expression we are selecting the coefficient of z^j in a finite sum. So, by starting the inserted formal power series at the b_1 term, we force the coefficient of z^j in the composition to be a finite sum of elements in R for all $j \geq 0$. This shows there are no problems with our definition of composition.

The derivative is a function $d/dz(\cdot)$ from $R[[z]]$ to $R[[z]]$ defined by

$$\frac{d}{dz}\left(\sum_{n=0}^{\infty} a_n z^n\right)=\sum_{n=0}^{\infty}(n+1)a_{n+1}z^n,$$

where $n+1$ is the element $1+\cdots+1$ in R. The integral is a function $\int \cdot\, dz$ from $R[[z]]$ to $R[[z]]$ defined by

$$\int\left(\sum_{n=0}^{\infty} a_n z^n\right)dz=\sum_{n=1}^{\infty}\frac{a_{n-1}}{n}z^n$$

provided the multiplicative inverse of n exists in R for $n \geq 1$. In our definition of integration, the coefficient of z^0 in the integral of any formal power series is taken to be 0. The derivative and integral for formal power series obey the usual differentiation and integration laws, such as the product rule, chain rule, and integration by parts.

As seen in Theorem 1.8, we may want to understand infinite products in the ring of formal power series such as $\prod_{i=1}^{\infty} 1/(1-z^i)$. To make sense of such products, we define the notion of convergence of formal power series.

If we are given formal power series $f^{(i)}(z) = \sum_{n\geq 0} f_n^{(i)} z^n$ for $i \geq 1$, then we say

$$\lim_{i\to\infty} f^{(i)}(z) = f(z) = \sum_{n\geq 0} f_n z^n$$

provided that for each $n \geq 0$ there is an m such $f_n^{(i)} = f_n$ for all $i \geq m$. Further, we define

$$\prod_{i=1}^{\infty} f^{(i)}(z) = \lim_{n\to\infty} \prod_{i=1}^{n} f^{(i)}(z) \qquad \text{and} \qquad \sum_{i=1}^{\infty} f^{(i)}(z) = \lim_{n\to\infty} \sum_{i=1}^{n} f^{(i)}(z).$$

For example, for any $0 \leq k < n$,

$$\left.\prod_{i=1}^{n} \frac{1}{1-z^i}\right|_{z^k} = \left.\prod_{i=1}^{k} \frac{1}{1-z^i}\right|_{z^k},$$

meaning that $\prod_{i=1}^{\infty} 1/(1-z^i)$ is well defined.

We also want to work over the ring of formal power series with infinitely many variables, $R[[x_1, x_2, \ldots]]$. To formally define such a ring, we establish some notation. We say that a function $\gamma\colon \{1,2,\ldots\} \to \{0,1,2,\ldots\}$ is a weak composition of n with m parts if

1. $\gamma(i) = \gamma_i = 0$ for all $i > m$,
2. $\gamma_m > 0$, and
3. $\sum_{i=1}^{m} \gamma_i = n$.

In such a situation, we shall simply write $\gamma = (\gamma_1, \gamma_2, \ldots, \gamma_m)$ with the understanding that $\gamma_i = 0$ for $i \geq m$. Let $x^\gamma = \prod_{i=1}^{m} x_i^{\gamma_i}$. With this notation, the 0s before the final nonzero element count as parts for a weak composition.

We refer to elements of the form x^γ as monomials and say that x^γ has degree n if γ is a weak composition of n.

For example, $\gamma = (2,0,1,0,3)$ is a weak composition of 6 with 5 parts which corresponds to the function $\gamma(1) = 2$, $\gamma(2) = 0$, $\gamma(3) = 1$, $\gamma(4) = 0$, $\gamma(5) = 3$, and $\gamma(i) = 0$ for $i > 5$ and $x^\gamma = x_1^2 x_2^0 x_3^1 x_4^0 x_5^3 = x_1^2 x_3 x_5^3$ is a monomial of degree 6.

We say that a function $\gamma\colon \{1,2,\ldots\} \to \{0,1,2,\ldots\}$ is a composition of n with m parts if

1. $\gamma(i) = \gamma_i = 0$ for all $i > m$,
2. $\gamma_i > 0$ for all $i \leq m$, and
3. $\sum_{i=1}^{m} \gamma_i = n$.

Thus compositions are not allowed to have 0 parts. In this situation we write $|\gamma| = n$.

For $n \geq 1$, let WC_n denote the set of all weak compositions of n, let $WC_0 = \{0\}$ where 0 denotes the zero function, and let $WC = \bigcup_{n \geq 0} WC_n$. Then we define

$$R[[x_1, x_2, \ldots]] = \left\{ \sum_{\gamma \in WC} a_\gamma x^\gamma : a_\gamma \in R \right\},$$

and for any $n \geq 1$ we define

$$R_n[[x_1, x_2, \ldots]] = \left\{ \sum_{\gamma \in WC_n} a_\gamma x^\gamma : a_\gamma \in R \right\}.$$

The ring of formal power series $R[[z]]$ can be recovered from $R[[x_1, x_2, \ldots]]$ by setting $x_1 = z$ and $0 = x_2 = x_3 = \cdots$. Similarly, the ring of formal power series in finitely many variables, denoted $R[[x_1, \ldots, x_n]]$, can be defined by taking $x_i = 0$ for $i \geq n$ in the ring $R[[x_1, x_2, \ldots]]$.

We can define infinite products and sums in $R[[x_1, x_2, \ldots]]$ in the same way we did for $R[[z]]$. Let $\mathbf{x} = (x_1, x_2, \ldots)$ and suppose that $f^{(i)}(\mathbf{x}) = \sum_{\gamma \in WC} f_\gamma^{(i)} x^\gamma$ are formal power series for $i \geq 1$. We define

$$\lim_{i \to \infty} f^{(i)}(\mathbf{x}) = f(\mathbf{x}) = \sum_{\gamma \in WC} f_\gamma x^\gamma$$

if for each $\gamma \in WC$ there is an m such $f_\gamma^{(i)} = f_\gamma$ for all $i \geq m$. Further, we define

$$\prod_{i=1}^{\infty} f^{(i)}(\mathbf{x}) = \lim_{n \to \infty} \prod_{i=1}^{n} f^{(i)}(\mathbf{x}) \qquad \text{and} \qquad \sum_{i=1}^{\infty} f^{(i)}(\mathbf{x}) = \lim_{n \to \infty} \sum_{i=1}^{n} f^{(i)}(\mathbf{x}).$$

For example, with these definitions, $\prod_{i \geq 1} 1/(1 - x_i t)$ is well defined in $R[[t, x_1, \ldots]]$.

We will work in the subring $BR[[x_1, x_2, \ldots]]$ of $R[[x_1, x_2, \ldots]]$ which consists of all those elements of $f = \sum_{\gamma \in WC} c_\gamma x^\gamma$ in $R[[x_1, x_2, \ldots]]$ such that there some $m \geq 0$ such that $c_\gamma \neq 0$ implies $|\gamma| \leq m$. These are the elements f of $R[[x_1, x_2, \ldots]]$ such that degrees of monomials that appear in F are bounded. It follows that

$$BR[[x_1, x_2, \ldots]] = \bigoplus_{n \geq 0} R_n[[x_1, x_2, \ldots]].$$

A function $\varphi : R \to R'$ is a ring homomorphism if $\varphi(1) = 1$ and $\varphi(r_1 r_2 + r_3) = \varphi(r_1)\varphi(r_2) + \varphi(r_3)$ for all $r_1, r_2, r_3 \in R$. Such a ring homomorphism φ may be extended to be a ring homomorphism $\varphi : R[[z]] \to R'[[z]]$ by defining

$$\varphi\left(\sum_{n=0}^{\infty} a_n z^n \right) = \sum_{n=0}^{\infty} \varphi(a_n) z^n.$$

If the formal power series $\sum_{n=0}^{\infty} a_n z^n$ and $\sum_{n=0}^{\infty} b_n z^n$ in $R[[z]]$ are reciprocals of one another, then

$$0 = \varphi(0) = \varphi\left(\sum_{m=0}^{n} a_m b_{n-m} \right) = \sum_{m=0}^{n} \varphi(a_m) \varphi(b_{n-m})$$

for $n \geq 1$. Therefore $\varphi\left(\sum_{n=0}^{\infty} a_n z^n\right)$ and $\varphi\left(\sum_{n=0}^{\infty} b_n z^n\right)$ are reciprocals of one another; in symbols,

$$\varphi\left(\sum_{n=0}^{\infty} a_n z^n\right)^{-1} = \left(\sum_{n=0}^{\infty} \varphi(a_n) z^n\right)^{-1}.$$

Many of the results in this book involve using ring homomorphisms to find generating functions for interesting sequences, and so the fact that ring homomorphisms interact nicely with operations on generating functions such as reciprocation will be used frequently.

Although elements in $R[[z]]$ are not defined to be functions of z, our definitions for coefficient, sum, product, reciprocal, composition, and derivative are the same definitions we could give for complex-valued functions. So, within an appropriate radius of convergence, it is safe to interpret generating functions as actual functions.

We end this section by showing how interpreting a generating function as a complex-valued function can help us understand the asymptotic growth of the coefficients of the power series, a_n. We state and use a few theorems from complex analysis, the proofs of which can be found in most textbooks on the subject.

Each series $f(z) = \sum_{n=0}^{\infty} a_n z^n$ has a radius of convergence R, which is either a nonnegative real number or ∞, such that the series converges for all $|z| < R$ and diverges for all $|z| > R$. Further, $f(z)$ is differentiable at all values of z with $|z| < R$ and, provided $R \neq \infty$, there is an a with $|a| = R$ such that $f(z)$ is not differentiable at a. This a is a singularity of f.

We can often identify the radius of convergence of the power series representation of $f(z)$ by finding the singularity closest to 0. For example, consider

$$f(z) = \frac{1}{1 - 2\sin z}.$$

Singularities occur when the denominator is zero, so the singularities of $f(z)$ are at $\pi/6 + 2k\pi$ and $5\pi/6 + 2k\pi$ for all integers k. The singularity closest to 0 has magnitude $\pi/6$, so this is the radius of convergence for the power series representation of $f(z)$.

Alternatively, the radius of convergence can be found using a limit supremum. The limit supremum of a sequence of real numbers $x_0, x_1, \ldots$, denoted $\limsup_{n\to\infty} x_n$, is defined to be

$$\limsup_{n\to\infty} x_n = \lim_{n\to\infty} \left(\sup\{x_n, x_{n+1}, \ldots\}\right)$$

where $\sup\{x_n, x_{n+1}, \ldots\}$ is either equal to the smallest real number larger than every term in $x_n, x_{n+1}, \ldots$ or equal to ∞ if there is no real number larger than every term in $x_n, x_{n+1}, \ldots$. The root test from complex analysis says that the radius of convergence R of $f(z) = \sum_{n=0}^{\infty} a_n z^n$ satisfies

$$\limsup_{n\to\infty} |a_n|^{1/n} = \frac{1}{R},$$

where $1/0 = \infty$ and $1/\infty = 0$ by convention. This statement gives us information about the asymptotics, or the limiting behavior, of the sequence a_n. In particular, the definition of limit superior gives that for every $\varepsilon > 0$ the inequality

$$|a_n| < \left(\frac{1}{R} + \varepsilon\right)^n \tag{1.1}$$

holds for large enough n. The radius of convergence R is the largest number for which (1.1) holds for all $\varepsilon > 0$.

For example, again consider $f(z) = 1/(1-2\sin z)$. Since the radius of convergence is $\pi/6$, there are coefficients $a_0, a_1, \ldots$ such that the equality $f(z) = \sum_{n=0}^{\infty} a_n z^n$ holds for all $|z| < \pi/6$. These coefficients satisfy

$$|a_n| < \left(\frac{6}{\pi} + 0.0001\right)^n < (1.91)^n$$

for large enough n. By identifying the radius of convergence we have found an excellent bound on the growth of a_n.

We can often do much better than finding bounds. Continuing our example, L'Hôpital's rule gives

$$\lim_{z\to\pi/6} \left(z - \frac{\pi}{6}\right) f(z) = \lim_{z\to\pi/6} \frac{z - \frac{\pi}{6}}{1 - 2\sin z} = \frac{-1}{\sqrt{3}},$$

and so multiplying $f(z)$ by $\left(z - \frac{\pi}{6}\right)$ removes the singularity at $\pi/6$. This means

$$f(z) = \frac{-1/\sqrt{3}}{z - \pi/6} + g(z)$$

for some function $g(z)$ which has the same singularities as $f(z)$ except for $\pi/6$. Since the power series expansion of the geometric series $1/(z-a)$ is $-\sum_{n=0}^{\infty} z^n/a^{n+1}$,

$$\begin{aligned} g(z) &= f(z) + \frac{1/\sqrt{3}}{z - \pi/6} \\ &= \sum_{n=0}^{\infty} a_n z^n - \frac{1}{\sqrt{3}} \sum_{n=0}^{\infty} \left(\frac{6}{\pi}\right)^{n+1} z^n \\ &= \sum_{n=0}^{\infty} \left(a_n - \frac{1}{\sqrt{3}} \left(\frac{6}{\pi}\right)^{n+1}\right) z^n. \end{aligned}$$

The radius of convergence of this series is $5\pi/6$ because $5\pi/6$ is the singularity of $g(z)$ closest to 0. Using (1.1),

$$\left| a_n - \frac{1}{\sqrt{3}} \left(\frac{6}{\pi}\right)^{n+1} \right| < \left(\frac{6}{5\pi} + 0.01\right)^n < (0.40)^n$$

for large enough n. We have found an outstanding approximation for a_n, namely $(6/\pi)^{n+1}/\sqrt{3}$, without computing a single term in the sequence.

More generally, let a be a singularity of $f(z) = \sum_{n=0}^{\infty} a_n z^n$ which is closest to 0. If a is the only singularity of f with magnitude $|a| = R$ and if there is an integer k for which

$$\lim_{z\to a}(z-a)^k f(z) = c_k < \infty,$$

then there are constants $c_1, \dots, c_{k-1}$ and a complex-valued function $g(z)$ with radius of convergence $R' > R$ such that

$$f(z) = \frac{c_k}{(z-a)^k} + \cdots + \frac{c_1}{(z-a)} + g(z).$$

Let b_n be the coefficient of z^n in the series expansion of $\frac{c_k}{(z-a)^k} + \cdots + \frac{c_1}{(z-a)}$; an explicit formula for b_n can be found without too much trouble (see Exercises 1.19 and 1.20). Then we have

$$g(z) = \sum_{n=0}^{\infty} (a_n - b_n) z^n$$

and so (1.1) tells us

$$|a_n - b_n| < \left(\frac{1}{R'} + \varepsilon\right)^n \tag{1.2}$$

for some fixed $\varepsilon > 0$ and large enough n. This means b_n is a good approximation for a_n, especially if R' is larger than 1. If more accurate approximations are wanted, this process can be repeated on the function g to enlarge R' even further.

Now that we have shown why it may be useful to permit generating functions to be interpreted as complex-valued functions, we are ready to begin our foray into the wonderful world of symmetric functions.

Exercises

1.1. Prove $\begin{bmatrix} n \\ k \end{bmatrix}_q = \begin{bmatrix} n \\ n-k \end{bmatrix}_q$ bijectively.

1.2. Let $n = k_1 + \cdots + k_\ell$. The q-multinomial coefficient $\begin{bmatrix} n \\ k_1,\dots,k_\ell \end{bmatrix}_q$ is $\frac{[n]_q!}{[k_1]_q!\cdots[k_\ell]_q!}$. Show that

$$\begin{bmatrix} n \\ k_1,\dots,k_\ell \end{bmatrix}_q = \sum_{r\in R(1^{k_1},\dots,\ell^{k_\ell})} q^{\mathrm{inv}(r)},$$

where $R(1^{k_1},\dots,\ell^{k_\ell})$ denotes the set of rearrangements of k_1 1s, k_2 2s, etc.

1.3. A rim hook is a sequence of connected cells in the Young diagram of an integer partition which begins in a cell on the northeast boundary and travels along the northeast edge such that its removal leaves the Young diagram of a smaller integer partition. For example, below we display a rim hook containing 6 cells:

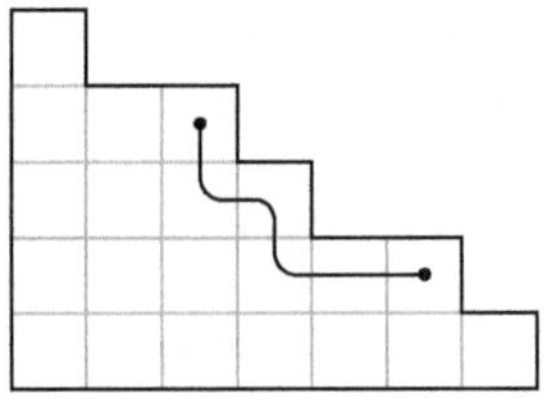

Let λ_a be the integer partition corresponding to the mathematical abacus a with beads $b_1, \ldots, b_k$ reading left to right. Show that moving bead b_i to an empty position j places to its left removes a rim hook with j cells from the Young diagram of λ_a.

In Exercises 1.4, 1.5, 1.6, and 1.7, prove the stated "q-analogued" version of a well-known identity involving binomial coefficients. Do not prove these identities by writing $\begin{bmatrix} n \\ k \end{bmatrix}_q$ as a fraction and manipulating powers of q; instead, use one of the combinatorial interpretations given in Theorems 1.4, 1.5, 1.6, 1.9 or use mathematical abaci.

1.4. The q-Pascal identity: $\begin{bmatrix} n \\ k \end{bmatrix}_q = q^k \begin{bmatrix} n-1 \\ k \end{bmatrix}_q + \begin{bmatrix} n-1 \\ k-1 \end{bmatrix}_q.$

1.5. The q-Cauchy identity: $x^n = \sum_{k=0}^{n} \begin{bmatrix} n \\ k \end{bmatrix}_q (x-q^0)\cdots(x-q^{k-1}).$

1.6. The q-binomial theorem: $(1+xq^0)\cdots(1+xq^{n-1}) = \sum_{k=0}^{n} q^{\binom{k}{2}} \begin{bmatrix} n \\ k \end{bmatrix}_q x^k.$

1.7. The q-Vandermonde identity: $\begin{bmatrix} a+b \\ c \end{bmatrix}_q = \sum_{k=0}^{n} q^{(a-k)(c-k)} \begin{bmatrix} a \\ k \end{bmatrix}_q \begin{bmatrix} b \\ c-k \end{bmatrix}_q.$

1.8. By expanding them as infinite products, show that the generating function for the number of integer partitions of n with distinct parts is equal to the generating function for the number of integer partitions of n with only odd parts.

1.9. Prove the identity in Exercise 1.8 bijectively.

1.10. Show that the number of integer partitions of $a-c$ with length $b-1$ and no parts larger than c is equal to the number of integer partitions of $a-b$ with length $c-1$ and no parts larger than b.

1.11. Show that the number of integer partitions with both odd and distinct parts with no parts greater than $2n-1$ is equal to the number of integer partitions λ which have a Young diagram which fits inside an $n \times n$ square and $\lambda = \lambda'$.

1.12. Remove a staircase (an integer partition of the form $(k, k-1, \ldots, 2, 1)$) from a Young diagram of an integer partition to show that

$$\prod_{n=1}^{\infty}(1+yz^n) = \sum_{k=0}^{\infty} \frac{y^k z^{\binom{k+1}{2}}}{(1-z)(1-z^2)\cdots(1-z^k)}.$$

1.13. Show that for any integer k the coefficient of y^k in

$$\prod_{n=0}^{\infty}(1+yz^n)\prod_{m=1}^{\infty}(1+y^{-1}z^m)$$

is equal to $z^{k(k-1)/2}\prod_{n=1}^{\infty}1/(1-z^n)$. This can be done by drawing a staircase next to an integer partition of n. Slice this picture along a diagonal line in order to find to two integer partitions with distinct parts.

1.14. From Exercise 1.13 deduce the Jacobi triple product identity:

$$\sum_{k=-\infty}^{\infty} y^k z^{k^2} = \prod_{n=1}^{\infty}(1-z^n)\left(1+yz^{2n-1}\right)\left(1+y^{-1}z^{2n-1}\right).$$

1.15. A pentagonal number is an integer of the form $(3k^2-k)/2$ for some integer k. Using Exercise 1.14, deduce Euler's pentagonal number theorem:

$$\prod_{n=1}^{\infty}(1-z^n) = \sum_{k=-\infty}^{\infty}(-1)^k z^{(3k^2-k)/2}.$$

Then show that

$$p(n) = p(n-1)+p(n-2)-p(n-5)-p(n-7)+p(n-12)+\cdots,$$

where $p(n)$ is the number of integer partitions of n and the integers $0,1,2,5,7,12,\ldots$ are the pentagonal numbers.

1.16. Prove Euler's pentagonal number theorem (given in Exercise 1.15) using a sign reversing involution on the set of integer partitions of n with distinct parts.

1.17. The Laplace transform is an integral transform used in a variety of science and engineering applications. It is defined on real valued functions $f(t)$ by

$$\mathscr{L}\{f(t)\} = F(s) = \int_0^{\infty} f(t)e^{-st}\,dt,$$

where s is positive real number. Show that $\mathscr{L}\{t^n\} = n!/s^{n+1}$ for integers n.

Formally extending this definition, we define the Laplace transform on formal power series by

$$\mathscr{L}\left\{\sum_{n=0}^{\infty} a_n t^n\right\} = \sum_{n=0}^{\infty} a_n n!/s^{n+1}.$$

These two definitions coincide when all integrals and sums converge.

Show that if $F(s)$ is the Laplace transform for the "exponential" generating function $\sum_{n=0}^{\infty} a_n t^n/n!$, then $F(1/s)/s$ is the "ordinary" generating function $\sum_{n=0}^{\infty} a_n s^n$. This allows us to convert exponential generating functions into ordinary generating functions or vice versa using the Laplace transform or its inverse.

1.18. Let $f(z) = \sum_{n=0}^{\infty} a_n z^n$. How can $f(z)$ be used to find a generating function for

$$\sum_{n=0}^{\infty} a_{jn} z^{jn}$$

for a positive integer j?

1.19. For any complex number a and nonnegative integer n, define the generalized binomial coefficient to be

$$\binom{a}{n} = \frac{a(a-1)\cdots(a-n+1)}{n!}.$$

Newton's generalized binomial theorem says that the power series for $(1+z)^a$ is

$$(1+z)^a = \sum_{n=0}^{\infty} \binom{a}{n} z^n.$$

Prove this theorem by taking derivatives of $(1+z)^a$ and evaluating them at $z = 0$.

1.20. Use Newton's binomial theorem to show that for any positive integer i,

$$\frac{1}{(1-z)^i} = \sum_{n=0}^{\infty} \binom{i+n-1}{n} z^n.$$

Then find an explicit formula for b_n, the coefficient of z^n in the series expansion of $\frac{c_k}{(z-a)^k} + \cdots + \frac{c_1}{(z-a)}$. This b_n approximates the coefficients a_n in (1.2).

Solutions

1.1 Using the interpretation of $\left[{n \atop k}\right]_q$ given in Theorem 1.9, conjugate an integer partition with a Young diagram which fits inside of an $k \times (n-k)$ rectangle to find another integer partition of the same size with a Young diagram which fits inside of a $(n-k) \times k$ rectangle.

1.2 Use the same ideas as in the proof of Theorem 1.4, define a bijection $\varphi : S_{k_1} \times \cdots \times S_{k_\ell} \times R(1^{k_1}, \ldots, \ell^{k_\ell}) \to S_n$ such that

$$\mathrm{inv}(\alpha_1) + \cdots + \mathrm{inv}(\alpha_\ell) + \mathrm{inv}(r) = \mathrm{inv}(\varphi((\alpha_1, \ldots, \alpha_\ell, r)))$$

for all $(\alpha_1, \ldots, \alpha_\ell, r) \in S_{k_1} \times \cdots \times S_{k_\ell} \times R(1^{k_1}, \ldots, \ell^{k_\ell})$.

The bijection φ can be defined by starting with $(\alpha_1, \ldots, \alpha_\ell, r)$. Write down r. Underneath the is in r, write down α_i where each integer in α_i is increased by $k_1 + \cdots + k_{i-1}$. We end up with a permutation in S_n with the correct number of inversions.

1.3 Suppose bead b_i passes over ℓ beads when moving to its new position in the empty place j places to its left. Then the number of empty spaces to the left of bead b_i is decreased by $j-\ell$ and the number of empty spaces to the left of each of the ℓ skipped over beads is decreased by 1. In the corresponding integer partition

$$\lambda_a = (\text{empty}(b_k), \ldots, \text{empty}(b_1)),$$

this means that $j-\ell$ is subtracted from part i and 1 is subtracted from the ℓ parts appearing after part i.

In terms of Young diagrams, cells in the shape of an "L" are removed from the Young diagram of λ_a. For example, suppose we move the rightmost bead in the mathematical abacus

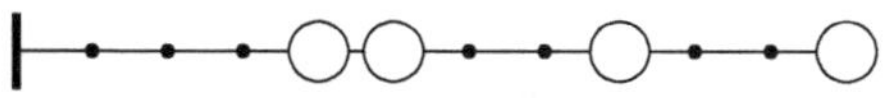

9 places to its left; thereby creating the abacus

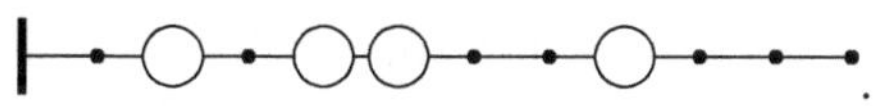

This corresponds to removing the shaded "L" shown below

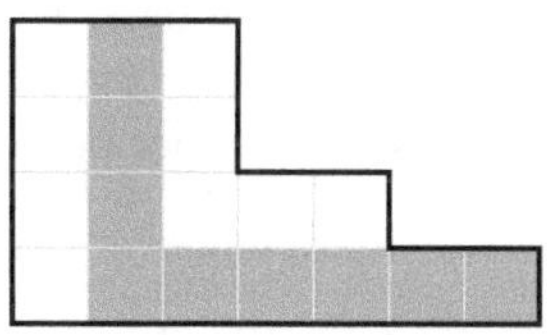

Removing such an "L" shape is equivalent to removing a rim hook which begins at the top of the "L," traces a path along the northeast boundary of the Young diagram, and ends at the bottom of the "L." Therefore moving a bead b_i to an empty place j places to its left corresponds to removing the rim hook from the Young diagram of λ_a which starts in row i reading bottom to top.

1.4 Although this proof uses rearrangements, the identity can be just as easily proved using other combinatorial interpretations.

A rearrangement $r \in R(0^k, 1^{n-k})$ can either begin with a 0 or a 1. If r begins with a 0, then removing this 0 will decrease the number of 0s by 1 and will not change the number of inversions in r. So the rearrangements which begin with a 0 correspond to $\begin{bmatrix} n-k \\ k-1 \end{bmatrix}_q$.

If r begins with a 1, then this 1 causes a total of k inversions. Rearrangements which begin with a 1 correspond to $q^k \begin{bmatrix} n-k \\ k \end{bmatrix}$. Since elements in $R(0^k, 1^{n-k})$ must start with either a 0 or a 1, the statement follows.

1.5 Let X be a vector space with a finite number of elements x and let $V_n(q)$ be a vector space of dimension n over the finite field with q elements for q prime. Consider all linear maps $L : V_n(q) \to X$. A linear transformation is determined by its action on the n basis vectors in a basis for $V_n(q)$. There are x linearly independent choices for each basis vector, so there are x^n total linear maps L.

Suppose the dimension of the null space of L is $n-k$. By Theorem 1.6 and exercise 1.1, there are $\begin{bmatrix} n \\ k \end{bmatrix}_q = \begin{bmatrix} n \\ n-k \end{bmatrix}_q$ possible choices for which subspace of $V_n(q)$ will serve as the null space. The remaining k basis vectors must be sent to k linearly independent vectors; there are $(x-q^0)\cdots(x-q^{k-1})$ choices here.

Therefore

$$\begin{aligned} x^n &= (\text{the number of linear maps } L : V_n(q) \to X) \\ &= \sum_{k=0}^{n} (\text{the number of } L : V_n(q) \to X \text{ with null space of dimension } n-k) \\ &= \sum_{k=0}^{n} \begin{bmatrix} n \\ k \end{bmatrix}_q (x-q^0)\cdots(x-q^{k-1}). \end{aligned}$$

We have verified this polynomial identity true for primes q, and thus the identity must hold for all q.

1.6 Let ℓ_i denote the number of 1s appearing to the left of the ith 0 in a rearrangement $r = r_1 \cdots r_n \in R(0^k, 1^{n-k})$. For example, if $r = 0\,1\,0\,1\,1\,0\,1\,0\,0\,0\,1\,1\,1\,1\,0$, then the sequence $\ell_1, \ldots, \ell_7$ is equal to $0,1,3,4,4,4,8$. From this definition, the ith occurrence of 0 in r is the character in position $\ell_i + (i-1)$, and so the sum of the positions of the 0s in r is $(\ell_1 + 0) + \cdots + (\ell_k + (k-1))$.

The number of inversions in r is the number of times a 1 appears to the left of a 0. Therefore we have

$$\begin{aligned} \operatorname{inv}(r) + \binom{k}{2} &= (\ell_1 + \cdots + \ell_k) + (0 + 1 + \cdots + (k-1)) \\ &= 0\chi(r_1 = 0) + 1\chi(r_2 = 0) + \cdots + (n-1)\chi(r_n = 0), \end{aligned}$$

where for any statement A, $\chi(A)$ is 1 if A is true and 0 if A is false.

To each term in the expansion of the product $(1+xq^0)\cdots(1+xq^{n-1})$ we can associate a rearrangement r in this way: If 1 is selected from the $(1+xq^i)$ term in the product, write down 1. If $q^i x$ is selected, write down 0. Therefore the coefficient of x^k in this product is equal to

$$\sum_{r = r_1 \cdots r_n \in R(0^k, 1^{n-k})} q^{0\chi(r_1=0) + 1\chi(r_2=0) + \cdots + (n-1)\chi(r_n=0)},$$

which simplifies to

$$\sum_{r = r_1 \cdots r_n \in R(0^k, 1^{n-k})} q^{\operatorname{inv}(r) + \binom{k}{2}} = q^{\binom{k}{2}} \begin{bmatrix} n \\ k \end{bmatrix}_q$$

by Theorem 1.4, as needed.

1.7 Consider an integer partition λ with a Young diagram which fits inside of a $c \times (a+b-c)$ rectangle created by a lattice path from $(0,c)$ to $(a+b-c,0)$. Identify the smallest value of k for which the Young diagram of λ contains $(n-k) \times (a-k)$ rectangle. Here is an example of what such a rectangle might look like when $a = 6, b = 16$, and $c = 7$:

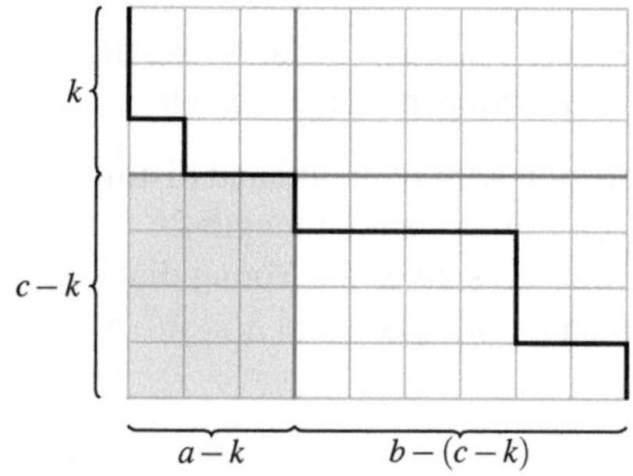

The lattice path can be broken into two pieces: a path from $(0,c)$ to $(a-k,c-k)$ and then a path from $(a-k,c-k)$ to $(a+b-c,0)$. These two lattice paths correspond to integer partitions which fit into a $k \times (a-k)$ and $(c-k) \times (b-(c-k))$ rectangles. Since the area of the rectangle inside λ has $(a-k)(c-k)$ cells, the identity follows; the right hand side sorts integer partitions according to k, keeping track of the total number of cells as the exponent of q.

1.8 Using the same logic as the proof of Theorem 1.8, the generating function for the number of integer partitions of n with distinct parts is

$$\begin{aligned}(1+z)(1+z^2)(1+z^3)\cdots &= \frac{1-z^2}{1-z}\frac{1-z^4}{1-z^2}\frac{1-z^6}{1-z^3}\cdots \\ &= \frac{1}{(1-z)(1-z^3)(1-z^5)}\cdots.\end{aligned}$$

This is the generating function for the number of integer partitions with only odd parts.

1.9 Given $\lambda = (\lambda_1, \ldots, \lambda_\ell) \vdash n$ with distinct parts, write each λ_i as $2^{a_i} d_i$ where d_i is an odd number. Then we have

$$\begin{aligned}n &= \lambda_1 + \cdots + \lambda_\ell \\ &= 2^{a_1} d_1 + \cdots + 2^{a_k} d_k \\ &= (\text{a sum of distinct powers of } 2) \cdot 1 + (\text{a sum of distinct powers of } 2) \cdot 3 + \cdots,\end{aligned}$$

where we have grouped the terms according to their odd components. This corresponds to an integer partition with only odd parts; the coefficient the odd number d tells us how many times the integer partition should contain d. This process is reversible since each integer has a unique base 2 representation.

1.10 We prove this bijectively, starting with $\lambda \vdash (a-c)$ with $\ell(\lambda) = b-1$ where no parts of λ are larger than c. Follow these three steps:

1. Add a row of c cells to the bottom to the Young diagram of λ.
2. Delete the first column.
3. Conjugate.

The result of these three reversible steps is the Young diagram of an integer partition of $a-b$ with length $c-1$ and no parts larger than b.

1.11 Take an integer partition with both odd and distinct parts with no parts greater than $2n-1$ and bend each row of cells of length $2k-1$ in the Young diagram into an "L" shape. Nesting these "L"s gives the Young diagram of an integer partition λ such that λ fits inside of an $n \times n$ square and $\lambda = \lambda'$. A picture:

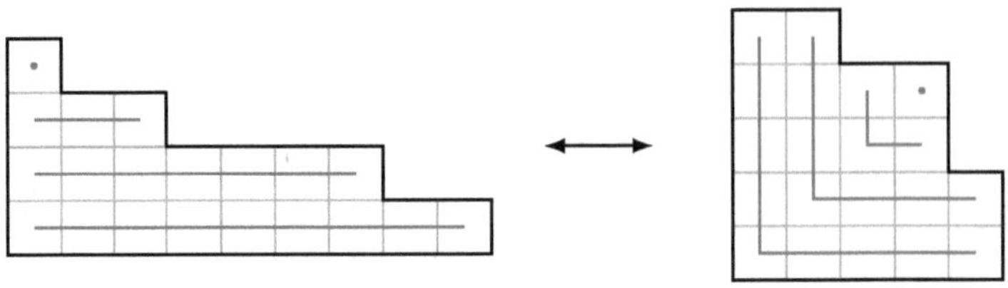

1.12 The left-hand side of the equation is equal to

$$\sum_{n=0}^{\infty} \sum_{\lambda \vdash n \text{ has distinct parts}} y^{\ell(\lambda)} z^n.$$

Since an integer partition λ in the above sum must have distinct parts, it must contain a staircase of height $\ell(\lambda) = k$. Remove this staircase and left justify the remaining cells in the Young diagram; an example is below:

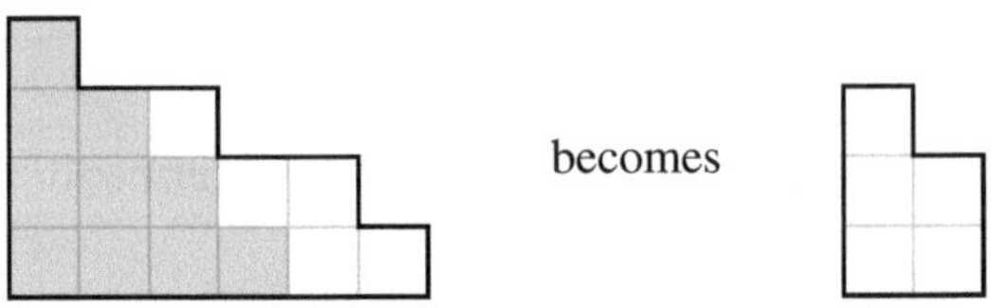

What remains must be an integer partition with $\leq k$ parts; these are in 1–1 correspondence by conjugation with integer partitions with parts of size $\leq k$. The generating function for these integer partitions is

$$\frac{1}{1-z}\frac{1}{1-z^2}\cdots\frac{1}{1-z^k}.$$

Sorting by the height of the removed staircase, which contains $1+2+\cdots+k = \binom{k+1}{2}$ cells, the generating function for the number of integer partitions with distinct parts is therefore equal to

$$\sum_{k=0}^{\infty} \frac{y^k z^{\binom{k+1}{2}}}{(1-z)(1-z^2)\cdots(1-z^k)}.$$

1.13 First assume that k is nonnegative. For the product $z^{k(k-1)/2}\prod_{n=1}^{\infty}1/(1-z^n)$, we place a staircase of $k-1$ cells (we use the mirror image of the Young diagram for the integer partition $(k-1,k-2,\ldots,1)$ since $1+\cdots+k-1=k(k-1)/2$) next to an integer partition of n, like the picture below which shows when $k=3$:

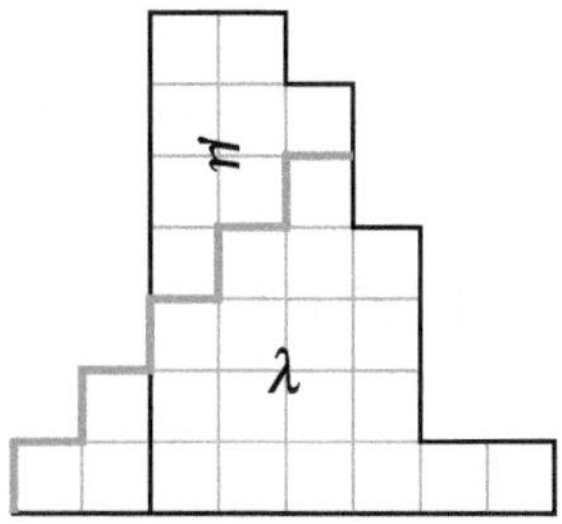

Cut these cells along the diagonal line following the top of the staircase to form two integer partitions. Let λ be the integer partition with Young diagram found by left justifying the cells below the diagonal and let μ be the integer partition found by conjugating and then left justifying the cells above the diagonal. The integer partitions λ and μ which come from the above diagram are shown below:

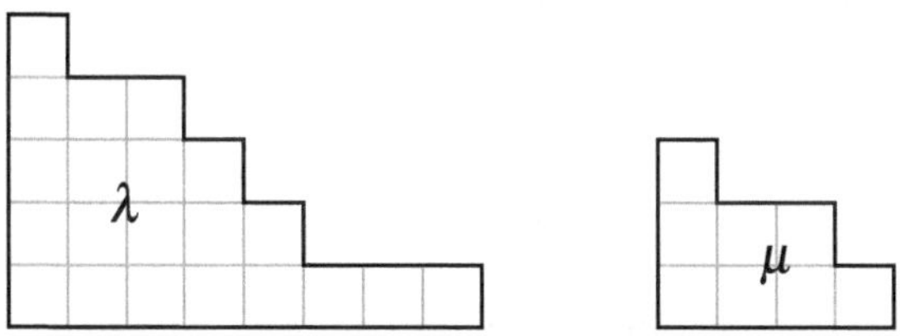

These integer partitions λ and μ must have distinct parts. Furthermore, depending on whether the last step of the diagonal line is vertical or horizontal, $\ell(\lambda)-\ell(\mu)$ is either k or $k-1$. If $\ell(\lambda)-\ell(\mu)=k-1$, add a 0 part to λ in order to make $\ell(\lambda)-\ell(\mu)=k$.

If k is negative, then $k(k-1)/2=|k|(|k|+1)/2$, so place a staircase of k cells next to an integer partition. Then, take μ to be the integer partition found below the diagonal line and let λ be the conjugate of the integer partition found above the diagonal line. We now have an ordered pair (λ,μ) with $\ell(\lambda)-\ell(\mu)=k$ or $k-1$; if the difference is $k-1$, then add a 0 part to λ to make the difference k.

We have shown the desired equality since the coefficient of y^k in the product

$$\prod_{n=0}^{\infty}(1+yz^n)\prod_{m=1}^{\infty}(1+y^{-1}z^m)$$

is equal to the number of pairs of integer partitions (λ,μ) where λ can have a 0 part and $\ell(\lambda)-\ell(\mu)=k$.

1.14 Sum the result in Exercise 1.13 over all possible integers k to find

$$\sum_{k=-\infty}^{\infty} y^k z^{k(k-1)/2} = \prod_{n=1}^{\infty} (1-z^n)\left(1+yz^{n-1}\right)\left(1+y^{-1}z^n\right).$$

Then find the result by taking "z" as "z^2" and "y" as "yz."

1.15 Take "z" as "$z^{3/2}$" and "y" as "$-z^{-1/2}$" in the Jacobi triple product identity to find

$$\sum_{k=-\infty}^{\infty} (-1)^k z^{(3k^2-k)/2} = \prod_{n=1}^{\infty} \left(1-z^{3n}\right)\left(1-z^{3n-2}\right)\left(1-z^{3n-1}\right) = \prod_{n=1}^{\infty} (1-z^n).$$

Therefore we have

$$1 = \left(\prod_{n=0}^{\infty} \frac{1}{1-z^n}\right)\left(\prod_{n=0}^{\infty} (1-z^n)\right) = \left(\sum_{n=0}^{\infty} p(n) z^n\right)\left(\sum_{n=0}^{\infty} (-1)^k z^{(3k^2-k)/2}\right).$$

Comparing coefficients of z^n on both sides of the equality gives

$$0 = p(n) - p(n-1) - p(n-2) + p(n-5) + p(n-7) - \cdots.$$

1.16 Since the product $\prod_{n=1}^{\infty}(1-z^n)$ is equal to

$$\sum_{\lambda \text{ has distinct parts}} (-1)^{\ell(\lambda)} z^{|\lambda|}, \tag{1.3}$$

we will consider integer partitions with distinct parts. The sign of such an integer partition λ is defined to be $(-1)^{\ell(\lambda)}$.

Given an integer partition λ with distinct parts, let $top(\lambda)$ be the number of cells in the top row of the Young diagram of λ and let $stair(\lambda)$ be the number of cells in the "staircase" which starts in the bottom right cell of the Young diagram of λ and travels along a diagonal with slope -1. For example, we depict $top(\lambda)$ and $stair(\lambda)$ for $\lambda = (8,7,6,4,3)$ below:

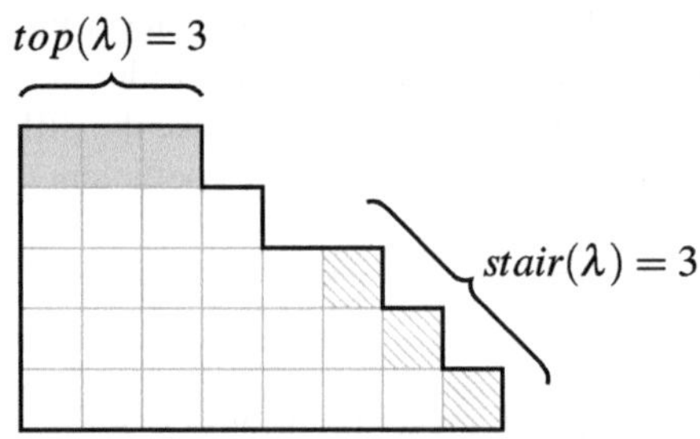

We define λ to be a fixed point under the involution φ if either

1. $\ell(\lambda) = k$ for some integer k, $top(\lambda) = k$, and $stair(\lambda) = k$, or
2. $\ell(\lambda) = k$ for some integer k, $top(\lambda) = k+1$, and $stair(\lambda) = k$.

Otherwise, define φ in the following way: if $top(\lambda) \leq stair(\lambda)$, then $\varphi(\lambda)$ is the integer partition found by removing the cells in the top section of the Young diagram of λ and placing them alongside of the staircase. For example, we would change the integer partition displayed above to this one:

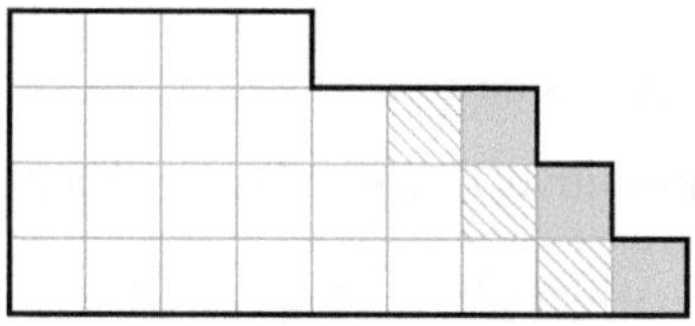

If $top(\lambda) > stair(\lambda)$, then define $\varphi(\lambda)$ to be the integer partition which undoes the above operation, that is, remove the cells in the staircase and place them on the top.

Unless λ is a fixed point, the sign of λ and $\varphi(\lambda)$ differs by -1. Therefore (1.3) is equal to the signed sum over all possible fixed points of the involution $\varphi(\lambda)$. An example of each of the two varieties of fixed points are shown below:

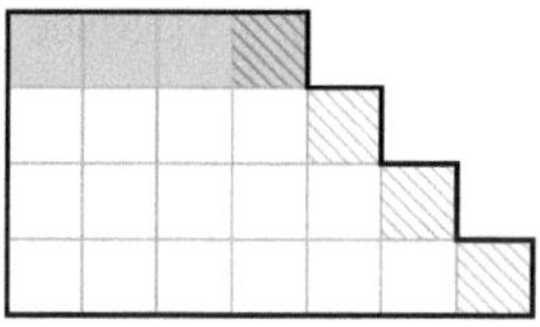
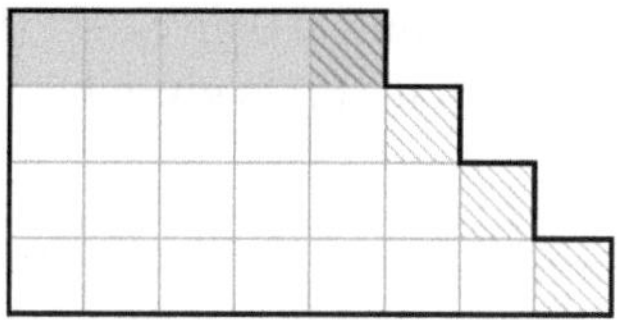

In the first case we count the number of cells in $|\lambda|$ by looking at the staircase of height $k-1$ next to the $k \times k$ square of cells. This gives

$$|\lambda| = k^2 + \binom{k}{2} = \frac{3k^2 - k}{2}.$$

In the second case, the staircase next to the $k \times (k+1)$ rectangle gives that

$$|\lambda| = k(k+1) + \binom{k}{2} = \frac{3k^2 + k}{2} = \frac{3(-k)^2 - (-k)}{2}.$$

These fixed points correspond to the pentagonal numbers, as desired.

1.17 We have

$$\mathscr{L}\{1\} = \int_0^\infty e^{-st}\, dt = \frac{1}{s}$$

and, using integration by parts,

$$\begin{aligned}\mathscr{L}\{t^n\} &= \int_0^\infty t^n e^{-st}\, dt \\ &= -\frac{t^n}{s} e^{-st}\Big|_0^\infty + \frac{n}{s}\int_0^\infty t^{n-1} e^{-st}\, dt \\ &= \frac{n}{s}\mathscr{L}\{t^{n-1}\}\end{aligned}$$

for all positive integers n. The statement $\mathscr{L}\{t^n\} = n!/s^{n+1}$ follows by induction.

The Laplace transform of the exponential generating function $\sum_{n=0}^{\infty} a_n t^n/n!$ is

$$F(s) = \mathscr{L}\left\{\sum_{n=0}^{\infty} \frac{a_n}{n!} t^n\right\} = \sum_{n=0}^{\infty} a_n \frac{1}{s^{n+1}},$$

and so $F(1/s)/s = \sum_{n=0}^{\infty} a_n s^n$, as desired.

1.18 Let $\zeta = e^{2\pi i/j}$ be a primitive j^{th} root of unity. Then

$$\frac{f(\zeta z) + \cdots + f(\zeta^j z)}{j} = \frac{1}{j} \sum_{n=0}^{\infty} a_n \frac{\zeta^n + \cdots + \zeta^{nj}}{j} z^n.$$

Since $\zeta^n + \cdots + \zeta^{nj}$ is equal to j if j divides n and 0 otherwise, this sum is equal to $\sum_{n=0}^{\infty} a_{nj} z^{nj}$, as desired.

1.19 If $f(z) = (1+z)^a$, then the nth derivative $f^{(n)}(z) = a(a-1)\cdots(a-n+1)$ $(1+z)^{a-n}$. Evaluating at 0 gives $f^{(n)}(0) = a(a-1)\cdots(a-n+1)$. Since the power series for a function $f(z)$ is given by

$$f(z) = \sum_{n=0}^{\infty} \frac{f^{(n)}(0)}{n!} z^n,$$

we have now proved Newton's binomial theorem.

1.20 Newton's binomial theorem gives

$$\begin{aligned}
(1-z)^{-i} &= \sum_{n=0}^{\infty} \frac{(-i)(-i-1)\cdots(-i-n+1)}{n!}(-z)^n \\
&= \sum_{n=0}^{\infty} \frac{i(i+1)\cdots(i+n-1)}{n!} z^n \\
&= \sum_{n=0}^{\infty} \binom{i+n-1}{n} z^n,
\end{aligned}$$

as desired. This says that the coefficient of z^n in $c_i/(z-a)^i = (-1)^i c_i a^{-i}/(1-z/a)^i$ is equal to $c_i(-1)^i\binom{i+n-1}{n}/a^{n+i}$ and therefore

$$b_n = (-1)^k \frac{c_k}{a^{n+k}} \binom{k+n-1}{n} + \cdots + (-1)^1 \frac{c_1}{a^{n+1}} \binom{1+n-1}{n}.$$

Notes

A delightful text introducing generating functions and their uses is Wilf's *Generatingfunctionology* [117]. Stanley's *Enumerative Combinatorics* [109] is an excellent text which introduces generating functions and permutation statistics. These two books deserve to be read by every student of combinatorics.

The systematic study of the permutation statistics des, exc, inv, and maj was begun by MacMahon [83]. He proved that des and exc where equidistributed over permutations and des and exc where equidistributed over permutations. The first bijective proofs of these fact was given by Foata. Our proof of the fact that des and exc are equidistributed over S_n can be found in [43]. The proof of Theorem 1.3 and the subsequent bijection is due to Carlitz [18]. The proof of Theorem 1.5 is a modification of a proof by Foata [44].

Those interested in integer partitions are referred to Andrew's *The Theory of Integer Partitions* [4] and Wilf's notes *Lectures and Integer Partitions* which can be found on his Web site. For instance, Andrew's book includes a proof of the direct formula for the number of integer partitions we display on our page 9, due to Hardy and Ramanujan. The interpretation of integer partitions as abaci can be found in Loehr's *Bijective Combinatorics* [81].

Chapter 2
Symmetric Functions

This chapter provides a lean but solid introduction to symmetric functions. All of the theory needed for our later chapters is carefully introduced while simultaneously giving the reader a firm hook on which to hang future studies. Our method is novel in that we emphasize the combinatorics of transition matrices and most of our proofs are combinatorial.

The subject is vast and an attempt to create an encyclopedic account would distract from our focus of using symmetric functions to solve enumeration problems. Therefore we have made heartbreaking choices on what topics to include or not include in this chapter, although we admit to succumbing a few interesting digressions which are not strictly needed in our development.

2.1 Standard Bases for Symmetric Functions

Let $x_1, x_2, \ldots$ be an infinite collection of indeterminates and, just as introduced in Section 1.3, let $B\mathbb{Q}[[x_1, x_2, \ldots]]$ be the subring of $\mathbb{Q}[[x_1, x_2, \ldots]]$ containing those monomials with bounded degree. Given a permutation $\sigma = \sigma_1 \ldots \sigma_N \in S_N$ and $P(x_1, x_2, \ldots) \in B\mathbf{Q}[[x_1, x_2, \ldots]]$, we define

$$\sigma P(x_1, x_2, \ldots x_N, x_{N+1}, x_{N+2}, \ldots) = P(x_{\sigma_1}, x_{\sigma_2}, \ldots x_{\sigma_N}, x_{N+1}, x_{N+2}, \ldots).$$

We say that $P(x_1, x_2, \ldots)$ is a symmetric function if for all $N \geq 1$ and all $\sigma \in S_N$,

$$\sigma P(x_1, x_2, \ldots) = P(x_1, x_2, \ldots).$$

Thus $P(x_1, x_2, \ldots)$ is a symmetric function if it is invariant under all finite permutations of the variables $x_1, x_2, \ldots$.

A. Mendes, J. Remmel, *Counting with Symmetric Functions*, Developments in Mathematics 43, DOI 10.1007/978-3-319-23618-6_2

We define $\Lambda(x_1,x_2,\ldots)$ to be the set of all symmetric functions in $B\mathbf{Q}[[x_1,x_2,\ldots]]$. Since the sum and the product of any two symmetric functions are again symmetric functions, it follows that $\Lambda(x_1,x_2,\ldots)$ is a ring. Further, we let

$$\Lambda_n(x_1,x_2,\ldots) = \Lambda(x_1,x_2,\ldots) \cap B\mathbf{Q}_n[[x_1,x_2,\ldots]]$$

and we will refer to $\Lambda_n(x_1,x_2,\ldots)$ as the vector space of symmetric functions of degree n. Our definitions ensure that we can write any symmetric function $P(x_1,x_2,\ldots)$ in the form

$$P(x_1,x_2,\ldots) = \sum_{n=0}^{N} P_n(x_1,x_2,\ldots),$$

where $P_n(x_1,x_2,\ldots) \in \Lambda_n[[x_1,x_2,\ldots]]$ for all n by breaking $P(x_1,x_2,\ldots)$ into its degree n components. In symbols, this means

$$\Lambda(x_1,x_2,\ldots) = \bigoplus_{n=0}^{\infty} \Lambda_n(x_1,x_2,\ldots).$$

By taking $x_i = 0$ for all $i \geq N+1$, the ring of symmetric functions $\Lambda(x_1,x_2,\ldots)$ specializes to the polynomial ring $\Lambda(x_1,\ldots,x_N)$. In this situation, an element $f \in \Lambda(x_1,\ldots,x_N)$ is called a symmetric polynomial in the variables $x_1,\ldots,x_N$ with coefficients in $\mathbb{Q}$. This means that for all permutations $\sigma = \sigma_1 \cdots \sigma_N \in S_N$,

$$f(x_1,\ldots,x_N) = f(x_{\sigma_1},\ldots,x_{\sigma_N}).$$

For example, one symmetric polynomial in the variables x_1, x_2, and x_3 is

$$2x_1 + 2x_2 + 2x_3 - x_1x_2 - x_1x_3 - x_2x_3 + 4x_1x_2x_3.$$

There are six standard bases for Λ_n: the monomial symmetric functions, the elementary symmetric functions, homogeneous symmetric functions, power symmetric functions, the Schur symmetric functions, and the forgotten symmetric functions. The main objective of this book is to exploit the relationships between these bases in order to solve counting problems.

The Monomial Symmetric Functions

If $\gamma = (\gamma_1,\ldots,\gamma_N)$ is a weak composition of n, then we let $\lambda(\gamma)$ the partition found by sorting γ in weakly decreasing order. For example, if $\gamma = (2,0,3,1,0,1,0,0,4)$, then $\lambda(\gamma) = (4,3,2,1,1)$.

Let $\lambda = (\lambda_1,\ldots,\lambda_\ell)$ be an integer partition of n. The monomial symmetric function $m_\lambda = m_\lambda(x_1,x_2,\ldots)$ is defined to be

$$m_\lambda = \sum_{\gamma \in WC_n, \lambda(\gamma)=\lambda} x^\gamma.$$

Put differently, m_λ is the sum of all the monomials whose exponents can be rearranged to give the partition λ. For example, the monomial symmetric polynomial $m_{(2,1)}(x_1,x_2,x_3)$ is

$$m_{(2,1)}(x_1,x_2,x_3) = x_1^2x_2^1 + x_1^2x_3^1 + x_1^1x_2^2 + x_2^2x_3^1 + x_1^1x_3^2 + x_2^1x_3^2.$$

Theorem 2.1. *The set $\{m_\lambda : \lambda \vdash n\}$ is a basis for Λ_n.*

Proof. If α and β are two weak compositions of n such that $\lambda(\alpha) = \lambda(\beta) = \lambda$, then α and β are rearrangements of one another. Thus the coefficients of x^α and x^β in any given symmetric function $P(x_1,x_2,\dots)$ are the same. This implies that we can write $P(x_1,x_2,\dots)$ in the form

$$P(x_1,x_2,\dots) = \sum_{\lambda \vdash n} c_\lambda m_\lambda$$

for constants c_λ, implying that $\{m_\lambda : \lambda \vdash n\}$ spans $\Lambda(x_1,x_2,\dots)$.

Since m_λ and m_μ have no monomials in common if $\lambda \neq \mu$, the set $\{m_\lambda : \lambda \vdash n\}$ is an independent set, thereby showing the theorem true. □

Theorem 2.1 tells us that the dimension of $\Lambda_n(x_1,x_2,\dots)$ is $p(n)$, the number of partitions of n.

The Elementary, Homogeneous, and Power Symmetric Functions

The nth elementary symmetric function e_n is defined using a generating function. Let $E(z)$ denote the generating function for the sequence $e_0, e_1, e_2, \dots$. Define e_n by

$$E(z) = \sum_{n=0}^{\infty} e_n z^n = \prod_{i=1}^{\infty}(1 + x_i z) = (1 + x_1 z)(1 + x_2 z)\cdots.$$

For example, if $0 = x_4 = x_5 = \cdots$, the generating function $E(z)$ becomes

$$(1+x_1z)(1+x_2z)(1+x_3z)$$
$$= 1 + (x_1+x_2+x_3)z + (x_1x_2+x_1x_3+x_2x_3)z^2 + x_1x_2x_3z^3$$

and so the first few elementary symmetric polynomials in three variables are $e_0 = 1$, $e_1 = x_1 + x_2 + x_3$, $e_2 = x_1x_2 + x_1x_3 + x_2x_3$, and $e_3 = x_1x_2x_3$. In general, we can employ similar logic as found in the proof of Theorem 1.8 to conclude each variable x_i can appear at most once in a given monomial in e_n. In other words, the elementary symmetric function e_n is the sum of all square-free monomials of degree n—this means that each monomial in e_n is not divisible by x_i^2 for any x_i. The symmetric function e_n is also equal to $m_{(1^n)}$.

The elementary symmetric function e_n can be expressed as a sum of column strict tableaux of shape 1^n. We have

$$e_n = \sum_{T \in CS_{(1^n)}} w(T).$$

For example, the terms in the symmetric polynomial

$$e_3(x_1,x_2,x_3,x_4) = x_1x_2x_3 + x_1x_2x_4 + x_1x_3x_4 + x_2x_3x_4$$

are the weights of the following column strict tableaux of shape 1^3 which are filled with integers no larger than 4:

3	4	4	4
2	2	3	3
1	1	1	2

For any integer partition $\lambda = (\lambda_1,\ldots,\lambda_\ell) \vdash n$, we define $e_\lambda = e_{\lambda_1}\cdots e_{\lambda_\ell}$. The fundamental theorem of symmetric functions says that the set $\{e_\lambda : \lambda \vdash n\}$ is a basis for Λ_n; this is our forthcoming Theorem 2.17.

The nth homogeneous symmetric function h_n is defined in a similar manner as e_n. Letting $H(z)$ denote the generating function for h_n, we define

$$H(z) = \sum_{n=0}^{\infty} h_n z^n = \prod_{i=1}^{\infty} \frac{1}{1-x_i z}.$$

For example, if we take $0 = x_4 = x_5 = \cdots$, then $H(z)$ becomes

$$\begin{aligned}
&\left(\frac{1}{1-x_1z}\right)\left(\frac{1}{1-x_2z}\right)\left(\frac{1}{1-x_3z}\right) \\
&\qquad = (1+x_1z+x_1^2z^2+\cdots)(1+x_2z+x_2^2z^2+\cdots)(1+x_3z+x_3^2z^2+\cdots) \\
&\qquad = 1+(x_1+x_2+x_3)z^1+(x_1^2+x_2^2+x_3^2+x_1x_2+x_1x_3+x_2x_3)z^2+\cdots
\end{aligned}$$

and so the first few homogeneous symmetric polynomials in three variables are $h_0 = 1$, $h_1 = x_1+x_2+x_3$, and $h_2 = x_1^2+x_2^2+x_3^2+x_1x_2+x_1x_3+x_2x_3$. In general, by writing each term in the infinite product as a geometric series and expanding, we see that h_n contains all possible degree n monomials, each with leading coefficient 1. In other words,

$$h_n = \sum_{\lambda \vdash n} m_\lambda.$$

The homogeneous symmetric function h_n can be expressed as a sum of column strict tableaux of shape n; we have

$$h_n = \sum_{T \in CS_{(n)}} w(T).$$

For example, the terms in the symmetric polynomial

$$h_2(x_1,x_2,x_3) = x_1^2 + x_2^2 + x_3^2 + x_1x_2 + x_1x_3 + x_2x_3$$

are the weights of the following column strict tableaux of shape 2 which are filled with integers no larger than 3:

$$\boxed{1\,|\,1} \quad \boxed{2\,|\,2} \quad \boxed{3\,|\,3} \quad \boxed{1\,|\,2} \quad \boxed{1\,|\,3} \quad \boxed{2\,|\,3}$$

For any integer partition $\lambda = (\lambda_1, \ldots, \lambda_\ell) \vdash n$, we define $h_\lambda = h_{\lambda_1} \cdots h_{\lambda_\ell}$. The set $\{h_\lambda : \lambda \vdash n\}$ is a basis for Λ_n; this is our Corollary 2.20.

The nth power symmetric function p_n is defined to be

$$p_n(x_1,x_2,x_3,\ldots) = x_1^n + x_2^n + x_3^n + \cdots$$

and so $p_n = m_{(n)}$. The power symmetric function p_n can be expressed as a weighted sum of tableaux if we require that every integer in a tableau of shape (n) be the same. Just like the elementary and homogeneous symmetric functions, we define $p_\lambda = p_{\lambda_1} \cdots p_{\lambda_\ell}$ for any integer partition $\lambda = (\lambda_1, \ldots, \lambda_\ell) \vdash n$. We will show that $\{p_\lambda : \lambda \vdash n\}$ is a basis for Λ_n in Corollary 2.24.

The Schur Symmetric Functions

The most important basis for Λ_n with respect to its relationship to other areas of mathematics is the Schur symmetric functions—they are crucial in understanding the representation theory of the symmetric group. Given an integer partition $\lambda \vdash n$, we define the Schur symmetric function s_λ by

$$s_\lambda = \sum_{T \in CS_\lambda} w(T).$$

For example, all possible column strict tableaux of shape $(2,1)$ which are filled with integers less than or equal to 3 are

$$\begin{array}{|c|c}\hline 3 & \\ \hline 1 & 2 \\ \hline\end{array} \quad \begin{array}{|c|c}\hline 2 & \\ \hline 1 & 3 \\ \hline\end{array} \quad \begin{array}{|c|c}\hline 2 & \\ \hline 1 & 1 \\ \hline\end{array} \quad \begin{array}{|c|c}\hline 3 & \\ \hline 1 & 1 \\ \hline\end{array} \quad \begin{array}{|c|c}\hline 2 & \\ \hline 1 & 2 \\ \hline\end{array} \quad \begin{array}{|c|c}\hline 3 & \\ \hline 1 & 3 \\ \hline\end{array} \quad \begin{array}{|c|c}\hline 3 & \\ \hline 2 & 2 \\ \hline\end{array} \quad \begin{array}{|c|c}\hline 3 & \\ \hline 2 & 3 \\ \hline\end{array}$$

and so

$$s_{(2,1)}(x_1,x_2,x_3) = 2x_1x_2x_3 + x_1^2x_2 + x_1^2x_3 + x_1x_2^2 + x_1x_3^2 + x_2^2x_3 + x_2x_3^2.$$

From this definition it may not be clear that the Schur symmetric function is even a symmetric function, much less a basis for Λ_n.

Theorem 2.2. *For any $\lambda \vdash n$, the Schur symmetric function s_λ is an element of Λ_n.*

Proof. Every element in S_n is a product of adjacent transpositions, that is, every element in S_n is the product of permutations of the form $(i\, i+1)$, so it is enough to show that $s_\lambda(x_1, x_2, \ldots)$ is unchanged under the action of switching x_i and x_{i+1} for all positive integers i. This means that we need to show that for every column strict tableau with k occurrences of i and j occurrences of $i+1$, there is a corresponding column strict tableau of the same shape with j occurrences of i and k occurrences of $i+1$.

Take $T \in CS_\lambda$. The appearances of i in relationship to the appearances of $i+1$ in T must look something like the appearances of the 3s and 4s below:

3	3	4	4	4	4										
			3	3	3	4	4	4							
							3	3	3	3	3	3	3	4	4

Each row in T may have a sequence of is followed by a sequence of $(i+1)$s. Rows are aligned so that an i cannot appear atop another i in the row below.

Suppose a given row of T contains a sequence of k is followed by j $(i+1)$s such that none of these is or $(i+1)$s are immediately above or below cells containing an i or $i+1$. Change this row of T so that it now contains j is followed by k $(i+1)$s. Make this change for every row of T to create the column strict tableau T'. For example, changing the 3s and 4s in the cells displayed above produces:

3	4	4	4	4	4										
			3	3	3	3	4	4							
							3	3	3	3	4	4	4	4	4

The new column strict tableau has the same shape as before with the number of is and $(i+1)$s switched, as needed. □

We end this section by using Vandermonde determinants to provide an alternative definition of the Schur symmetric function in the case where there are finitely many variables $x_1, \ldots, x_N$.

Given an integer partition $\lambda \vdash n$ where $n \leq N$, write λ as $(\lambda_1, \ldots, \lambda_N)$ so that λ has N parts (the last parts of λ can be 0). We define $\Delta_\lambda(x_1, \ldots, x_N)$ to be the determinant of the matrix with i, j entry equal to $x_i^{\lambda_j+N-j}$. For example,

$$\Delta_{(3,1,0,0)}(x_1,x_2,x_3,x_4) = \begin{vmatrix} x_1^{3+3} & x_1^{1+2} & x_1^{0+1} & x_1^{0+0} \\ x_2^{3+3} & x_2^{1+2} & x_2^{0+1} & x_2^{0+0} \\ x_3^{3+3} & x_3^{1+2} & x_3^{0+1} & x_3^{0+0} \\ x_4^{3+3} & x_4^{1+2} & x_4^{0+1} & x_4^{0+0} \end{vmatrix} = \begin{vmatrix} x_1^6 & x_1^3 & x_1 & 1 \\ x_2^6 & x_2^3 & x_2 & 1 \\ x_3^6 & x_3^3 & x_3 & 1 \\ x_4^6 & x_4^3 & x_4 & 1 \end{vmatrix}.$$

Using the expansion of the determinant of a matrix as a signed sum over the symmetric group S_n, we have

$$\Delta_\lambda(x_1,\ldots,x_N) = \sum_{\sigma=\sigma_1\cdots\sigma_N\in S_N} \operatorname{sign}(\sigma) x_{\sigma_1}^{\lambda_1+N-1} x_{\sigma_2}^{\lambda_2+N-2} \cdots x_{\sigma_N}^{\lambda_N+0},$$

where $\operatorname{sign}(\sigma)$ is the sign of the permutation σ. This implies that Δ_λ is a polynomial in $x_1,\ldots,x_N$. Theorem 2.3 says that the polynomial $\Delta_{(0,\ldots,0)}(x_1,\ldots,x_N)$, known as the Vandermonde determinant, factors nicely.

Theorem 2.3. *We have* $\Delta_{(0,\ldots,0)}(x_1,\ldots,x_N) = \prod_{1\le i<j\le N}(x_i - x_j)$.

Proof. Switching x_i and x_j interchanges two rows in the determinant

$$\Delta_{(0,\ldots,0)} = \begin{vmatrix} x_1^{N-1} & x_1^{N-2} & \cdots & 1 \\ x_2^{N-1} & x_2^{N-2} & \cdots & 1 \\ & & \vdots & \\ x_N^{N-1} & x_N^{N-2} & \cdots & 1 \end{vmatrix},$$

which has the net effect of changing the sign of $\Delta_{(0,\ldots,0)}$ by a factor of -1. This implies that $\Delta_{(0,\ldots,0)}(x_1,\ldots,x_N)$ is divisible by $(x_i - x_j)$ for every $i < j$.

More generally, this shows $\prod_{1\le i<j\le N}(x_i - x_j)$ divides $\Delta_{(0,\ldots,0)}$. Since these two polynomials are sums of monomials of degree $(N-1)+(N-2)+\cdots+0$, they must be equal up to some constant factor. By considering the main diagonal of the determinant, the coefficient of $x_1^{N-1}x_2^{N-2}\cdots x_N^0$ is 1 in both the determinant and the product, so this constant factor is 1. □

Slight modifications of the proof of Theorem 2.3 show that $\Delta_{(0,\ldots,0)}(x_1,\ldots,x_N)$ divides $\Delta_\lambda(x_1,\ldots,x_N)$ for all $\lambda \vdash n$. Furthermore, since switching x_i and x_j changes $\Delta_\lambda(x_1,\ldots,x_N)$ by a factor of -1, $\Delta_\lambda(x_1,\ldots,x_N)/\Delta_{(0,\ldots,0)}(x_1,\ldots,x_N)$ is a symmetric polynomial. For example,

$$\frac{\Delta_{(2,1,0)}(x_1,x_2,x_3)}{\Delta_{(0,0,0)}(x_1,x_2,x_3)} = \frac{\begin{vmatrix} x_1^4 & x_1^2 & 1 \\ x_2^4 & x_2^2 & 1 \\ x_3^4 & x_3^2 & 1 \end{vmatrix}}{(x_1-x_2)(x_1-x_3)(x_2-x_3)}$$

which, when expanded and simplified, is equal to

$$2x_1x_2x_3 + x_1^2x_2 + x_1^2x_3 + x_1x_2^2 + x_1x_3^2 + x_2^2x_3 + x_2x_3^2.$$

Theorem 2.4 explains why this calculation gives the Schur symmetric polynomial $s_{(2,1)}(x_1,x_2,x_3)$.

Theorem 2.4. *For any $\lambda \vdash n \leq N$, we have* $\dfrac{\Delta_\lambda(x_1,\dots,x_N)}{\Delta_{(0,\dots,0)}(x_1,\dots,x_N)} = s_\lambda(x_1,\dots,x_N)$.

Proof. Expanding the determinant $\Delta_\lambda(x_1,\dots,x_N)$ as a sum over permutations in S_N and using Theorem 2.3, the identity in the statement of the theorem is the same as

$$\prod_{1\leq i<j\leq N} \frac{1}{x_i - x_j} \sum_{\sigma=\sigma_1\cdots\sigma_N \in S_N} \operatorname{sign}(\sigma) x_{\sigma_1}^{\lambda_1+N-1} x_{\sigma_2}^{\lambda_2+N-2} \cdots x_{\sigma_N}^{\lambda_N+0} = \sum_{T\in CS_\lambda} w(T).$$

Multiply both sides of this equation by $x_1^N x_2^{N-1} \cdots x_N^1$. With this term, we factor out the first term in each of the parentheses in $\prod_{i<j} 1/(x_i - x_j)$ and use $x_1^{N-1} x_2^{N-2} \cdots x_N^0$ to turn this product into $\prod_{i<j} 1/(1 - x_j/x_i)$. The remaining $x_1 \cdots x_N$ is used to increase each exponent in the sum on the left by 1. Our equation becomes

$$\prod_{1\leq i<j\leq N} \frac{1}{1-\frac{x_i}{x_j}} \sum_{\sigma\in S_N} \operatorname{sign}(\sigma) x_{\sigma_1}^{\lambda_1+N} x_{\sigma_2}^{\lambda_2+N-1} \cdots x_{\sigma_N}^{\lambda_N+1} = x_1^N x_2^{N-1} \cdots x_N^1 \sum_{T\in CS_\lambda} w(T). \tag{2.1}$$

We will prove this formulation of the identity with a sign reversing involution.

Looking at the left-hand side of this equality, we begin by constructing combinatorial objects in the following manner:

1. Affix an additional $N-j+1$ cells to the left of the jth row of the Young diagram of λ, counting rows from bottom to top.
2. Select a permutation $\sigma = \sigma_1 \cdots \sigma_N \in S_N$ and write σ vertically to the right of the Young diagram, reading bottom to top.
3. Starting from the bottom, place the integer σ_i into each cell in row i of our picture.

For example, if $\lambda = (3,1,0,0)$, the choice of $\sigma = 2\ 1\ 4\ 3$ gives

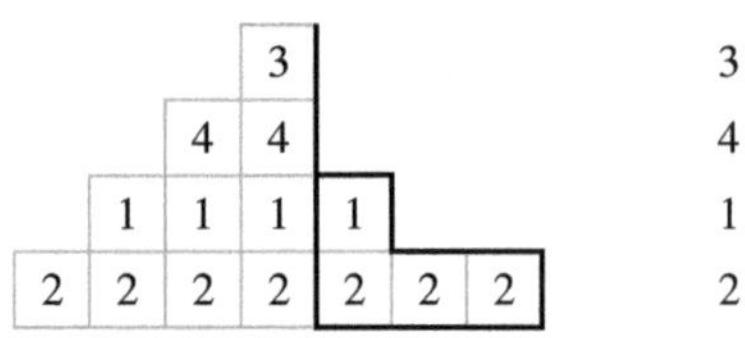

by following steps 1, 2, and 3. These three steps account for the sum on the left-hand side of (2.1). To account for the product

$$\prod_{i<j} \frac{1}{1-\frac{x_j}{x_i}} = \prod_{i<j} \left(1 + \left(\frac{x_j}{x_i}\right) + \left(\frac{x_j}{x_i}\right)^2 + \left(\frac{x_j}{x_i}\right)^3 + \cdots\right),$$

we finish creating our combinatorial in step 4:

4. In each row i, change any number of σ_is to an integer larger than σ_i. If every σ_i is changed, select any number of integers larger than σ_i to write down to the left of row i. Arrange the integers in each row so as to form a nondecreasing sequence.

For example, we can choose to change the above object into the one below:

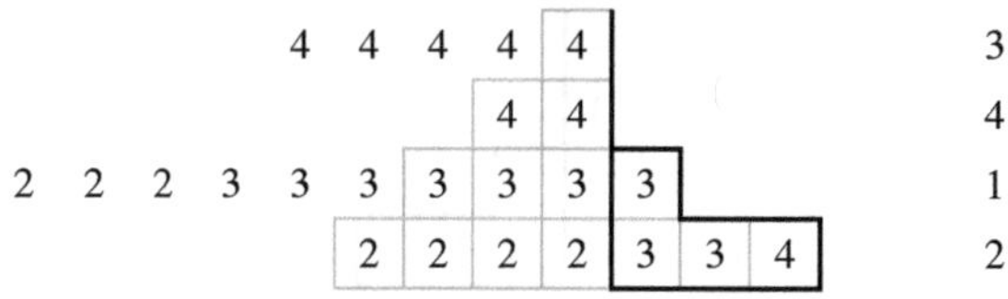

We define the sign of such an object T to be $\operatorname{sign}(\sigma)$ and we define the weight of the object to be

$$\left(\frac{x_1^{\text{the number of 1's in } T}}{x_1^{\text{the number of integers not in a cell in row } \sigma_1}}\right)\cdots\left(\frac{x_n^{\text{the number of } n\text{'s in } T}}{x_n^{\text{the number of integers not in a cell in row } \sigma_n}}\right).$$

For instance, the sign of the object displayed above is $\operatorname{sign}(2\ 1\ 4\ 3) = +1$ and the weight is $x_1^{-6}x_2^7x_3^5x_4^9$. By construction, the signed, weighted sum over all possible objects T is equal to the left-hand side of (2.1).

Let T be an object under consideration. We now describe how to create a new object $\varphi(T)$ with the same weight as T but with opposite sign. Starting from the most north cell in the most east column, scan the columns of T from top to bottom, moving right to left, looking for the first violation of column strictness. In the sample object displayed above, this violation occurs at the place where a 3 appears above another 3.

If T has no violations of column strictness, define $\varphi(T) = T$. Otherwise, let c be this first violating cell—this means the integer in c is not greater than the integer in the cell immediately below c. Define $\varphi(T)$ to be T with c and every cell in the same row and to the left of c switched with the cell kitty-corner to its south west. Additionally, if c is in the ith row, switch the positions of σ_i and σ_{i-1} in σ. Below we show the image of the object T displayed above together with added diagonal lines to help the reader more readily identify how cells have been changed:

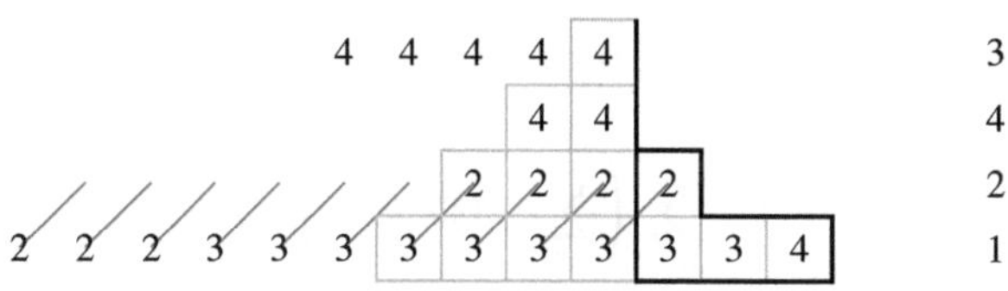

Since integers increase within rows, the integer in c is switched with a cell containing an integer no greater than the integer in c. Therefore the first violating cell in $\varphi(T)$ must be in the same position as the first violating cell in T, that is, φ is an involution. Introducing the transposition (σ_i, σ_{i+1}) changes the sign of σ by a factor of -1 and, since the integers both inside and outside of the cells in T and $\varphi(T)$ are the same, the weights of T and $\varphi(T)$ are also the same. In conclusion, φ is an involution which is weight preserving and, unless T is a fixed point, sign reversing.

The fixed points under the involution φ must look something like below:

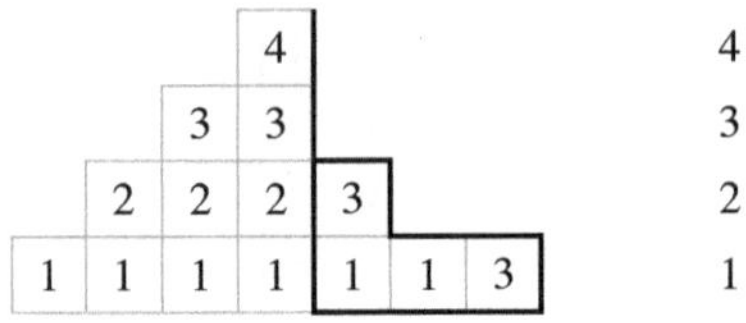

There can be no violations of column strictness and so the column immediately to the left of the Young diagram of shape λ must contain the integers $1,\ldots,n$ reading bottom to top. Therefore every fixed point must have σ equal to $1\ 2\ \cdots\ n$, every integer to the left of the Young diagram in row i containing i, and there cannot be any integers appearing outside of a cell.

These fixed points correspond to column strict tableaux of shape λ with an additional weight of $x_1^N x_2^{N-1} \cdots x_N^1$ coming from the cells to the left of the tableau; in other words, we have found the right-hand side of equation (2.1). □

2.2 Relationships Between Bases for Symmetric Functions

Our first relationship between symmetric functions, Theorem 2.5, follows immediately from our definitions of e_n and h_n. However, although simple, we will reap an incredible amount of information about generating functions for permutation statistics from Theorem 2.5.

Theorem 2.5. *The generating functions $E(z)$ and $H(z)$ for the elementary and homogeneous symmetric polynomials satisfy $H(z) = 1/E(-z)$.*

Proof. By definition, $H(z) = \prod_{i=1}^{\infty} \frac{1}{1-x_i z} = \frac{1}{\prod_{i=1}^{\infty}(1+x_i(-z))} = 1/E(-z)$. □

Rewriting Theorem 2.5 as $1 = H(z)E(-z)$, we find

$$1 = \sum_{n=0}^{\infty} \left(\sum_{i=0}^{n} (-1)^i e_i h_{n-i} \right) z^n.$$

Comparing coefficients of z^n shows that $\sum_{i=0}^{n}(-1)^i e_i h_{n-i}$ is equal to 0 for all $n \geq 1$. In following the philosophy of providing simple combinatorial proofs whenever reasonable, we will prove this fact with a sign reversing involution on pairs of column strict tableaux.

Proof (A second proof of Theorem 2.5). Consider ordered pairs (S,T) where S is a column strict tableau of shape 1^i and T a column strict tableau of shape $(n-i)$ for some $i \leq n$. Define the sign of (S,T) to be $(-1)^i$ and the weight to be $w(S)w(T)$. Then the signed, weighted sum over all possible pairs (S,T) is equal to $\sum_{i=0}^{n}(-1)^i e_i h_{n-i}$.

If the topmost integer in S is not smaller than the rightmost in T, move this integer from S to T. Otherwise, move the rightmost integer in T to the top of S. An example:

$$\begin{array}{|c|}\hline 6\\\hline 4\\\hline 2\\\hline 1\\\hline\end{array}\;,\;\begin{array}{|c|c|c|c|}\hline 1&1&4&4\\\hline\end{array}\longleftrightarrow\begin{array}{|c|}\hline 4\\\hline 2\\\hline 1\\\hline\end{array}\;,\;\begin{array}{|c|c|c|c|c|}\hline 1&1&4&4&6\\\hline\end{array}$$

This process is the desired sign reversing involution. □

The next theorem nicely illustrates a common theme in our work: after a theorem is proved combinatorially (that is, proved with a bijection or a sign reversing involution), we can usually modify the proof to arrive at new, related results.

Theorem 2.6. *For $k \geq 1$ and $n \geq 1$, $\sum_{i=0}^{n-1}(-1)^i e_i s_{(1^k,n-i)} = (-1)^{n-1} e_{n+k}$.*

Proof. The left hand side of this equation is the signed, weighted sum over all pairs of the form (S,T) where S is a column strict tableau of shape 1^i, T is a column strict tableau of shape $(1^k, n-i)$ with $n-i \geq 1$, the sign is $(-1)^i$, and the weight is $w(S)w(T)$. Apply the same sign reversing and weight preserving involution as in the second proof of Theorem 2.5: if the topmost integer in S is not smaller than the rightmost in T, move this integer from S to T. Otherwise, undo this operation.

Since we require that $n-i \geq 1$, there are fixed points which cannot be changed by this involution. Such a fixed point (S,T) must have the topmost integer in S smaller than the single element on the bottom row of T. For example, one fixed point when $k=3$ and $n=5$ is

$$\begin{array}{|c|}\hline 5\\\hline 4\\\hline 3\\\hline 1\\\hline\end{array}\;,\;\begin{array}{|c|}\hline 9\\\hline 8\\\hline 7\\\hline 6\\\hline\end{array}$$

Since the sign of such a fixed point is $(-1)^{n-1}$, the weighted sum over all fixed points (S,T) corresponds to $(-1)^{n-1}e_{n+k}$; this can be seen by affixing T atop S. □

Corollary 2.7. *For $k \geq 1$,*

$$\sum_{n=1}^{\infty} s_{(1^k,n)} z^{n+k} = \frac{\sum_{n=k+1}^{\infty}(-1)^{n-k-1} e_n z^n}{E(-z)}.$$

Proof. Multiplying both sides of this equation by $E(-z) = \sum_{n=0}^{\infty}(-1)^n e_n z^n$ and expanding, we find this statement:

$$\sum_{n=1}^{\infty}\left(\sum_{i=0}^{n-1}(-1)^i e_i s_{(1^k,n-i)}\right) z^{n+k} = \sum_{n=k+1}^{\infty}(-1)^{n-k-1} e_n z^n.$$

The result follows by replacing the inner summand on the left-hand side with $(-1)^{n-k-1}e_{n+k}$ as allowed by Theorem 2.6 and reindexing. □

Simple bijections and involutions can give relationships between the elementary, homogeneous, and power symmetric functions, as we show in Theorems 2.8 and 2.9. These two theorems are commonly attributed to Isaac Newton or Albert Girard.

Theorem 2.8. *For* $n \geq 1$, $\sum_{i=0}^{n-1} h_i p_{n-i} = nh_n$.

Proof. The right-hand side corresponds to the weighted sum over all column strict tableaux of shape n where one of the n cells is shaded. Define a bijection on such objects in this way: if a cell c containing i is marked, remove c and all of the cells right of c that also contain i to create two-column strict tableau. This process is depicted below:

The result is a pair (S,T) where S is a column strict tableau of shape i and T is a column strict tableau of shape $n-i$ where every cell in T contains the same integer. The weighted sum over all such pairs corresponds to the left-hand side of the equation. □

Theorem 2.9. *For* $n \geq 1$, $\sum_{i=0}^{n-1} (-1)^i e_i p_{n-i} = (-1)^{n-1} ne_n$.

Proof. The left-hand side corresponds to the signed and weighted sum over all pairs of the form (S,T) where S is a column strict tableau of shape 1^i, T is a column strict tableau of shape $n-i$ where every cell in T contains the same integer, the sign of (S,T) is $(-1)^i$, and the weight is $w(S)w(T)$.

Define a weight preserving and sign reversing involution on such pairs (S,T) in the following way. If the integer in T also appears in S, move that integer from S to T. If the integer in T does not appear in S and T contains more than one cell, then move one cell from T to S. Otherwise, declare (S,T) to be a fixed point. This operation is displayed below:

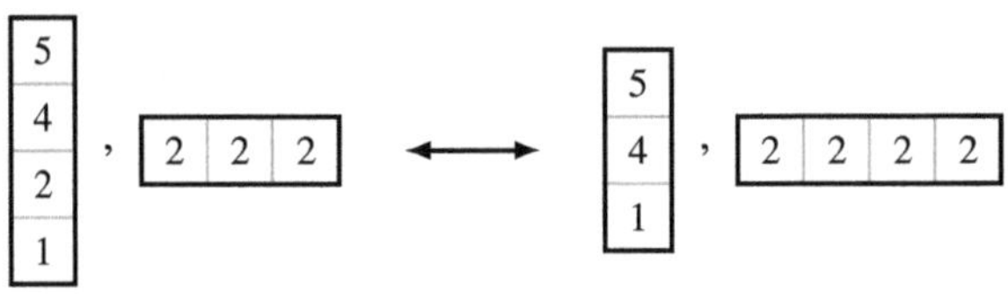

The fixed points (S,T) under this operation have sign $(-1)^{n-1}$, have only one cell in T, and the integer in that cell does not appear in S. If we place the single cell in T into S and shade it gray, these fixed points correspond to $(-1)^{n-1}ne_n$ since there are ne_n ways to form a column strict tableau of shape 1^n with one cell shaded gray. □

Corollary 2.10. *We have*

$$\sum_{n=1}^{\infty} p_n z^n = \frac{\sum_{n=1}^{\infty} (-1)^{n-1} ne_n z^n}{E(-z)}.$$

Proof. Multiplying both sides of this equation by $E(-z) = \sum_{n=0}^{\infty}(-1)^n e_n z^n$ and expanding, we find this statement:

$$\sum_{n=1}^{\infty}\left(\sum_{i=0}^{n-1}(-1)^i e_i p_{n-i}\right)z^n = \sum_{n=1}^{\infty}(-1)^{n-1} n e_n z^n.$$

This follows immediately by replacing the inner summand on the left-hand side with $(-1)^{n-1}e_n$ as allowed by Theorem 2.9. □

Theorem 2.11. *For $n \geq 1$, $\sum_{\lambda\vdash n} n!p_\lambda/z_\lambda = n!h_n$.*

Proof. Write a permutation $\sigma \in S_n$ above the cells of a standard tableau of shape n. The weighted sum over all possible objects is equal to $n!h_n$.

Suppose that the largest integer inside a cell in an object T is i. Locate the largest integer in σ atop an i in T, say σ_j. Cut the σ_j cell and all cells to the right off of T, creating two objects. Repeat this procedure on the remaining portion of T until there are no more cuts to be made. For example, if the object T is shown below,

8	1	5	2	6	10	11	4	12	3	7	9
1	1	1	1	2	2	2	3	3	3	3	3

then we would change T into

8	1	5	2		6		10		11		4		12	3	7	9
1	1	1	1		2		2		2		3		3	3	3	3

Even if many components which result from these cuts were rearranged, the process could be reversed in order to reconstruct T.

The integers on the top of the object created by cutting T can be considered a permutation of n written in cyclic notation. If the cycle type of this permutation is the integer partition $\lambda = (\lambda_1, \ldots, \lambda_\ell)$, then each part λ_i corresponds to a column strict tableau of shape (λ_i) where every integer is the same. Since Theorem 1.10 gives that the number of permutations with cycle type λ is $n!/z_\lambda$, these objects are counted by $\sum_{\lambda\vdash n} n!p_\lambda/z_\lambda$. □

Theorem 2.12. *For $n \geq 1$, $\sum_{\lambda\vdash n}(-1)^{n-\ell(\lambda)}n!p_\lambda/z_\lambda = n!e_n$.*

Proof. Take a permutation $\sigma \in S_n$ written in cyclic notation and, underneath each cycle of length λ_i, write a column strict tableau of shape (λ_i) where each cell contains the same integer. If σ has cycle type λ, then we define the sign of such an object to be $(-1)^{n-\ell(\lambda)}$. The signed, weighed sum over all such objects is equal to the left-hand side of the equation.

Momentarily ignoring the sign of an object T, apply the inverse to the bijection found in the proof of Theorem 2.11. We will now define a sign reversing involution in order to cancel any terms with a sign of -1.

If no two integers appearing in the cells of T are the same, define T as a fixed point. Otherwise, scan the cells of T from left to right looking for the first occurrence of two consecutive cells containing the same integer, say i. When this happens, find the largest two integers in the permutation σ which appear above an i and switch them. As an example, our involution pairs these objects:

$$\begin{array}{ccccccccc} 5 & 1 & 2 & 3 & 6 & 4 & 8 & 9 & 7 \\ \hline \multicolumn{1}{|c|}{1} & \multicolumn{1}{c|}{1} & \multicolumn{1}{c|}{1} & \multicolumn{1}{c|}{2} & \multicolumn{1}{c|}{2} & \multicolumn{1}{c|}{4} & \multicolumn{1}{c|}{4} & \multicolumn{1}{c|}{4} & \multicolumn{1}{c|}{5} \\ \hline \end{array} \longleftrightarrow \begin{array}{ccccccccc} 2 & 1 & 5 & 3 & 6 & 4 & 8 & 9 & 7 \\ \hline \multicolumn{1}{|c|}{1} & \multicolumn{1}{c|}{1} & \multicolumn{1}{c|}{1} & \multicolumn{1}{c|}{2} & \multicolumn{1}{c|}{2} & \multicolumn{1}{c|}{4} & \multicolumn{1}{c|}{4} & \multicolumn{1}{c|}{4} & \multicolumn{1}{c|}{5} \\ \hline \end{array}$$

This is a sign reversing involution because we have introduced exactly one transposition into the permutation σ. Fixed points correspond to a permutation atop a column strict tableau of shape (n) where no two cells contain the same integer. These fixed points, which have sign $(-1)^{n-n} = 1$, naturally correspond to $n!e_n$. □

We end this section by showing that some of these relationships between symmetric functions can be rephrased in terms of matrix determinants.

Theorem 2.13. *For all $n \geq 1$,*

$$e_n = \begin{vmatrix} h_1 & h_2 & h_3 & \cdots & h_n \\ 1 & h_1 & h_2 & \cdots & h_{n-1} \\ 0 & 1 & h_1 & \cdots & h_{n-2} \\ \vdots & \vdots & \vdots & & \vdots \\ 0 & 0 & 0 & \cdots & h_1 \end{vmatrix}.$$

Proof. The assertion is true when $n = 1$ because $e_1 = h_1$. We proceed by induction.

Removing the ith row and last column of the $n \times n$ determinant leaves a determinant of the form

$$\begin{vmatrix} A & B \\ 0 & C \end{vmatrix}.$$

where A is an $(i-1) \times (i-1)$ matrix of the same form as the original $n \times n$ matrix, B is an $(i-1) \times (n-i)$ matrix, 0 is the $(i-1) \times (n-i)$ zero matrix, and C is an $(n-i) \times (n-i)$ upper triangular matrix with 1s along the diagonal. By the induction hypothesis, the determinant of this matrix is e_{i-1}.

Expanding the determinant of the original $n \times n$ matrix along the last column, we find

$$\sum_{i=0}^{n-1} (-1)^{n+i-1} h_{n-i} e_i = e_n + (-1)^{n-1} \sum_{i=0}^{n} (-1)^i e_i h_{n-i}$$

which, by theorem 2.5, is equal to e_n. □

Theorem 2.14. *For all $n \geq 1$,*

$$p_n = \begin{vmatrix} e_1 & 1 & 0 & \cdots & 0 \\ 2e_2 & e_1 & 1 & \cdots & 0 \\ 3e_3 & e_2 & e_1 & \cdots & 0 \\ \vdots & \vdots & \vdots & & \vdots \\ ne_n & e_{n-1} & e_{n-2} & \cdots & e_1 \end{vmatrix}.$$

Proof. The assertion is true when $n = 1$ because $p_1 = e_1$. We proceed by induction.

Removing the ith column and last row of the $n \times n$ determinant leaves a determinant of the form

$$\begin{vmatrix} A & 0 \\ B & C \end{vmatrix}.$$

where A is an $(i-1) \times (i-1)$ matrix of the same form as the original $n \times n$ matrix, 0 is the $(i-1) \times (n-i)$ zero matrix, B is an $(n-i) \times (i-1)$ matrix, and C is an $(n-i) \times (n-i)$ lower triangular matrix with 1s along the diagonal. By the induction hypothesis, the determinant of this matrix is p_{i-1}.

Expanding the determinant of the original $n \times n$ matrix along the last row, we find

$$(-1)^{n-1} n e_n p_0 - \sum_{i=1}^{n-1} (-1)^i e_i p_{n-i} = (-1)^{n-1} n e_n - \sum_{i=0}^{n-1} (-1)^i e_i p_{n-i} + p_n$$

which, by Theorem 2.9, is equal to p_n. □

2.3 Transition Matrices

Let $\{a_\lambda : \lambda \vdash n\}$ and $\{b_\lambda : \lambda \vdash n\}$ be two bases for Λ_n. There is a $p(n) \times p(n)$ change of basis matrix A with entries indexed by partitions λ and μ such that

$$a_\mu = \sum_{\lambda \vdash n} A_{\lambda,\mu} b_\lambda, \tag{2.2}$$

where $A_{\lambda,\mu}$ is the λ,μ entry of A. The matrix A is the a-to-b transition matrix.

Let $\lambda^{(1)}, \ldots, \lambda^{(p(n))}$ be the integer partitions of n listed in the reverse lexicographic order, say, and take $f \in \Lambda_n$. Since $\{a_\lambda : \lambda \vdash n\}$ and $\{b_\lambda : \lambda \vdash n\}$ are bases, there are constants $c_{\lambda^{(1)}}, \ldots, c_{\lambda^{(p(n))}}$ and $d_{\lambda^{(1)}}, \ldots, d_{\lambda^{(p(n))}}$ such that

$$\begin{aligned} f &= c_{\lambda^{(1)}} a_{\lambda^{(1)}} + \cdots + c_{\lambda^{(p(n))}} a_{\lambda^{(p(n))}} \\ &= d_{\lambda^{(1)}} b_{\lambda^{(1)}} + \cdots + d_{\lambda^{(p(n))}} b_{\lambda^{(p(n))}}. \end{aligned}$$

Using standard matrix notation, equation (2.2) is equivalent to

$$\begin{bmatrix} A_{\lambda^{(1)},\lambda^{(1)}} & \cdots & A_{\lambda^{(1)},\lambda^{(p(n))}} \\ \vdots & \ddots & \vdots \\ A_{\lambda^{(p(n))},\lambda^{(1)}} & \cdots & A_{\lambda^{(p(n))},\lambda^{(p(n))}} \end{bmatrix} \begin{bmatrix} c_{\lambda^{(1)}} \\ \vdots \\ c_{\lambda^{(p(n))}} \end{bmatrix} = \begin{bmatrix} d_{\lambda^{(1)}} \\ \vdots \\ d_{\lambda^{(p(n))}} \end{bmatrix}.$$

Thus multiplying by the a-to-b transition matrix A allows us to take a symmetric function f expressed in terms of the a basis and write f in terms of the b basis.

This section is devoted to providing combinatorial interpretations for the entries of the transition matrices between various symmetric functions.

We do not have to use infinitely many variables to find such transition matrices. To see this, notice that for the monomial symmetric function $m_\lambda(x_1,x_2,\dots,x_N)$ to be nonzero, it must be the case that $N \geq n$—otherwise there may not be enough variables to create a monomial with the needed exponents. Thus $\{m_\lambda(x_1,\dots,x_N) : \lambda \vdash n\}$ is a basis for $\Lambda_n(x_1,\dots,x_N)$.

This means that the a-to-m transition matrix A is the same in $\Lambda_n(x_1,\dots,x_N)$ as it is in $\Lambda_n(x_1,x_2,\dots)$. If the b-to-m transition matrix is B, then it follows that the a-to-b transition matrix is $B^{-1}A$. Since A and B are the same in $\Lambda_n(x_1,\dots,x_N)$ and $\Lambda_n(x_1,x_2,\dots)$, the a-to-b transition matrix is also the same in $\Lambda_n(x_1,\dots,x_N)$ and $\Lambda_n(x_1,x_2,\dots)$. Thus when we are studying the transition matrices between bases of $\Lambda_n(x_1,x_2,\dots)$, it is enough to only consider symmetric polynomials in N variables $x_1,x_2,\dots,x_N$ for some $N \geq n$.

The s-to-m Transition Matrix

If we let $K_{\lambda,\mu}$ equal the number of column strict tableau of shape λ and content μ, then definition of the Schur symmetric function says that the coefficient of m_λ in s_μ is $K_{\mu,\lambda}$. This coefficient is called a Kostka number.

The Kostka matrix is the square matrix indexed by integer partitions of n written in reverse lexicographic order with λ,μ entry equal to $K_{\mu,\lambda}$. For example, the Kostka matrix with rows indexed by λ (the content) and columns indexed by μ (the shape) when $n = 4$ is

$$\begin{array}{c} \\ (4) \\ (3,1) \\ (2^2) \\ (2,1^2) \\ (1^4) \end{array}
\begin{array}{c} \begin{array}{ccccc} (4) & (3,1) & (2^2) & (2,1^2) & (1^4) \end{array} \\
\left[\begin{array}{ccccc}
1 & 0 & 0 & 0 & 0 \\
1 & 1 & 0 & 0 & 0 \\
1 & 1 & 1 & 0 & 0 \\
1 & 2 & 1 & 1 & 0 \\
1 & 3 & 2 & 3 & 1
\end{array}\right] \end{array}.$$

The $(2,1^2),(3,1)$ entry is 2 because there are two-column strict tableau of shape $(3,1)$ and type $(2,1^2)$:

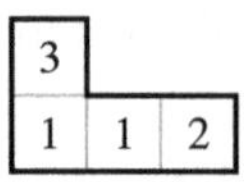

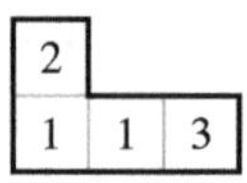

The Kostka matrix is the s-to-m transition matrix, or the change of basis matrix, which turns a linear combination of Schur functions into a linear combination of monomial symmetric functions by matrix multiplication. In other words, the coefficient of m_λ in $a_1 s_{(4)} + a_2 s_{(3,1)} + a_3 s_{(2^2)} + a_4 s_{(2,1^2)} + a_5 s_{(1^4)}$ can be found by performing the matrix multiplication

$$\begin{bmatrix} 1&0&0&0&0 \\ 1&1&0&0&0 \\ 1&1&1&0&0 \\ 1&2&1&1&0 \\ 1&3&2&3&1 \end{bmatrix} \begin{bmatrix} a_1 \\ a_2 \\ a_3 \\ a_4 \\ a_5 \end{bmatrix}.$$

Theorem 2.15. *The set $\{s_\lambda : \lambda \vdash n\}$ is a basis for Λ_n.*

Proof. If $\lambda < \mu$ in the reverse lexicographic order, then the first part in which λ and μ disagree is larger in λ than in μ. In this case there are no column strict Young tableau of shape μ and type λ. Further, $K_{\lambda,\lambda} = 1$ for all $\lambda \vdash n$. This tells us that the Kostka matrix is invertible because it is lower triangular with ones along the diagonal. Since $\{m_\lambda : \lambda \vdash n\}$ is a basis for Λ_n, so is $\{s_\lambda : \lambda \vdash n\}$. □

The e-to-m Transition Matrix

Given integer partitions $\lambda = (\lambda_1, \ldots, \lambda_\ell)$ and $\mu = (\mu_1, \ldots, \mu_k)$, let $\mathbf{Z}_2 M_{\lambda,\mu}$ be the number of $\ell \times k$ matrices with entries either 0 or 1 such that the sum of the ith row is λ_i and the sum of the jth column is μ_j. For example, if $\lambda = (3,2,1)$ and $\mu = (2,2,2)$, then one possible matrix is

$$\begin{bmatrix} 1&1&1 \\ 0&1&1 \\ 1&0&0 \end{bmatrix}$$

because the row sums are λ and the column sums are μ.

Theorem 2.16. *The coefficient of m_λ in e_μ is $\mathbf{Z}_2 M_{\lambda,\mu}$.*

Proof. Given $\lambda \vdash n$, we will count the number of ways we can form the monomial $x_1^{\lambda_1} \cdots x_k^{\lambda_k}$ by multiplying out $e_\mu = e_{\mu_1} \cdots e_{\mu_\ell}$ by organizing our work into a table where rows are indexed by $x_1, \ldots, x_k$ and columns are indexed by $e_{\mu_1}, \ldots, e_{\mu_\ell}$. Place a 1 in the x_i row and e_{μ_j} column entry of the table if the monomial selected from e_{μ_j} to contribute to a final product contains x_i and place a 0 in the table otherwise.

For example, when $\lambda = (3^2, 2, 1^2)$ and $\mu = (3^2, 2^2)$, one possible table is

$$\begin{array}{c} \\ x_1 \\ x_2 \\ x_3 \\ x_4 \\ x_5 \end{array} \begin{array}{c} \begin{array}{cccc} e_3 & e_3 & e_2 & e_2 \end{array} \\ \begin{bmatrix} 1&1&0&1 \\ 1&1&1&0 \\ 0&1&0&1 \\ 0&0&1&0 \\ 1&0&0&0 \end{bmatrix} \end{array}.$$

This table corresponds to the terms in each parenthesis in

$$e_{(3^2,2^2)}(x_1,x_2,\ldots) = (x_1x_2x_3 + x_1x_2x_4 + \cdots)^2(x_1x_2 + x_1x_3 + \cdots)^2$$

which are selected to form the monomial $x_1^3x_2^3x_3^2x_4^1x_5^1$.

The number of ways to form such a table is the coefficient of m_λ in e_μ. Each table is an element in $\mathbf{Z}_2M_{\lambda,\mu}$ and so the theorem is proved. □

Theorem 2.16 gives a combinatorial interpretation for the entries in the e-to-m transition matrix. This matrix in the case $n = 4$ is shown below:

$$\begin{array}{c} \\ (4) \\ (3,1) \\ (2^2) \\ (2,1^2) \\ (1^4) \end{array} \begin{array}{c} \begin{array}{ccccc} (4) & (3,1) & (2^2) & (2,1^2) & (1^4) \end{array} \\ \begin{bmatrix} 0 & 0 & 0 & 0 & 1 \\ 0 & 0 & 0 & 1 & 4 \\ 0 & 0 & 1 & 2 & 6 \\ 0 & 1 & 2 & 5 & 12 \\ 1 & 4 & 6 & 12 & 24 \end{bmatrix} \end{array}.$$

This is a symmetric matrix because the number of matrices with row sum λ and column sum μ is the same as the number of matrices with column sum λ and row sum μ by transposition.

The next theorem is known as the fundamental theorem of symmetric functions.

Theorem 2.17. *The set* $\{e_\lambda : \lambda \vdash n\}$ *is a basis for* Λ_n.

Proof. The only possible 0-1 matrix with row sum λ and column sum λ' is the matrix with the upside-down Young diagram of λ displayed in 1s in the matrix. For instance, the only 0-1 matrix with row sum $(4,2,1)$ and column sum $(3,2,1,1)$ is

$$\begin{bmatrix} 1 & 1 & 1 & 1 \\ 1 & 1 & 0 & 0 \\ 1 & 0 & 0 & 0 \end{bmatrix}$$

The same argument will show that if $\lambda' < \mu'$, then there are no possible 0-1 matrices with row sum μ and column sum λ' because there are not enough parts in μ to account for the first part of λ. Therefore a reordering of the rows and columns of the e-to-m transition matrix results in a triangular matrix with 1s along the diagonal. This transition matrix is therefore invertible, implying that $\{e_\lambda : \lambda \vdash n\}$ is a basis for Λ_n. □

The h-to-e and e-to-h transition matrices

Let $B_{\lambda,\mu}$ be the set of all possible Young diagrams of μ where the rows of μ are partitioned into "bricks" of lengths giving the integer partition λ. The four $T \in B_{\lambda,\mu}$ when $\lambda = (4,2,2,1,1)$ and $\mu = (5,3,2)$ are here:

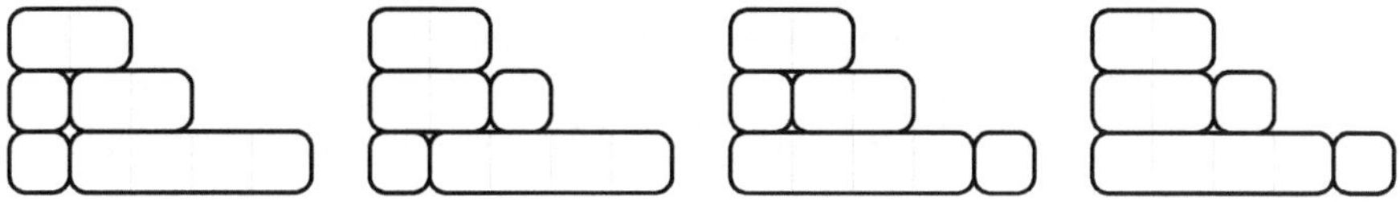

These elements in $B_{\lambda,\mu}$ are called brick tabloids of content λ and shape μ.

Theorem 2.18. *The coefficient of e_λ in h_μ is $(-1)^{n-\ell(\lambda)}\left|B_{\lambda,\mu}\right|$. In other words,*

$$h_\mu = \sum_{\lambda \vdash n} (-1)^{n-\ell(\lambda)} \left|B_{\lambda,\mu}\right| e_\lambda.$$

Proof. The right-hand side of this identity can be interpreted combinatorially. Use the summand and the $\left|B_{\lambda,\mu}\right|$ term to select a brick tabloid of content λ and shape μ for some $\lambda \vdash n$. Using the e_λ term, fill each brick with a decreasing sequence of distinct positive integers. Define the weight of such a brick tabloid to be the usual weight of a tableau. Finally, define the sign of such an object to be $(-1)^{n-\ell(\lambda)}$ (this power is the total number of cells in brick tabloid plus the number of bricks in the tabloid). The signed sum over all such combinatorial objects is equal to the right-hand side of the identity in the statement of the theorem.

For example, one such combinatorial object created in this way is shown below:

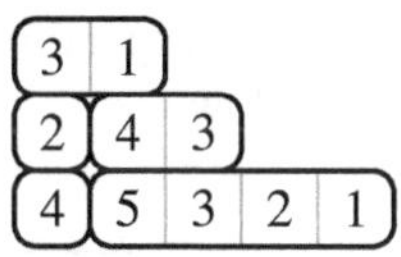

The weight of this object is $x_1^2x_2^2x_3^3x_4^2x_5$ and the sign is $(-1)^{10-5}$.

Let $\mathcal{B}$ be the set of combinatorial objects created in this way. We now define a sign reversing weight preserving involution φ on $\mathcal{B}$. Starting in the top row and scanning the bricks in $B \in \mathcal{B}$ from left to right, locate the first time if there is either a brick of length ≥ 2 or there is a brick of length 1 followed by another brick in the same row such that the integer labels between the two consecutive bricks decrease.

If there is a brick of length ≥ 2, then let $\varphi(B)$ be the object found by chopping the first cell off the brick of length ≥ 2, thereby creating two bricks. If there is a brick of length 1 followed by another brick in the same row such that the integer labels between the two consecutive bricks decrease, then let $\varphi(B)$ be the object found by combining the bricks. If neither situation is found after scanning all the rows of B, let $\varphi(B) = B$.

For example, the image of the combinatorial object shown above is here:

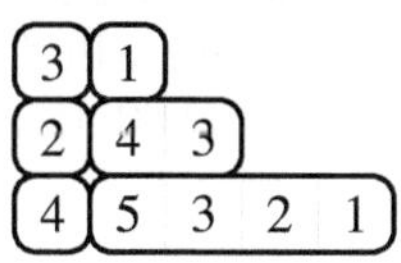

Fixed points under this involution must have every brick of length 1 and the integer labels within each row must weakly increase. The sign of such an object is $(-1)^{n-n} = 1$ and the weights give rise to exactly the homogeneous symmetric function h_μ. This proves the desired identity. □

Theorem 2.18 gives a combinatorial interpretation for the entries of the h-to-e transition matrix. This matrix in the case $n = 4$ is shown below; the entries are $(-1)^{n-\ell(\lambda)}|B_{\lambda,\mu}|$ with λ (the content) indexing the rows and μ (the shape) indexing the columns:

$$\begin{array}{c} \\ (4) \\ (3,1) \\ (2^2) \\ (2,1^2) \\ (1^4) \end{array} \begin{array}{c} \begin{array}{ccccc} (4) & (3,1) & (2^2) & (2,1^2) & (1^4) \end{array} \\ \left[\begin{array}{ccccc} -1 & 0 & 0 & 0 & 0 \\ 2 & 1 & 0 & 0 & 0 \\ 1 & 0 & 1 & 0 & 0 \\ -3 & -2 & -2 & -1 & 0 \\ 1 & 1 & 1 & 1 & 1 \end{array}\right] \end{array}.$$

Theorem 2.19. *The h-to-e transition matrix is its own inverse.*

Proof. Writing down the matrix multiplication explicitly, we wish to show that

$$\sum_{\alpha \vdash n} (-1)^{n-\ell(\lambda)+n-\ell(\alpha)} |B_{\lambda,\alpha}| |B_{\alpha,\mu}| = \begin{cases} 1 & \text{if } \lambda = \mu, \\ 0 & \text{if } \lambda \neq \mu. \end{cases} \tag{2.3}$$

Given an element $T_1 \in B_{\lambda,\alpha}$ and $T_2 \in B_{\alpha,\mu}$, form a "double brick tabloid" by placing the bricks in each row of T_1 into the corresponding brick in T_2. For example, if T_1 and T_2 are the brick tabloids shown below

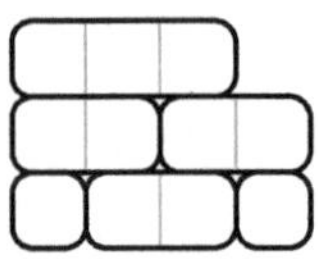

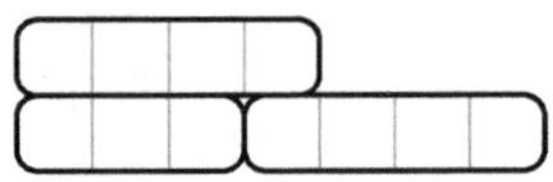

then we would combine T_1 and T_2 to create the double brick tabloid shown here:

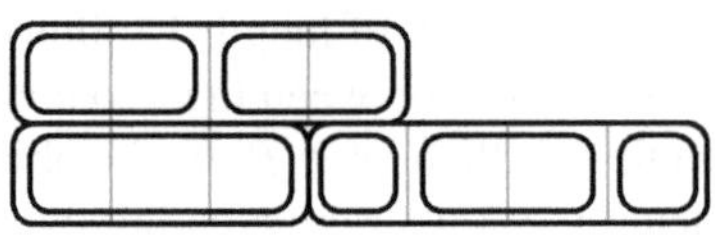

Given a double brick tabloid created from T_1 and T_2, call the bricks in the rows of T_1 "big bricks" and call the bricks inside the big bricks "little bricks." If we define the sign of a double brick tabloid to be $(-1)^{\text{the number of big and little bricks}}$, then the signed, weighted sum of all possible double brick tabloids of shape μ filled with little bricks of content λ is equal to the left-hand side of (2.3).

To find the right-hand side of (2.3), we will define a sign reversing involution. Scan the double brick tabloid from top to bottom and then from left to right looking for the first time there are either

1. two consecutive big bricks within a row, or
2. two little bricks inside of one big brick.

The double brick tabloid is a fixed point if there are no instances of either situation 1 or situation 2. Otherwise, if we encounter 1 first, then combine the two big bricks into one. If we encounter 2 first, then split the violating big brick b into two big bricks immediately after the first little brick in b. These are inverse operations which reverse the sign of the double brick tabloid.

For example, the double brick tabloid shown earlier in this proof would be changed to this double brick tabloid:

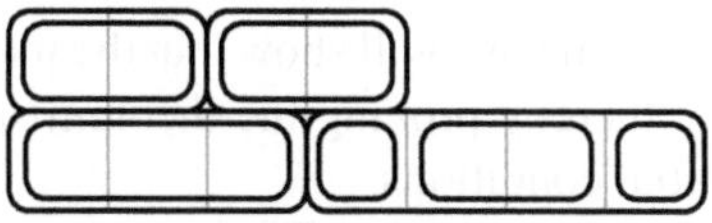

Each row in a fixed point must contain exactly one big brick containing exactly one little brick. Therefore only one fixed point of positive sign exists exactly when $\lambda = \mu$, which is the right-hand side of (2.3). □

Corollary 2.20. *The set* $\{h_\lambda : \lambda \vdash n\}$ *is a basis for* Λ_n.

Proof. The h-to-e transition matrix is invertible and the elementary symmetric functions are a basis, so the homogeneous symmetric functions are also a basis. □

Corollary 2.21. *The coefficient of* h_λ *in* e_μ *is* $(-1)^{n-\ell(\lambda)} |B_{\lambda,\mu}|$. *In other words,*

$$e_\mu = \sum_{\lambda \vdash n} (-1)^{n-\ell(\lambda)} |B_{\lambda,\mu}| h_\lambda.$$

The p-to-e and p-to-m Transition Matrices

Small modifications to brick tabloids can help us describe the elements in both the p-to-e and the p-to-m transition matrices. Define the weight of $T \in B_{\lambda,\mu}$, denoted $w(T)$, to be the product of the lengths of the bricks ending each row in T and let

$$w(B_{\lambda,\mu}) = \sum_{T \in B_{\lambda,\mu}} w(T).$$

For example, the weights of the four brick tabloids displayed after the proof of Theorem 2.17 on page 51 are 16, 8, 4, and 2, showing that $w(B_{(4,2^2,1^2),(5,3,2)}) = 30$. Theorem 2.22 below tells us that the λ,μ entry of the p-to-e transition matrix is equal to $(-1)^{n-\ell(\lambda)} w(B_{\lambda,\mu})$.

Theorem 2.22. *The coefficient of* e_λ *in* p_μ *is* $(-1)^{n-\ell(\lambda)} w(B_{\lambda,\mu})$.

Proof. Let $c_{\lambda,\mu}$ be the coefficient of e_λ in p_μ. We will show the following facts:

1. $c_{(n),(n)} = (-1)^{n-1} n$.
2. If $\lambda \vdash n$ has more than one part, then $c_{\lambda,(n)} = \sum_{i=1}^{n-1} (-1)^{i-1} c_{\lambda \setminus i,(n-i)}$ where $\lambda \setminus i$ denotes the integer partition λ with one part of size i removed with $c_{\lambda \setminus i,\mu} = 0$ if λ does not have a part of size i.

3. If $\alpha+\beta$ denotes the partition created by the multiset union of α and β where $\alpha \vdash \mu_1$ and $\beta \vdash n-\mu_1$, then
$$c_{\lambda,\mu} = \sum_{\alpha+\beta=\lambda} c_{\alpha,(\mu_i)} c_{\beta,\mu\backslash\mu_i}.$$

After proving these identities true, we will show that the integers $(-1)^{n-\ell(\lambda)}w(B_{\lambda,\mu})$ satisfy the same identities, thereby proving the theorem since both integers satisfy the same recursion and initial conditions.

Theorem 2.9 tells us that $(-1)^{n-1}ne_n = \sum_{i=0}^{n-1}(-1)^i e_i p_{n-i}$. Rewriting this,

$$\begin{aligned}
p_n &= (-1)^{n-1}ne_n + \sum_{i=1}^{n-1}(-1)^{i-1}e_i p_{n-i} \\
&= (-1)^{n-1}ne_n + \sum_{i=1}^{n-1}(-1)^{i-1}e_i \left(\sum_{\alpha\vdash n-i} c_{\alpha,(n-i)}e_\alpha\right) \\
&= (-1)^{n-1}ne_n + \sum_{\lambda\vdash n}\left(\sum_{i=1}^{n-1}(-1)^{i-1}c_{\lambda\backslash i,(n-i)}\right)e_\lambda
\end{aligned}$$

where in the last line we have combined the e_i and the e_α terms to create e_λ where λ is an integer partition with more than one part. Looking at the coefficients of e_n and e_λ in this expression verifies items 1 and 2. As for the third item,

$$\begin{aligned}
\sum_{\lambda\vdash n} c_{\lambda,\mu}e_\lambda &= p_\mu = p_{\mu_1}p_{\mu\backslash\mu_1} \\
&= \left(\sum_{\alpha\vdash n} c_{\alpha,(\mu_1)}e_\alpha\right)\left(\sum_{\beta\vdash n} c_{\beta,\mu\backslash\mu_1}e_\beta\right) \\
&= \sum_{\alpha+\beta=\lambda} c_{\alpha,(\mu_1)}c_{\beta,\mu\backslash\mu_1}e_\lambda.
\end{aligned}$$

Comparing coefficients of e_λ on the extremes gives item 3.

Now we show that $(-1)^{n-\ell(\lambda)}w(B_{\lambda,\mu})$ satisfies the same recursions. Item 1 follows since $(-1)^{n-\ell((n))}w(B_{(n),(n)}) = (-1)^{n-1}n$.

Item 2 follows by sorting the bricks appearing in the one row of (n) by the length of the first brick. Suppose that $\lambda \neq (n)$ and i is a part of λ. Then there are $w(B_{\lambda\backslash i,(n-i)})$ ways to create weighted brick tabloid of shape (n) after starting with a brick of length i. Therefore we have

$$\begin{aligned}
(-1)^{n-\ell(\lambda)}w(B_{\lambda,(n)}) &= (-1)^{n-\ell(\lambda)}\sum_{i=1}^{n-1}w(B_{\lambda\backslash i,(n-i)}) \\
&= \sum_{i=1}^{n-1}(-1)^{i-1}\left((-1)^{(n-i)-(\ell(\lambda)-1)}w(B_{\lambda\backslash i,(n-i)})\right),
\end{aligned}$$

which verifies item 2.

Item 3 follows by sorting the bricks according to the bricks appearing in the top row. The number of weighted brick tabloids with bricks in the lengths appearing in α in the top row is equal to $w(B_{\alpha,(\mu_1)})w(B_{\beta,\mu\backslash\mu_1})$ where β is the integer partition for which $\alpha+\beta=\lambda$. Therefore $(-1)^{n-\ell(\lambda)}w(B_{\lambda,\mu})$ is equal to

$$\sum_{\alpha+\beta=\lambda}\left((-1)^{\mu_1-\ell(\alpha)}w(B_{\alpha,(\mu_1)})\right)\left((-1)^{(n-\mu_1)-\ell(\beta)}w(B_{\beta,\mu\backslash\mu_1})\right),$$

which verifies item 3 and completes the proof. □

An ordered brick tabloid of content $\mu=(\mu_1,\ldots,\mu_\ell)$ and shape λ is a brick tabloid in $B_{\mu,\lambda}$ such that the bricks of length $\mu_1,\ldots,\mu_\ell$ are labeled with $1,\ldots,\ell$ such that brick labels decrease within rows. For example, all four possible ordered brick tabloids of content $(3,2^2,1^2)$ and shape $(4^2,1)$ are shown below:

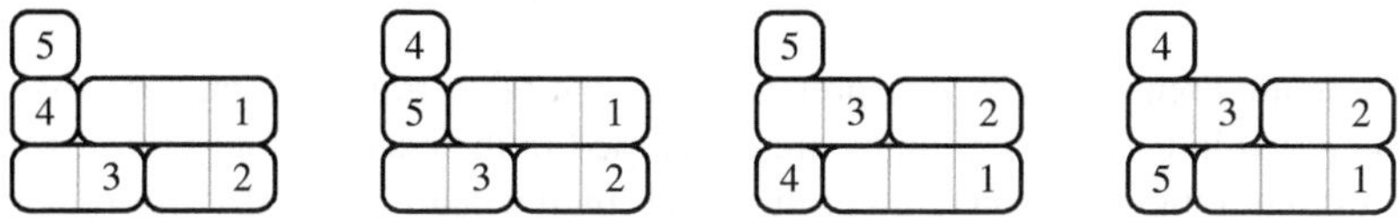

Let $OB_{\mu,\lambda}$ be the number of ordered brick tabloids of content μ and shape λ.

Theorem 2.23. *The coefficient of m_λ in p_μ is $OB_{\mu,\lambda}$.*

Proof. The number of ordered brick tabloids of content μ and shape λ corresponds directly to the number of times the monomial $x_1^{\lambda_1}\cdots x_k^{\lambda_k}$ appears in the expansion of the product

$$p_\mu=p_{\mu_1}\cdots p_{\mu_\ell}=\left(x_1^{\mu_1}+x_2^{\mu_1}+\cdots\right)\cdots\left(x_1^{\mu_\ell}+x_2^{\mu_\ell}+\cdots\right).$$

Specifically, if row λ_i in an ordered brick tabloid contains bricks labeled $\mu_{i_1},\ldots,\mu_{i_k}$, then this ordered brick tabloid corresponds to selecting the x_i term from each of $p_{\mu_1},\ldots,p_{\mu_{i_k}}$ to contribute to the final monomial. □

Theorem 2.23 tells us the coefficient in the p-to-m transition matrix is $OB_{\mu,\lambda}$. For clarity and for reference in section 6, we display this matrix in the case $n=4$ with rows indexed by λ (the shape) and columns indexed by μ (the content):

$$\begin{array}{c} \\ (4) \\ (3,1) \\ (2^2) \\ (2,1^2) \\ (1^4) \end{array}\begin{array}{c} \begin{array}{ccccc} (4) & (3,1) & (2^2) & (2,1^2) & (1^4) \end{array} \\ \begin{bmatrix} 1 & 1 & 1 & 1 & 1 \\ 0 & 1 & 0 & 2 & 4 \\ 0 & 0 & 2 & 2 & 6 \\ 0 & 0 & 0 & 2 & 12 \\ 0 & 0 & 0 & 0 & 24 \end{bmatrix} \end{array}.$$

Corollary 2.24. *The set $\{p_\lambda:\lambda\vdash n\}$ is a basis for Λ_n.*

Proof. The p-to-m transition matrix is invertible since it is upper triangular with nonzero diagonal entries. □

At this point we have a number of combinatorial descriptions for the entries of the transition matrices between standard bases of the ring of symmetric functions. We have recorded what we have done so far by labeling the edge on the directed graph below with the λ, μ entry of the corresponding transition matrix:

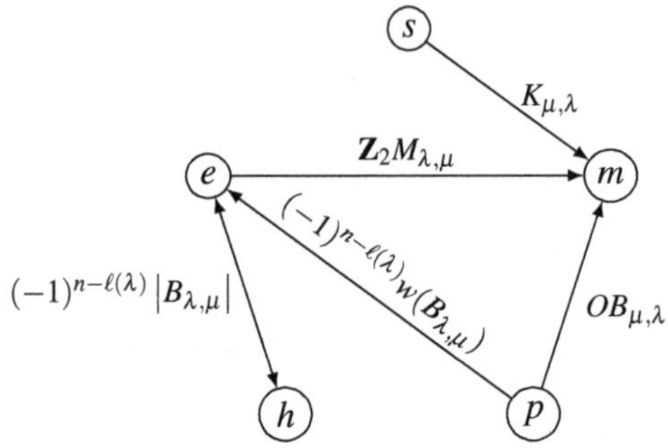

There are combinatorial interpretations for the other transition matrices we have not included in this section; many are developed in the exercises. For reference, we have drawn a more complete diagram which includes all of the transition matrices introduced in this text in Appendix A.

This graph is connected if edge directions are ignored, so we can now combine transition matrices by matrix inversion or multiplication to turn any one basis into another. For instance, to find the m-to-h transition matrix, we multiply the inverse of the e-to-m matrix and the e-to-h matrix; in the case of $n = 4$ this is

$$\begin{bmatrix} -1 & 0 & 0 & 0 & 0 \\ 2 & 1 & 0 & 0 & 0 \\ 1 & 0 & 1 & 0 & 0 \\ -3 & -2 & -2 & -1 & 0 \\ 1 & 1 & 1 & 1 & 1 \end{bmatrix} \begin{bmatrix} 0 & 0 & 0 & 0 & 1 \\ 0 & 0 & 0 & 1 & 4 \\ 0 & 0 & 1 & 2 & 6 \\ 0 & 1 & 2 & 5 & 12 \\ 1 & 4 & 6 & 12 & 24 \end{bmatrix}^{-1} = \begin{bmatrix} 4 & -4 & -2 & 4 & -1 \\ -4 & 7 & 2 & -7 & 2 \\ -2 & 2 & 3 & -4 & 1 \\ 4 & -7 & -4 & 10 & -3 \\ -1 & 2 & 1 & -3 & 1 \end{bmatrix}.$$

2.4 A Scalar Product

This section defines a scalar product on Λ_n. This scalar product has a relationship to some of the results in Chapter 5. Although not discussed in this book, the scalar product is also closely related to an inner product in the representation theory of the symmetric group, see [104] for more details on that connection.

We define a scalar product on Λ_n by declaring that $\{p_\lambda/\sqrt{z_\lambda} : \lambda \vdash n\}$ is an orthonormal basis. In other words, we define our scalar product so that

$$\left\langle \frac{p_\lambda}{\sqrt{z_\lambda}}, \frac{p_\mu}{\sqrt{z_\mu}} \right\rangle = \begin{cases} 1 & \text{if } \lambda = \mu, \\ 0 & \text{if } \lambda \neq \mu, \end{cases}$$

for all $\lambda, \mu \vdash n$ and then extend the definition by linearity.

We say that two bases $\{a_\lambda : \lambda \vdash n\}$ and $\{b_\lambda : \lambda \vdash n\}$ of Λ_n are dual bases if

$$\langle a_\lambda, b_\mu \rangle = \begin{cases} 1 & \text{if } \lambda = \mu, \\ 0 & \text{if } \lambda \neq \mu, \end{cases}$$

for all $\lambda, \mu \vdash n$. This means that the basis $\{p_\lambda/\sqrt{z_\lambda} : \lambda \vdash n\}$ is dual with itself. The next theorem provides a useful characterization of dual bases in Λ_n.

Theorem 2.25. *Bases $\{a_\lambda : \lambda \vdash n\}$ and $\{b_\lambda : \lambda \vdash n\}$ of Λ_n are dual if and only if*

$$\sum_{\lambda \vdash n} a_\lambda(X) b_\lambda(Y) = \sum_{\lambda \vdash n} \frac{p_\lambda(X) p_\lambda(Y)}{z_\lambda}$$

where $X = (x_1, x_2, \ldots)$ and $Y = (y_1, y_2, \ldots)$.

Proof. Let A be the a-to-$p_\lambda/\sqrt{z_\lambda}$ transition matrix and let B be the b-to-$p_\lambda/\sqrt{z_\lambda}$ transition matrix. This means

$$a_\lambda = \sum_{\alpha \vdash n} A_{\alpha,\lambda} \frac{p_\alpha}{\sqrt{z_\alpha}} \qquad \text{and} \qquad b_\mu = \sum_{\beta \vdash n} B_{\beta,\mu} \frac{p_\beta}{\sqrt{z_\beta}}.$$

Then we have

$$\begin{aligned} \langle a_\lambda, b_\mu \rangle &= \left\langle \sum_{\alpha \vdash n} A_{\alpha,\lambda} \frac{p_\alpha}{\sqrt{z_\alpha}}, \sum_{\beta \vdash n} B_{\beta,\mu} \frac{p_\beta}{\sqrt{z_\beta}} \right\rangle \\ &= \sum_{\alpha,\beta \vdash n} A_{\alpha,\lambda} B_{\beta,\mu} \left\langle \frac{p_\alpha}{\sqrt{z_\alpha}}, \frac{p_\beta}{\sqrt{z_\beta}} \right\rangle \\ &= \sum_{\alpha \vdash n} A_{\alpha,\lambda} B_{\alpha,\mu}. \end{aligned}$$

This last sum is the μ, λ entry in the matrix multiplication $B^{\mathrm{T}}A$. Therefore the bases $\{a_\lambda : \lambda \vdash n\}$ and $\{b_\lambda : \lambda \vdash n\}$ of Λ_n are dual if and only if $B^{\mathrm{T}}A = I$.

On the other hand, we have

$$\begin{aligned} \sum_{\lambda \vdash n} a_\lambda(X) b_\lambda(Y) &= \sum_{\lambda \vdash n} \left(\sum_{\alpha \vdash n} A_{\alpha,\lambda} \frac{p_\alpha(X)}{\sqrt{z_\alpha}} \right) \left(\sum_{\beta \vdash n} B_{\beta,\lambda} \frac{p_\beta(Y)}{\sqrt{z_\beta}} \right) \\ &= \sum_{\alpha,\beta,\lambda \vdash n} A_{\alpha,\lambda} B_{\beta,\lambda} \frac{p_\alpha(X)}{\sqrt{z_\alpha}} \frac{p_\beta(Y)}{\sqrt{z_\beta}} \end{aligned}$$

The coefficient of $p_\alpha(X) p_\beta(Y)/\sqrt{z_\alpha z_\beta}$ in this last line is

$$\sum_{\lambda \vdash n} A_{\alpha,\lambda} B_{\beta,\lambda},$$

which is the β, α entry in the matrix multiplication BA^{T}. Therefore identity in the statement of the theorem is true if and only if $BA^{\mathrm{T}} = I$.

The theorem follows since $B^{\mathrm{T}}A = I$ if and only if $BA^{\mathrm{T}} = I$. □

The next theorem gives an alternative expression for the sums in Theorem 2.25.

Theorem 2.26. *We have* $\prod_{i,j\geq 1} \frac{1}{1-x_iy_j} = \sum_{\lambda} \frac{p_\lambda(X)p_\lambda(Y)}{z_\lambda}$.

Proof. Starting with the left-hand side of the identity, we have

$$\prod_{i,j\geq 1} \frac{1}{1-x_iy_j} = \exp\left(\ln \prod_{i,j\geq 1} \frac{1}{1-x_iy_j}\right) = \exp\left(\sum_{i,j\geq 1} \ln \frac{1}{1-x_iy_j}\right).$$

Using $\ln 1/(1-x) = \sum_{k\geq 1} x^k/k$ and $\exp x = \sum_{m\geq 0} x^m/m!$, the above expression is

$$\exp\left(\sum_{i,j,k\geq 1} \frac{x_i^k y_j^k}{k}\right) = \exp\left(\sum_{k\geq 1} \frac{p_k(X)p_k(y)}{k}\right) = \sum_{m\geq 0}\left(\sum_{k\geq 1} \frac{p_k(X)p_k(Y)}{k}\right)^m \frac{1}{m!}.$$

Let $\cdot|_{2n}$ denote the degree $2n$ terms for $n \geq 1$ in a sum or product. Applying degree $2n$ extraction on both sides of our string of inequalities gives

$$\begin{aligned}\prod_{i,j\geq 1} \frac{1}{1-x_iy_j}\Bigg|_{2n} &= \sum_{m\geq 0}\left(\sum_{k\geq 1} \frac{p_k(X)p_k(Y)}{k}\right)^m \frac{1}{m!}\Bigg|_{2n}\\ &= \sum_{m=1}^{n}\left(\sum_{k=1}^{n} \frac{p_k(X)p_k(Y)}{k}\right)^m \frac{1}{m!}\Bigg|_{2n}\end{aligned}$$

where we are able to truncate the infinite sums since the tail end of the series cannot contribute to a degree $2n$ term. Using the multinomial theorem $(x_1+\cdots+x_n)^m = \sum_{a_1+\cdots+a_n=m}\binom{m}{a_1,\ldots,a_n} x_1^{a_1}\cdots x_n^{a_n}$, this expression is equal to

$$\sum_{m=1}^{n} \frac{1}{m!} \sum_{a_1+\cdots+a_n=m} \frac{m!}{a_1!\cdots a_n!} \prod_{k=1}^{n}\left(\frac{p_k(X)p_k(Y)}{k}\right)^{a_k}\Bigg|_{2n}.$$

The degree of the terms in $\prod_{i=1}^{n}(p_k(X)p_k(Y)/k)^{a_k}$ are $2(a_1+2a_2+\cdots+na_n)$. Furthermore, if $\lambda = (1^{a_1}\cdots n^{a_n})$ is a partition of n with m parts, then $a_1+\cdots+a_n = m$, $a_1+2a_2+\cdots+na_n = n$, and $p_\lambda(X)p_\lambda(Y)/z_\lambda = \prod_{k=1}^{n}(p_k(X)p_k(Y))^{a_k}/(k^{a_i}a_i!)$. We now have

$$\prod_{i,j\geq 1} \frac{1}{1-x_iy_j}\Bigg|_{2n} = \sum_{m=1}^{n} \sum_{\substack{a_1+\cdots+a_n=m\\ a_1+2a_2+\cdots+na_n=n}} \prod_{k=1}^{n} \frac{p_k(X)^{a_k}p_k(Y)^{a_k}}{k^{a_k}a_k!} = \sum_{\lambda\vdash n} \frac{p_\lambda(X)p_\lambda(Y)}{z_\lambda}.$$

The theorem follows by summing this identity over all nonnegative integers n. □

Theorem 2.27. *The homogeneous symmetric functions $\{h_\lambda : \lambda \vdash n\}$ and the monomial symmetric functions $\{m_\lambda : \lambda \vdash n\}$ are dual bases in Λ_n.*

Proof. The definition of the homogeneous symmetric function says

$$\prod_{i\geq 1}\frac{1}{1-x_iy_j}=\sum_{n\geq 0}h_n(X)y_j^n$$

for any $j\geq 1$, and so

$$\prod_{i,j\geq 1}\frac{1}{1-x_iy_j}=\prod_{j=1}^{\infty}\sum_{n\geq 0}h_n(X)y_j^n.$$

The left-hand side of this identity is symmetric in the variables $y_1,y_2,\ldots$. If we take the coefficient of $m_\lambda(Y)$ on the right-hand side of the equation, the coefficient is $h_\lambda(X)$. This proves

$$\prod_{i,j\geq 1}\frac{1}{1-x_iy_j}=\sum_{\lambda}h_\lambda(X)m_\lambda(Y),$$

which, by Theorems 2.25 and 2.26, is enough to prove the theorem. □

Theorem 5.6, found in Chapter 5, will provide a combinatorial proof that the Schur symmetric functions $\{s_\lambda:\lambda\vdash n\}$ are an orthonormal basis for Λ_n.

Theorem 2.28. *Let $\{a_\lambda:\lambda\vdash n\}$ and $\{b_\lambda:\lambda\vdash n\}$ be one pair of dual bases in Λ_n and let $\{a'_\lambda:\lambda\vdash n\}$ and $\{b'_\lambda:\lambda\vdash n\}$ be second pair of dual bases. If A is the a-to-a' transition matrix and B is the b-to-b' transition matrix, then $A=\left(B^{-1}\right)^{\mathrm{T}}$.*

Proof. Since A is the a-to-a' transition matrix, $a_\mu=\sum_{\lambda\vdash n}A_{\lambda,\mu}a'_\lambda$. The b'-to-b transition matrix is B^{-1}, and so $b'_\lambda=\sum_{\mu\vdash n}B^{-1}_{\mu,\lambda}b_\mu$ where $B^{-1}_{\mu,\lambda}$ is the μ,λ entry of B^{-1}. We now have

$$A_{\lambda,\mu}=\left\langle\sum_{\lambda\vdash n}A_{\lambda,\mu}a'_\lambda,b'_\lambda\right\rangle=\langle a_\mu,b'_\lambda\rangle=\left\langle a_\mu,\sum_{\mu\vdash n}B^{-1}_{\mu,\lambda}b_\mu\right\rangle=B^{-1}_{\mu,\lambda},$$

which is the same as $A=\left(B^{-1}\right)^{\mathrm{T}}$. □

At this point we know the dual basis for the monomial symmetric functions (the homogeneous), the homogeneous symmetric functions (the monomials), the power symmetric functions (the power symmetric functions, divided by a factor of z_λ), and the Schur symmetric functions (the Schur symmetric functions). But what is dual to the elementary symmetric functions? We define the forgotten symmetric functions $\{f_\lambda:\lambda\vdash n\}\subseteq\Lambda_n$ to be dual to the basis $\{e_\lambda:\lambda\vdash n\}$.

The forgotten symmetric functions can be found using Theorem 2.28. The homogeneous and the monomial symmetric functions are dual, and, by Theorem 2.18, the h-to-e transition matrix has μ,λ entry $(-1)^{n-\ell(\lambda)}\left|B_{\lambda,\mu}\right|$. By Theorem 2.28, the f-to-m transition matrix has μ,λ entry $(-1)^{n-\ell(\mu)}\left|B_{\mu,\lambda}\right|$. Put differently,

$$f_\mu=\sum_{\lambda\vdash n}(-1)^{n-\ell(\mu)}\left|B_{\mu,\lambda}\right|m_\lambda.$$

Thus the forgotten symmetric functions can be expanded into monomials by counting brick tabloids.

2.5 The ω Transformation

In this section we define a ring homomorphism ω on Λ. This function will expand our understanding of fundamental relationships between standard bases for Λ and will allow us to explain why the transition matrices between certain bases in Λ_n are the same as previously described transition matrices.

Since the elementary symmetric functions $\{e_\lambda : \lambda \vdash n\}$ are basis for Λ_n for all n, the functions $e_0, e_1, \ldots$ are algebraically independent and generate Λ. This means every element of Λ can be uniquely expressed as a polynomial in the functions $e_1, \ldots, e_N$ for some N.

This means we can define a ring homomorphism ω on Λ by defining $\omega(e_n)$ for each $n \geq 1$ and then extending ω by linearity. Defining various ring homomorphisms on Λ can reveal many combinatorial identities; this is one of our major themes.

For this section we will take ω to be the ring homomorphism defined by setting $\omega(e_n) = h_n$ for all $n \geq 1$. It follows that $\omega(e_\lambda) = h_\lambda$ for all $\lambda \vdash n$.

Theorem 2.29. *The function ω is an involution.*

Proof. Using Theorem 2.18 to expand h_n in terms of the elementary symmetric functions, we have

$$\begin{aligned}\omega(h_n) &= \sum_{\lambda \vdash n} (-1)^{n-\ell(\lambda)} |B_{\lambda,(n)}| \omega(e_\lambda) \\ &= \sum_{\lambda \vdash n} (-1)^{n-\ell(\lambda)} |B_{\lambda,(n)}| h_\lambda.\end{aligned}$$

Corollary 2.21 says this sum is equal to e_n. Therefore $\omega^2(e_n) = \omega(h_n) = e_n$, showing that ω is an involution. □

Theorem 2.30. *For all $n \geq 1$, $\omega(p_n) = (-1)^{n-1} p_n$.*

Proof. We show this by induction on n, with the case $n = 1$ being true since $p_1 = e_1 = h_1$ and thus $\omega(p_1) = \omega(e_1) = h_1 = (-1)^{1-1} p_1$.

Assume by induction that $\omega(p_k) = (-1)^{k-1} p_k$ for $k \leq n$. By Theorem 2.8,

$$p_n = nh_n - \sum_{i=1}^{n-1} h_i p_{n-i}.$$

Applying ω to both sides and using the induction hypothesis, we find

$$\omega(p_n) = ne_n - \sum_{i=1}^{n-1} (-1)^{n-i-1} e_i p_{n-i},$$

which, by Theorem 2.9, we can conclude is equal to $(-1)^{n-1} p_n$. □

Theorem 2.30 implies that $\omega(p_\lambda) = (-1)^{n-\ell(\lambda)} p_\lambda$ for all $\lambda \vdash n$.

Theorem 2.31. *For any symmetric functions $f,g \in \Lambda_n$, $\langle \omega(f), \omega(g) \rangle = \langle f, g \rangle$.*

Proof. For any $\lambda, \mu \vdash n$,

$$\begin{aligned}\left\langle \omega\left(\frac{p_\lambda}{\sqrt{z_\lambda}}\right), \omega\left(\frac{p_\mu}{\sqrt{z_\mu}}\right)\right\rangle &= \left\langle (-1)^{n-\ell(\lambda)}\frac{p_\lambda}{\sqrt{z_\lambda}}, (-1)^{n-\ell(\mu)}\frac{p_\mu}{\sqrt{z_\mu}}\right\rangle \\ &= \begin{cases} 1 & \text{if } \lambda = \mu, \\ 0 & \text{if } \lambda \neq \mu, \end{cases} \\ &= \left\langle \frac{p_\lambda}{\sqrt{z_\lambda}}, \frac{p_\mu}{\sqrt{z_\mu}}\right\rangle.\end{aligned}$$

It is enough to prove the theorem true for a basis, like we just did for the basis $\{p_\lambda/\sqrt{z_\lambda} : \lambda \vdash n\}$. □

Theorem 2.31 allows us to find the image of the monomial symmetric functions under the ring homomorphism ω. For any $\lambda, \mu \vdash n$, we have

$$\langle e_\lambda, \omega(m_\mu)\rangle = \langle \omega(e_\lambda), m_\mu\rangle = \langle h_\lambda, m_\mu\rangle = \begin{cases} 1 & \text{if } \lambda = \mu, \\ 0 & \text{if } \lambda \neq \mu. \end{cases}$$

This says that the bases $\{e_\lambda : \lambda \vdash n\}$ and $\{\omega(m_\lambda) : \lambda \vdash n\}$ are dual. Since the forgotten symmetric functions are the functions which are dual to the elementary symmetric functions, it must be the case that $\omega(m_\lambda) = f_\lambda$.

At this point we know the values of ω on the elementary, homogeneous, power, monomial, and forgotten bases. What about the Schur symmetric functions? We will use Theorem 2.32 to prove that $\omega(s_\lambda) = s_{\lambda'}$ where λ' is the conjugate partition to λ. The identities in Theorem 2.32 are known as the Jacobi–Trudi identities and are of interest in their own right. The proof we have chosen to include is due to Ira Gessel and Xavier Viennot.

Theorem 2.32. *Let $\lambda \vdash n$ be an integer partition with ℓ parts $\lambda_1 \leq \lambda_2 \leq \cdots \leq \lambda_\ell$ written in nondecreasing order. Then*

$$s_\lambda = \det\left(h_{\lambda_i+i-j}\right)_{i,j=1,\ldots,\ell} \qquad \textit{and} \qquad s_{\lambda'} = \det\left(e_{\lambda_i+i-j}\right)_{i,j=1,\ldots,\ell}$$

where we set $h_k = 0$ and $e_k = 0$ if $k < 0$.

Proof. We first prove the identity involving the homogeneous symmetric functions.

Each homogeneous symmetric function h_{λ_i+i-j} is the weighted sum over all column strict tableaux of shape $(\lambda_i + i - j)$. By interpreting the integers appearing in the column strict tableaux as the x-coordinates of the north steps, each choice of such a column strict tableaux corresponds to a weighted path p in the plane which starts at $(1, j)$, makes unit steps either north or east, and ends in an infinite number of east steps at height $\lambda_i + i$.

For example, suppose that $\lambda = (4,4,4,4)$, $i = 3$, and $j = 1$. Looking at $h_{\lambda_i+i-j} = h_{4+3-1} = h_6$, the column strict tableau on the left is interpreted as the lattice path on the right in the figure below:

5	7	7	10	10	10

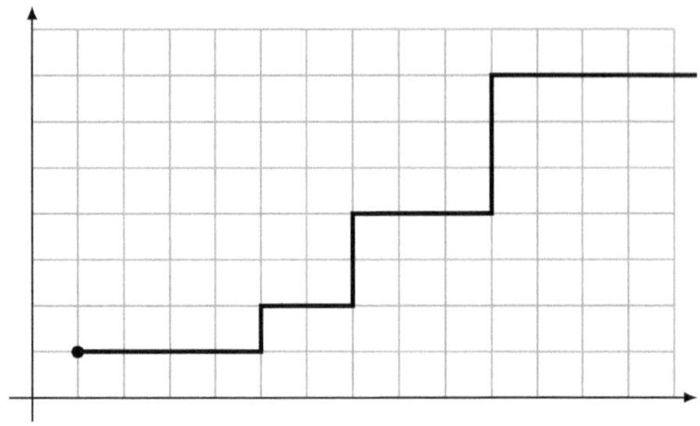

The column strict tableau on the left corresponds to the path on the right because the path has north steps at x-coordinates $5,7,7,10,10$, and 10. From such a path it is easy to find i (since the maximum height is $\lambda_i + i$) and j (since the starting point is at height j).

Let $\mathcal{P}_{j,\lambda_i+i}$ be the set of such lattice paths which begin at $(1, j)$ and end with an infinite sequence of east steps at height $\lambda_i + i$. If we define the weight of $p \in \mathcal{P}_{j,\lambda_i+i}$ to be the weight of the corresponding column strict tableau, then h_{λ_i+i-j} is the weighted sum over all $p \in \mathcal{P}_{j,\lambda_i+i}$.

Expanding the determinant as a signed sum over permutations $\sigma \in S_n$, we have

$$\det\left(h_{\lambda_i+i-j}\right)_{i,j=1,\ldots,\ell} = \sum_{\sigma\in S_n} \operatorname{sign}(\sigma) h_{\lambda_1+1-\sigma(1)} \cdots h_{\lambda_\ell+\ell-\sigma(\ell)}.$$

The terms in this sum can be considered collections of paths $(p_1,\ldots,p_\ell)$ where $p_i \in \mathcal{P}_{\sigma(i),\lambda_i+i}$ for $i = 1,\ldots,\ell$. The ordered ℓ-tuple of lattice paths $(p_1,\ldots,p_\ell)$ will be called a lattice path family. For example, if $\lambda = (4,4,4,4)$ and $\sigma = 3\ 2\ 1\ 4$, one such lattice path family is represented below:

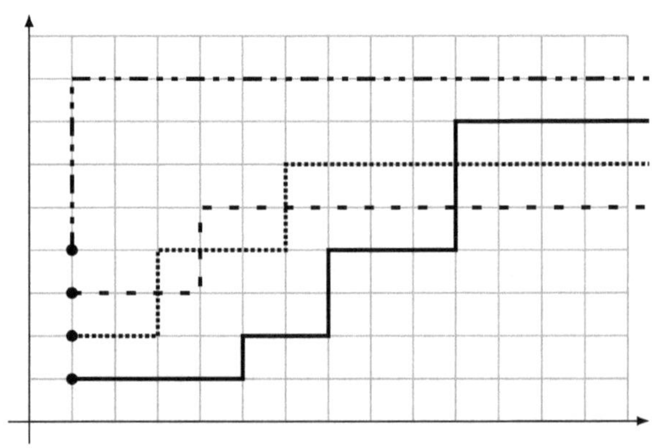

If we define the weight of the lattice path family $(p_1,\ldots,p_\ell)$ to be the product of the weights of the paths $p_1,\ldots,p_\ell$ and if we define the sign of the family to be the sign of the underlying permutation σ (which can be deduced from the lattice path family since the integer $\sigma(i)$ can be found for each i), then by construction, $\det\left(h_{\lambda_i+i-j}\right)_{i,j=1,\ldots,\ell}$ is the weighted, signed sum over all lattice path families.

Because we ordered the parts of λ in nondecreasing order, the maximum heights of the paths in a lattice path family $(p_1,\ldots,p_\ell)$, namely $\lambda_1+1,\ldots\lambda_\ell+\ell$, are distinct. Moreover, the ith highest path reading bottom to top on the right side of a lattice path family is the path p_i. This path must appear as the $\sigma(i)$th path reading bottom to top on the left. This means that the permutation σ can be found easily: the ith highest path on the right ends up as the $\sigma(i)$th highest path on the right.

To prove the identity involving the homogeneous symmetric functions in the statement of the theorem, we will describe a weight preserving, sign reversing involution on lattice path families which will leave fixed points corresponding to column strict tableau of shape λ.

The involution φ is as follows. If there is no place in the lattice path family $(p_1,\ldots,p_\ell)$ where two paths intersect, define the lattice path family to be fixed under the involution φ. Otherwise, find the most south and then most west coordinate where two paths intersect. Exactly two paths must intersect here, for if three lattice paths intersect at the same point, then two of these paths must have intersected at a more south or more west coordinate, contradicting our choice of intersection. Furthermore, by our choice of intersection, the paths involved must begin at consecutive coordinates.

Suppose this intersection involves the paths p_i and p_{i+1}. Define φ to be the lattice path family found by switching the tail ends of p_i and p_{i+1} after this point of intersection and leaving all other paths alone.

For example, considering the lattice path family displayed earlier in the proof, the first point of intersection is at the $(3,-3)$ coordinate and involves the second and the third paths. Applying the involution φ to this lattice path family gives the picture below:

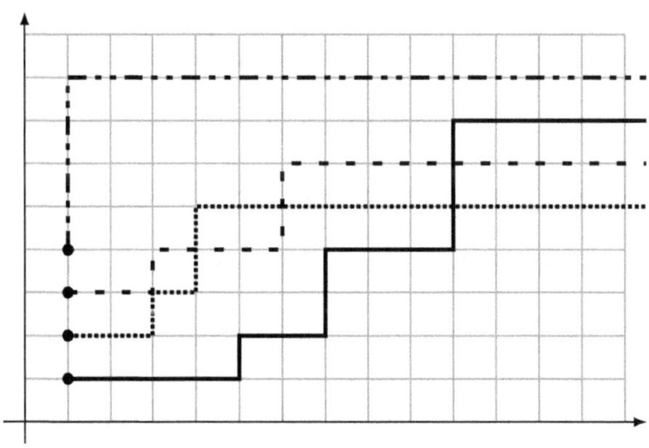

The function φ is weight preserving and is an involution because $(p_1,\ldots,p_\ell)$ and $\varphi(p_1,\ldots,p_\ell)$ have the same set of north steps and any coordinates of intersection remain unchanged; in particular, the most south and most west coordinate of intersection is preserved. Furthermore, since we have switched the ending positions of exactly two paths, the permutation σ is changed by one transposition, changing the sign of σ by -1.

Thus $\det\left(h_{\lambda_i+i-j}\right)_{i,j=1,\dots,\ell}$ is equal the weighted sum over all lattice path families where no two paths intersect. Since each path p_i lies below the path p_{i+1} for all i, the underlying permutation in such a nonintersecting lattice path family must be the identity permutation, which has sign $+1$.

Each nonintersecting lattice path family $(p_1,\dots,p_\ell)$ naturally corresponds to a column strict tableaux of shape λ. Starting from the top path and working downwards, fill the rows in a tableau of shape λ working bottom up with the x-coordinates of the north steps in each path. For example, below we display one nonintersecting lattice path family together with the corresponding tableau:

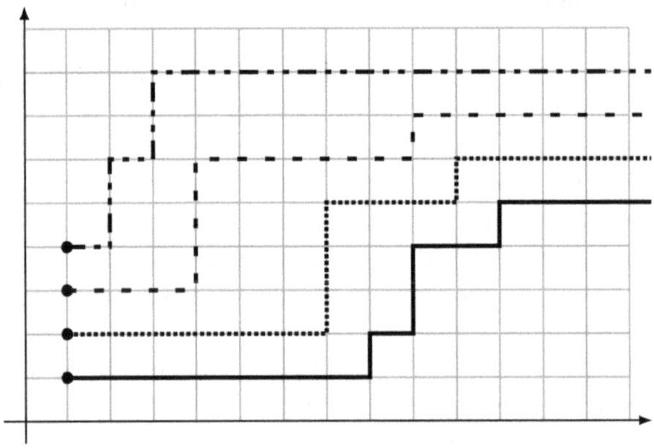

8	9	9	11
7	7	7	10
4	4	4	9
2	2	3	3

Since each path in the nonintersecting lattice path family moves north and east only, each row of the tableau is weakly increasing. Furthermore, by construction, the kth column in the tableau is strictly increasing since the kth north step in path p must appear higher than the kth north step in any path below p.

Since the Schur symmetric function is the weighted sum over all column strict tableaux of shape λ, at this point we have proved the identity

$$s_\lambda = \det\left(h_{\lambda_i+i-j}\right)_{i,j=1,\dots,\ell}.$$

The proof for the elementary symmetric function determinant is the same as that for the homogeneous symmetric functions with a few small modifications. We outline the main points but leave some of the finer details for the reader to verify.

The key difference is that each elementary symmetric function e_{λ_i+i-j} is the sum over column strict tableaux of shape $1^{(\lambda_i+i-j)}$, meaning that, unlike the homogeneous symmetric function, we cannot have repeated integers in a tableau. Thus the corresponding lattice paths cannot have two consecutive north steps—every north step must be immediately followed by an east step.

To adjust for this difference, we will associate each column strict tableau coming from e_{λ_i+i-j} with a lattice path p in the plane which starts at $(1,j)$, takes steps of the form $(1,0)$ or $(1,1)$, and ends in an infinite number of $(1,0)$ steps at height λ_i+i. We create the path p such that if the integer k appears in the column strict tableaux, then p has a diagonal step beginning at x-coordinate k.

Therefore the determinant

$$\det\left(e_{\lambda_i+i-j}\right)_{i,j=1,\ldots,\ell} = \sum_{\sigma\in S_n} \operatorname{sign}(\sigma) e_{\lambda_1+1-\sigma(1)} \cdots e_{\lambda_\ell+\ell-\sigma(\ell)}$$

can be interpreted as a signed, weighted sum over lattice path families with east and diagonal steps instead of east and north steps. For example, one such lattice path family when $\lambda = (2,2,4)$ is shown below:

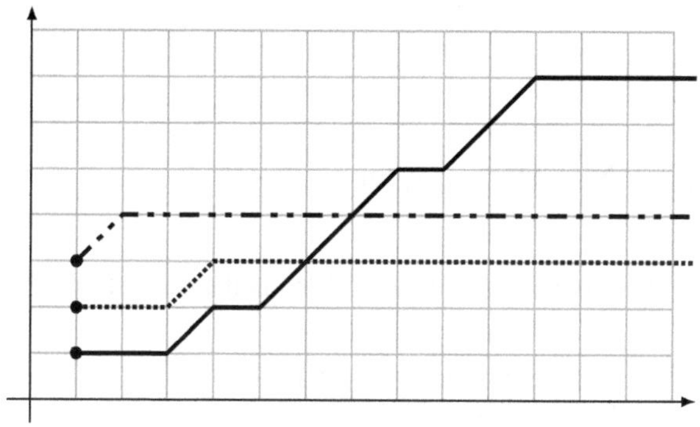

The weight of this lattice path family shown above is $x_1x_3^2x_5x_6x_7x_9x_{10}$ (since the diagonal steps begin at x-coordinates $1,3,3,5,6,7,9$, and 10) and the underlying permutation is $\sigma = 2\ 3\ 1$ with sign $+1$.

We can now apply the same involution as described for the homogeneous symmetric functions; find the most south and most west coordinate where two paths intersect and switch their tails.

The positions of the diagonal steps in a fixed point can be used to fill the rows of a tableau of shape λ in the same way as we did for the homogeneous symmetric functions. Fixed points naturally correspond to tableau with strictly increasing rows and weakly increasing columns. For example, the fixed point shows on the left corresponds to the tableau on the right:

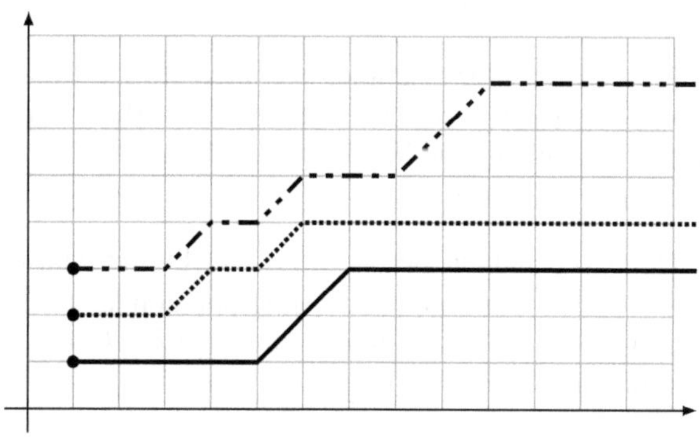

5	6		
3	5		
3	5	8	9

This proves

$$s_{\lambda'} = \det\left(e_{\lambda_i+i-j}\right)_{i,j=1,\ldots,\ell}$$

since conjugating these tableaux which correspond to fixed points gives the necessary column strict tableaux. □

Corollary 2.33. *For all $\lambda \vdash n$, $\omega(s_\lambda) = s_{\lambda'}$.*

Proof. Apply ω to the first identity in 2.32 to find the second identity. □

Exercises

2.1. Show that $s_{(1^k,n)} = \sum_{i=0}^{k} (-1)^{k-i} e_i h_{n+k-i}$.

2.2. Show that $p_r = \sum_{k=0}^{r-1} (-1)^k s_{(r-k,1^k)}$ for $r \geq 1$.

2.3. Show $\sum_{n=1}^{\infty} p_n z^n = zH'(z)/H(z)$ where $H'(z)$ is the derivative of $H(z) = \sum_{n=0}^{\infty} h_n z^n$.

2.4. Prove that the coefficient of h_λ in e_μ is $(-1)^{n-\ell(\lambda)}\left|B_{\lambda,\mu}\right|$ using an involution similar to that found in the proof of Theorem 2.18.

2.5. Show that for integer partitions $\lambda, \mu \vdash n$,

$$\sum_{\alpha \vdash n} (-1)^{\ell(\lambda)+\ell(\alpha)} OB_{\lambda,\alpha} w(B_{\alpha,\mu}) = \begin{cases} 0 & \text{if } \lambda \neq \mu, \\ z_\lambda & \text{if } \lambda = \mu. \end{cases}$$

2.6. Using Exercise 2.5, show that the coefficient of p_λ in e_μ is $(-1)^{n-\ell(\lambda)} OB_{\lambda,\mu}/z_\lambda$ and the coefficient of p_λ in m_μ is $(-1)^{\ell(\lambda)+\ell(\mu)} w(B_{\mu,\lambda})/z_\lambda$.

2.7. Define an alternating polynomial f in the variables $x_1,\ldots,x_n$ to be a polynomial with coefficients in $\mathbb{Q}$ such that

$$f(x_1,\ldots,x_n) = \operatorname{sign}(\sigma) f(x_{\sigma_1},\ldots,x_{\sigma_n})$$

for all permutations $\sigma = \sigma_1 \cdots \sigma_n \in S_n$. For example, one alternating polynomial in the variables x_1, x_2, and x_3 is $x_1x_2x_3(x_1-x_2)(x_1-x_3)(x_2-x_3)$. Show that any alternating polynomial must be divisible by the Vandermonde determinant $\Delta_{(0,\ldots,0)}$ and therefore must have minimum degree $\binom{n}{2}$.

2.8. Show that Δ_λ is an alternating polynomial (see Exercise 2.7). Further, show that if a monomial $x_1^{\lambda_1+n-1} \cdots x_n^{\lambda_n+n-n}$ is a term in an alternating polynomial f, then f must have all terms present in Δ_λ. These two facts imply that $\{\Delta_\lambda : \lambda \vdash k\}$ is a basis for the set of alternating polynomials of degree $k + \binom{n}{2}$.

2.9. Let RCS_λ denote the set of reverse column strict tableaux, that is, all tableaux where the integer labeling weakly decreases in rows and strictly decreases in columns. For example, here are all elements in $RCS_{(2,1)}$ that are filled with integers ≤ 3:

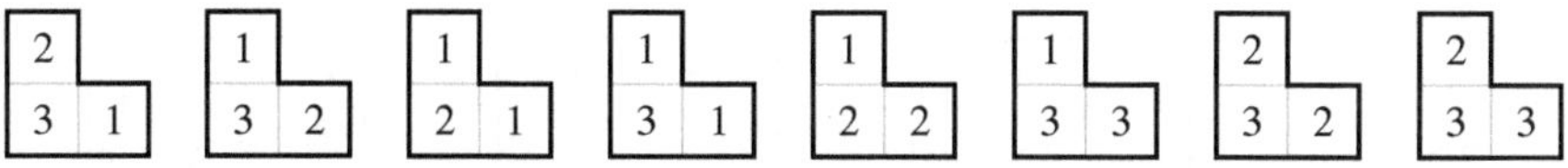

Show that $s_\lambda = \sum_{RCS_\lambda} w(T)$ for any $\lambda \vdash n$.

2.10. A labeling of the mathematical abacus a is a filling of the k beads in a with a permutation in S_k. Below we display a labeled abacus of length 10 with 5 beads filled with the permutation $3\ 4\ 2\ 1\ 5 \in S_5$:

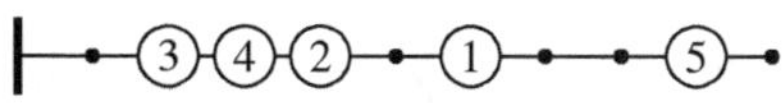

Let $b_1, \ldots, b_k$ be the beads in a labeled abacus a when read left to right, let $\text{label}(b_i)$ be the integer in bead b_i, and let $\text{position}(b_i)$ be the position of bead b_i. We define the weight of a to be $x_{\text{label}(b_1)}^{\text{position}(b_1)} \cdots x_{\text{label}(b_k)}^{\text{position}(b_k)}$ and we define the sign of a to be the sign of the permutation $\text{label}(b_k) \cdots \text{label}(b_1)$. For instance, the weight of the labeled abacus shown above is $x_3^2 x_4^3 x_2^4 x_1^6 x_5^9$ and the sign is $\text{sign}(5\ 1\ 2\ 4\ 3) = -1$.

Let λ be the integer partition corresponding to the abacus a. Show that

$$x_1 \cdots x_k \Delta_\lambda(x_1, \ldots, x_k) = \sum \text{sign}(\ell)\text{weight}(\ell)$$

where the sum runs over all possible labelings ℓ of the abacus a.

2.11. Let $\lambda \vdash n$. Using Exercise 2.10, show that $e_j \Delta_\alpha = \sum \Delta_\lambda$ where the sum runs over the integer partitions $\lambda \vdash (n+j)$ found by adding 1 to j distinct parts of α.

2.12. Using Exercise 2.11 to expand $e_\mu \Delta_{(0,\ldots,0)}$ into a sum of terms of the form Δ_λ, show that the λ, μ entry of the e-to-s transition matrix is equal to $K_{\lambda',\mu}$.

2.13. Let ν be a rim hook (see Exercise 1.3). The sign of ν, denoted $\text{sign}(\nu)$, is defined to be $(-1)^{\text{the number of rows spanned by } \nu - 1}$. For instance, the sign of the rim hook of length 6 pictured below is $(-1)^{3-1} = +1$:

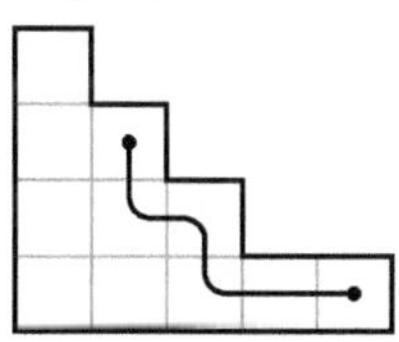

Using Exercise 2.10, show that

$$p_j \Delta_\alpha = \sum \text{sign}(\nu) \Delta_\lambda$$

where the sum runs over the integer partitions λ which can be found by adding a rim hook ν of length j to α.

2.14. A rim hook tableau of shape λ and content $\mu = (\mu_1, \ldots, \mu_\ell)$ is a filling of the cells of the Young diagram of λ with rim hooks of lengths $\mu_1, \ldots, \mu_\ell$ (see Exercise 1.3) labeled with $1, \ldots, \ell$ such that the removal of the last i rim hooks leaves the Young diagram of a smaller integer partition for all i. For example, below we display all possible rim hook tableau of shape $(5,4,4)$ and content $(5,5,1,1,1)$:

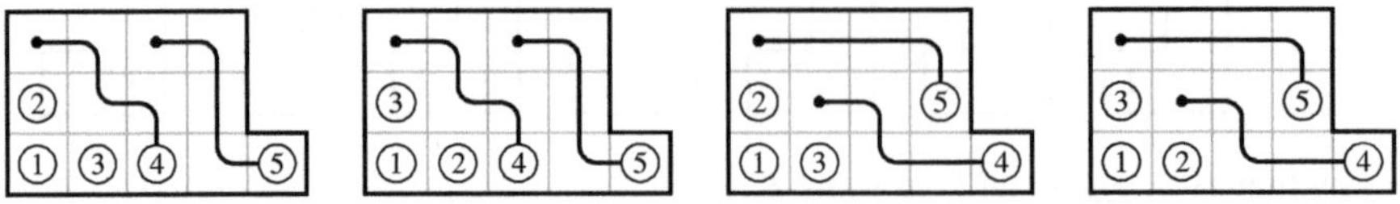

The sign of a rim hook tableau T is defined to be the product of the signs of the rim hooks in T (see Exercise 2.13). The four rim hook tableaux pictured above all have sign $+1$. We define

$$\chi^\lambda_\mu = \sum_{\substack{\text{rim hook tableaux } T \text{ with}\\ \text{shape } \lambda \text{ and content } \mu}} \operatorname{sign}(T).$$

Using Exercise 2.13 to expand $p_\mu \Delta_{(0,\ldots,0)}$ into a sum of terms of the form Δ_λ, show that the λ, μ entry of the p-to-s transition matrix is χ^λ_μ.

2.15. A special rim hook tabloid of shape λ and content μ is a rim hook tableau of shape λ and content μ (see Exercise 2.14) such that the labels on the rim hooks are erased and every rim hook contains at least one cell in the first column of the Young diagram of λ. The change of nomenclature from "tableau" to "tabloid" indicates that the rim hooks within a special rim hook tabloid are unordered.

For example, here are the only two possible special rim hook tabloids of content $(6,6,4,2)$ and shape $(5,5,4,3,1)$:

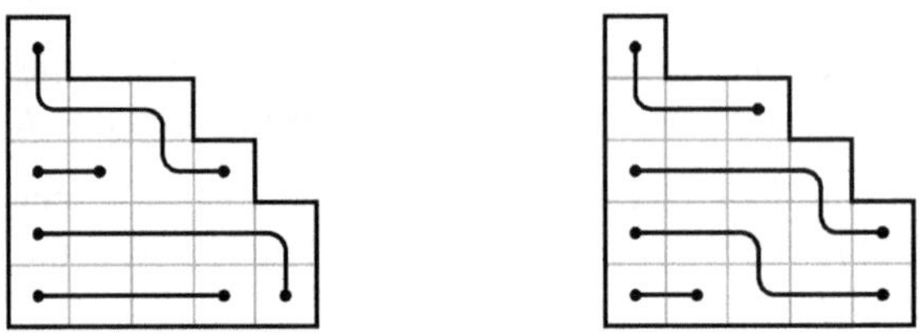

Let $K^{-1}_{\mu,\lambda}$ to be the integer defined by

$$K^{-1}_{\mu,\lambda} = \sum_{\substack{\text{special rim hook tabloids } T\\ \text{of shape } \lambda \text{ and content } \mu}} \operatorname{sign}(T)$$

where $\operatorname{sign}(T)$ is defined in Exercise 2.14. The goal of this exercise is to show that the inverse to the Kostka matrix has λ, μ entry equal $K^{-1}_{\mu,\lambda}$, that is, we want to show

$$\sum_{\alpha \vdash n} K_{\mu,\alpha} K^{-1}_{\alpha,\lambda} = \begin{cases} 1 & \text{if } \mu = \lambda, \\ 0 & \text{if } \mu \neq \lambda. \end{cases} \tag{2.4}$$

for all λ, μ. This can be done by considering pairs (C,S) where C is a column strict tableau and S is a special rim hook tabloid such that:

1. The special rim hook tabloid S of shape λ and content α is chosen first. Let a_i be the length of the special rim hook which begins in row i of S reading bottom to top. This number might be 0.
2. The column strict tableau C has shape μ and contains a_1 1s, a_2 2s, etc. By Theorem 2.2, the number of ways to form C is independent of this specification of number of 1's, 2's, etc.
3. The sign of (C,S) is equal to sign(S).

For example, one pair when $\mu = (7,6,5)$ and $\lambda = (5,5,4,3,1)$ is

3	3	5	5	5		
2	2	3	3	3	5	
1	1	2	2	2	2	3

,

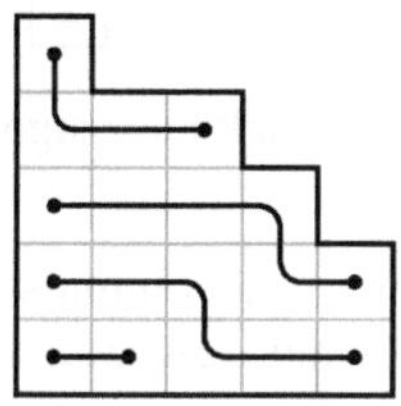

.

There is a unique way to switch the tail ends of two consecutive special rim hooks in S. Use this "tail-switching" idea to define a sign reversing involution on such pairs (C,S) in order to verify (2.4).

2.16. Given integer partitions $\lambda = (\lambda_1, \ldots, \lambda_\ell)$ and $\mu = (\mu_1, \ldots, \mu_k)$, let $NM_{\lambda,\mu}$ be the number of $\ell \times k$ matrices with nonnegative integer entries such that the sum of the ith row is λ_i and the sum of the jth column is μ_j. Show that the coefficient of m_λ in h_μ is $NM_{\lambda,\mu}$.

2.17. Show that

$$n!e_n = \begin{vmatrix} p_1 & 1 & 0 & \cdots & 0 & 0 \\ p_2 & p_1 & 2 & \cdots & 0 & 0 \\ \vdots & \vdots & \vdots & & \vdots & \vdots \\ p_{n-1} & p_{n-2} & p_{n-3} & \cdots & p_1 & n-1 \\ p_n & p_{n-1} & p_{n-2} & \cdots & p_2 & p_1 \end{vmatrix}.$$

Using the ω transformation, find a similar relationship involving the homogeneous and power symmetric functions.

Solutions

2.1 Looking at the right-hand side of the identity, we can consider ordered pairs (S,T) where S is a column strict tableau of shape 1^i for $0 \le i \le k$ and T is a column strict tableau of shape $(n+k-i)$. The sign is $(-1)^{k-i}$. We now apply a sign reversing involution similar to that found in the second proof of Theorem 2.5.

If the bottommost integer in S is not larger than the leftmost integer in T, then move this integer from S to T. Otherwise, if S has height smaller than k and if the bottommost integer in S is larger than the leftmost integer in T, then move this integer from T to S.

Fixed points under this sign reversing involution must have S with height k and the bottommost integer in S is larger than the leftmost integer in T, like the picture below if $k = 4$:

$$\begin{array}{|c|}\hline 6\\\hline 4\\\hline 3\\\hline 2\\\hline\end{array}\,,\;\begin{array}{|c|c|c|c|}\hline 1&1&4&4\\\hline\end{array}$$

These fixed points, which have sign $+1$, correspond to tableaux of shape $(1^k, n)$ by gluing S atop T.

2.2 Define the sign of a column strict tableau T of shape $(r-k, 1^k)$ to be $(-1)^k$. Consider the following sign reversing involution: locate the largest integer m appearing in T. If m appears in the first column of T and this first column has more than one cell, then move m to the right of the bottom row of T. If m appears in bottom row of T and m is larger than the largest cell in the first column of T, then move m to the first column.

Fixed points under this sign reversing involution must have the largest integer m appearing in both the bottom row of T and the first column of T. Furthermore, the first column of T must have only one cell. It follows that T must contain exactly one row, and that every cell in that row contains the same integer m. These fixed points correspond to p_r, as desired.

2.3 Using Theorem 2.8,

$$H(z)\sum_{n=1}^{\infty} p_n z^n = \sum_{n=1}^{\infty}\left(\sum_{i=0}^{n-1} h_i p_{n-i}\right) z^n = \sum_{n=1}^{\infty} n h_n z^n = z\sum_{n=0}^{\infty} n h_n z^{n-1} = zH'(z).$$

2.4 The proof is similar to the proof of Theorem 2.18 except that we allow weakly decreasing sequences in the bricks instead of strictly decreasing sequences. In particular, the desired identity is

$$e_\mu = \sum_{\lambda \vdash n} (-1)^{n-\ell(\lambda)} \left|B_{\lambda,\mu}\right| h_\lambda. \tag{2.5}$$

The right-hand side of (2.5) can be interpreted combinatorially. Use the summand and the $\left|B_{\lambda,\mu}\right|$ term to select a brick tabloid of content λ and shape μ for some $\lambda \vdash n$. Using the h_λ term, fill each brick with a weakly decreasing sequence of positive integers. Define the weight and sign in the same way as in the proof of theorem 2.18. The signed sum over all such combinatorial objects is equal to the right-hand side of (2.5).

Define an involution φ by starting in the top row and scanning the bricks from left to right, locating the first time there is either a brick of length ≥ 2 or there is a brick of length 1 followed by another brick in the same row such that the integer labels between the two consecutive bricks weakly decrease.

If there is a brick of length ≥ 2, change the object by chopping the first cell off the brick of length ≥ 2, thereby creating two bricks. If there is a brick of length 1 followed by another brick in the same row such that the integer labels between the two consecutive bricks weakly decrease, then change the object by combining the bricks. Do nothing if neither situation is found.

Fixed points must consist of only bricks of length 1 (and thus must have sign $+1$) and must have strictly increasing sequences of integers within each row, corresponding directly with e_μ, as desired.

2.5 Construct a set of combinatorial objects by following these steps:

1. Select an ordered brick tabloid of content λ and shape α for some $\alpha \vdash n$.
2. Select a brick tabloid of content α and shape μ. Select one cell in the last brick in each row and shade it gray. This shading accounts for the weight in a weighted brick tabloid.
3. Combine the brick tabloids selected in step 1 and step 2 by placing the bricks in each row of the ordered brick tabloid into the corresponding brick in the weighted brick tabloid.

For example, if the brick tabloids

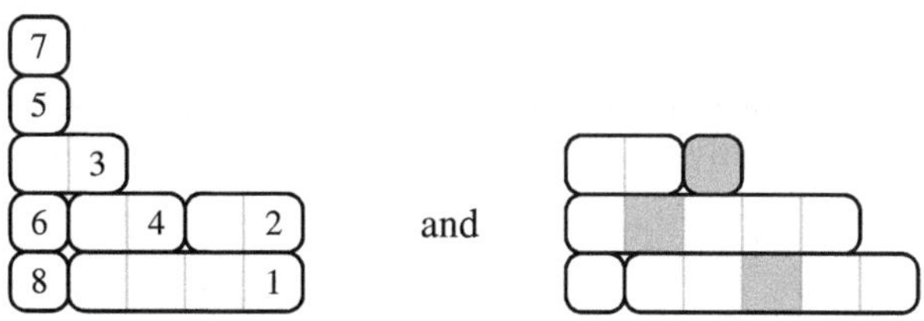

are selected in steps 1 and 2, then combining them in step 3 would create

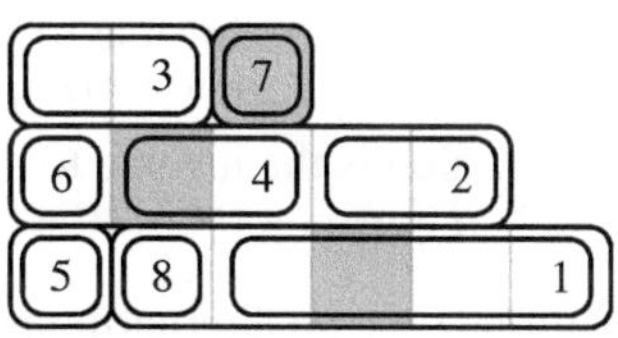

Let $\mathcal{T}$ be the set of all objects created by following steps 1, 2, and 3. Call a smaller brick appearing inside of another brick a "little brick" and the larger bricks "big bricks." Define the sign of $T \in \mathcal{T}$ to be $(-1)^{\text{the number of big and little bricks in } T}$. By construction, the signed, weighted sum of $T \in \mathcal{T}$ is the sum in the statement of this exercise.

Define a sign reversing involution by examining the last big brick in each row of T, starting from top to bottom, looking for either

1. a last big brick which contains more than one little brick or

2. a row with more than one big brick such that the last big brick contains only one little brick.

If case 1 is found, break the big brick into smaller big bricks by moving the little brick containing the shaded cell into its own big brick at the end of the row. If case 2 is found, combine the two big bricks into one big brick, sorting the little bricks so that the little brick labels decrease within the big brick. For example, the image of the $T \in \mathcal{T}$ displayed earlier is shown below:

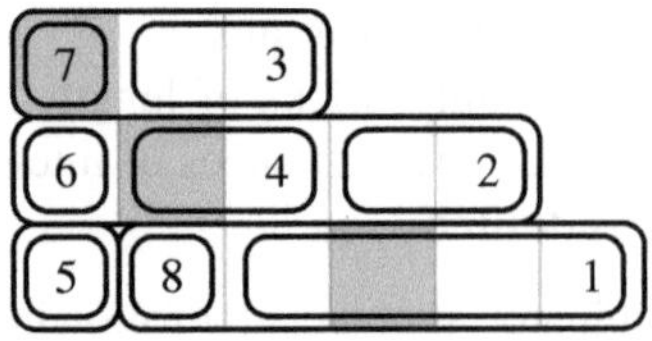

Fixed points must have exactly one big brick in each row, and that big brick must contain exactly one little brick. These fixed points, which all have sign $+1$, occur exactly when $\lambda = \mu$. If $\lambda = 1^{m_1}2^{m_2}\cdots$, then the total number of fixed points is $z_\lambda = 1^{m_1}2^{m_2}\cdots m_1!m_2!\cdots$ since $1^{m_1}2^{m_2}\cdots$ accounts for the placement of the shaded cell in each row and $m_1!m_2!\cdots$ accounts for the ways to rearrange the labels on little bricks of the same length.

2.6 Exercise 2.5 gives the λ, μ entry in the matrix multiplication

$$\|(-1)^{n-\ell(\lambda)}OB_{\lambda,\mu}\|_{\lambda,\mu\vdash n}\|(-1)^{n-\ell(\lambda)}w(B_{\lambda,\mu})\|_{\lambda,\mu\vdash n}.$$

The product of these two matrices is the diagonal matrix with λ^{th} diagonal entry equal to z_λ, which is nearly the identity matrix. From this we can say two things:

1. the inverse to $\|(-1)^{n-\ell(\lambda)}w(B_{\lambda,\mu})\|_{\lambda,\mu\vdash n}$ is $\|(-1)^{n-\ell(\lambda)}OB_{\lambda,\mu}/z_\lambda\|_{\lambda,\mu\vdash n}$ and
2. the inverse to $\|OB_{\lambda,\mu}\|_{\lambda,\mu\vdash n}$ is $\|(-1)^{\ell(\lambda)+\ell(\mu)}w(B_{\mu,\lambda})/z_\lambda\|_{\lambda,\mu\vdash n}$.

Stated differently,

1. the e-to-p transition matrix, which is the inverse to the p-to-e transition matrix given in Theorem 2.22, has λ,μ entry $(-1)^{n-\ell(\lambda)}OB_{\lambda,\mu}/z_\lambda$ and
2. the m-to-p transition matrix, which is the inverse to the p-to-m transition matrix given in Theorem 2.23, has λ,μ entry $(-1)^{\ell(\lambda)+\ell(\mu)}w(B_{\mu,\lambda})/z_\lambda$, as desired.

2.7 Considering the transposition $(i\ \ j)$, an alternating polynomial f must satisfy

$$f(x_1,\ldots,x_i,\ldots,x_j,\ldots,x_N) = -f(x_1,\ldots,x_j,\ldots,x_i,\ldots,x_N)$$

for all integers i, j. Therefore f must be divisible by $(x_i - x_j)$ for all i, j. This means that f must be divisible by the Vandermonde determinant $\prod_{i<j}(x_i - x_j)$ and that the degree of f must be at least $(N-1)+\cdots+0 = \binom{N}{2}$.

2.8 The determinant of a matrix is changed by a factor of -1 when two rows are interchanged. Since Δ_λ is the determinant

$$\begin{vmatrix} x_1^{\lambda_1+n-1} & x_1^{\lambda_2+n-2} & \cdots & x_1^{\lambda_n+0} \\ x_2^{\lambda_1+n-1} & x_2^{\lambda_2+n-2} & \cdots & x_2^{\lambda_n+0} \\ \vdots & \vdots & & \vdots \\ x_n^{\lambda_1+n-1} & x_n^{\lambda_2+n-2} & \cdots & x_n^{\lambda_n+0} \end{vmatrix},$$

switching the roles of x_i and x_j changes the sign of Δ_λ by -1. It is therefore an alternating polynomial.

If $x_1^{\lambda_1+n-1}x_2^{\lambda_2+n-2}\cdots x_n^{\lambda_n+0}$ is a term in an alternating polynomial f, then f must contain all terms of the form $\text{sign}(\sigma)x_{\sigma_1}^{\lambda_1+n-1}x_{\sigma_2}^{\lambda_2+n-2}\cdots x_{\sigma_n}^{\lambda_n+0}$ for σ a permutation of S_n. Therefore f contains

$$\Delta_\lambda(x_1,\ldots,x_n) = \sum_{\sigma=\sigma_1\cdots\sigma_n\in S_n} \text{sign}(\sigma)x_{\sigma_1}^{\lambda_1+n-1}x_{\sigma_2}^{\lambda_2+n-2}\cdots x_{\sigma_n}^{\lambda_n+0},$$

as desired.

2.9 We will describe a weight preserving function which turns any $T \in CS_\lambda$ into a $T' \in RCS_\lambda$.

Take $T \in CS_\lambda$. If a 1 appears in the same column as a 2 in T, switch their positions. Then if a sequence of 1s appears in the same row as a sequence of 2s in T, switch the appearances of these sequences. Now every 1 appears above or to the right of every 2. Repeat this procedure with the 1s and 3s in T, then the 1s and 4s, and so on, until every 1 is appears above or to the right of every larger integer in T.

Inductively repeat this process with 2, moving all appearances of 2 above or to the right of all larger integers. Then repeat this process with 3, 4, and so on. The result is the desired reverse column strict tableau T'.

2.10 If $b_1,\ldots,b_k$ are the beads in the abacus a, then the corresponding integer partition λ is equal to $(\text{empty}(b_k),\ldots,\text{empty}(b_1))$ where $\text{empty}(b_i)$ denotes the number of empty places to the left of b_i. We have

$$\begin{aligned} x_1\cdots x_k\Delta_\lambda(x_1,\ldots,x_k) &= \sum_{\sigma=\sigma_1\cdots\sigma_k\in S_k} \text{sign}(\sigma)x_{\sigma_1}^{\lambda_1+k}\cdots x_{\sigma_k}^{\lambda_k+1} \\ &= \sum_{\sigma=\sigma_1\cdots\sigma_k\in S_k} \text{sign}(\sigma)x_{\sigma_1}^{\text{empty}(b_k)+k}\cdots x_{\sigma_k}^{\text{empty}(b_1)+1}. \end{aligned}$$

Since there are exactly $i-1$ beads to the left of bead b_i, we know $\text{position}(b_i) = \text{empty}(b_i)+i$. Therefore our sum is equal to

$$\sum_{\sigma=\sigma_1\cdots\sigma_k\in S_k} \text{sign}(\sigma)x_{\sigma_1}^{\text{position}(b_k)}\cdots x_{\sigma_k}^{\text{position}(b_1)} = \sum_{\text{labeling } \ell \text{ of } a} \text{sign}(\ell)\text{weight}(\ell).$$

2.11 Let a be the mathematical abacus corresponding to the integer partition α. Since e_j is the sum of square-free monomials of degree j, each monomial in the product

$$e_j\Delta_\alpha = e_j \sum_{\ell \text{ is a labeling of } a} \text{sign}(\ell)\text{weight}(\ell)$$

can be associated with an ordered pair of the form $(x_{i_1}\cdots x_{i_j},\ell)$ where $i_1 < \cdots < i_j$ are j distinct positive integers and ℓ is a labeling of a. By defining the sign of such a pair to be $\text{sign}(\ell)$ and the weight to be $x_{i_1}\cdots x_{i_j}\text{weight}(\ell)$, it follows that $e_j\Delta_\alpha$ is the signed, weighted sum over all possible pairs of the form $(x_{i_1}\cdots x_{i_j},\ell)$.

Given $(x_{i_1}\cdots x_{i_j},\ell)$, starting with the leftmost possible bead and working rightward, move the beads with labels given by $i_1,\ldots,i_j$ one space to the right. If we cannot move bead b one space to the right because that space is occupied by another bead b', match $(x_{i_1}\cdots x_{i_j},\ell)$ with the pair found by interchanging the labels on b and b' in both $x_{i_1}\cdots x_{i_j}$ and ℓ. Since the permutations in ℓ and ℓ' differ by a transposition, these two objects have opposite signs. They have the same weight, and so their pairing will cancel them from the sum.

Each time we move one of the k beads to the right, we are increasing one part in the corresponding integer partition by 1. Therefore $e_j\Delta_\alpha$ corresponds to the signed sum over all possible labelings of abaci which correspond to an integer partition $\lambda \vdash (n+j)$ which can be found by adding 1 to j distinct parts of α, as desired.

2.12 Let $\mu = (\mu_1,\ldots,\mu_\ell) \vdash n$. By Exercise 2.11, $e_\mu\Delta_{(0,\ldots,0)} = e_{\mu_1}\cdots e_{\mu_\ell}\Delta_{(0,\ldots,0)}$ is the sum of terms of the form Δ_λ where λ is an integer partition created by adding 1 to μ_1 distinct parts in the integer partition $(0,\ldots,0)$, then adding 1 to μ_2 distinct parts to the result, then adding 1 to μ_3 distinct parts to the result, and so on. Place an i in the cell of the Young diagram for the integer partition λ if the cell was created by adding 1 in step i of this process.

For example, we can create $\Delta_{(6,3,3,2)}$ from $e_{(3,3,3,2,2,1)}\Delta_{(0,\ldots,0)}$ by starting with $(0,0,0,0)$, then successively adding 1 to distinct parts to create $(1,1,1,0)$, $(2,2,1,1)$, $(3,3,2,1)$, $(4,3,3,1)$, $(5,3,3,2)$, and then $(6,3,3,2)$. Recording these steps by placing $1,\ldots,6$ in the cells of the Young diagram gives

2	5				
1	3	4			
1	2	3			
1	2	3	4	5	6

.

The integers in this tableau of shape λ must weakly increase within columns and strictly increase within rows. Therefore the number of terms of the form Δ_λ in the expansion of $e_\mu\Delta_{(0,\ldots,0)}$ is equal to the number of column strict tableau of shape λ' and content μ, namely $K_{\lambda',\mu}$.

We now have that $e_\mu\Delta_{(0,\ldots,0)} = \sum_{\lambda\vdash n} K_{\lambda',\mu}\Delta_\lambda$. Dividing both sides of this equation by $\Delta_{(0,\ldots,0)}$ and using Theorem 2.4, we have $e_\mu = \sum_{\lambda\vdash n} K_{\lambda',\mu}s_\lambda$, as desired.

2.13 Let a be the mathematical abacus corresponding to the integer partition α. Since $p_j = x_1^j + x_2^j + \cdots$, each monomial in the product

$$p_j\Delta_\alpha = p_j \sum_{\ell \text{ is a labeling of } a} \text{sign}(\ell)\text{weight}(\ell)$$

can be associated with an ordered pair of the form (x_i^j, ℓ) where ℓ is a labeling of a. If the sign of such a pair is sign(ℓ) and the weight is x_i^jweight(ℓ), then $p_j\Delta_\alpha$ is the signed, weighted sum over all possible pairs of the form (x_i^j, ℓ).

Given (x_i^j, ℓ), move the bead with label i to the right j spaces. If we cannot do this because that space is occupied by another bead b', match (x_i^j, ℓ) with the pair found by interchanging the labels on b and b' in both x_i^j and ℓ. Since the permutations in ℓ and ℓ' differ by a transposition, these two objects have opposite signs. They have the same weight, and so their pairing will cancel them from the sum.

If we happen to move this bead b over another bead b', then we are multiplying the permutation giving the labels on ℓ by the transposition (label(b) label(b')). For example, if we move the bead with label 4

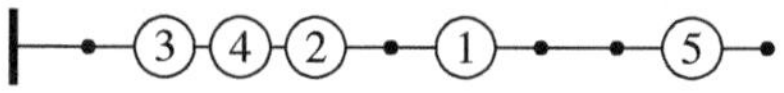

7 spaces to the right to form

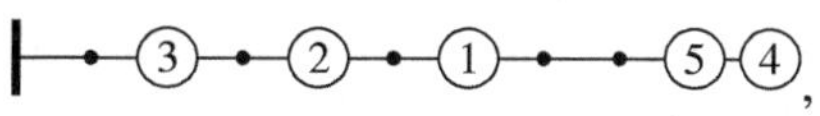

then we introduce the transpositions (2 4), (1 4), and (5 4) to the underlying permutation.

Exercise 1.3 tells us that we are adding one rim hook of size j to the corresponding integer partition each time we move a bead b to the right j spaces. The number of beads b passes is one less than the number of rows in the corresponding rim hook ν, and so this move changes the sign by sign(ν). Therefore, by Exercise 1.3, $p_j\Delta_\alpha$ corresponds to the signed sum over all possible labelings of abaci which correspond to an integer partition $\lambda \vdash (n+j)$ which can be found by adding a rim hook ν of length j to α.

2.14 Let $\mu = (\mu_1, \ldots, \mu_\ell) \vdash n$. By Exercise 2.13, $p_\mu\Delta_{(0,\ldots,0)} = p_{\mu_1} \cdots p_{\mu_\ell}\Delta_{(0,\ldots,0)}$ is the sum of terms of the form $\pm\Delta_\lambda$ where λ is an integer partition created by adding a first rim hook μ of size μ_ℓ to $(0,\ldots,0)$, then adding a rim hook of length $\mu_{\ell-1}$ to the result, then adding a rim hook of length to $\mu_{\ell-2}$ to the result, and so on. The $\pm$ sign on $\pm\Delta_\lambda$ is determined by the product of the rim hooks. Label the order the rim hooks were placed with the numbers $1,\ldots,\ell$ to find a rim hook tableau of shape λ and content μ.

This shows that $p_\mu\Delta_{(0,\ldots,0)} = \sum_{\lambda\vdash n}\chi_\mu^\lambda\Delta_\lambda$. Dividing both sides of this equation by $\Delta_{(0,\ldots,0)}$ and using Theorem 2.4, we have $p_\mu = \sum_{\lambda\vdash n}\chi_\mu^\lambda s_\lambda$, as desired.

2.15 If row i contains only the integer i for each row i in C reading bottom to top, then define (C,S) to be a fixed point of the involution. Otherwise, find the least i such that row i contains an integer larger than i, and let j be the maximum integer in this row.

Let ν_j be the special rim hook which begins in row j of S. By switching their tail ends, there is a unique way to change the special rim hooks ν_j and ν_{j-1} to two other special rim hooks which occupy the same cells as ν_j and ν_{j-1}. Change S into the special rim hook tabloid found by making this switch.

Performing this "tail-switching" operation on S changes the lengths of the special rim hooks v_j and v_{j-1}, and so to be a valid pair of the form (C,S) we need to change the frequencies of the js and the $(j-1)$s in C accordingly. Make this switch using the involution in the proof of Theorem 2.2.

For example, the image of the pair (C,S) displayed in Exercise 2.15 is shown below:

3	3	5	5	5		
2	2	3	3	3	5	
1	1	2	2	2	3	3

,

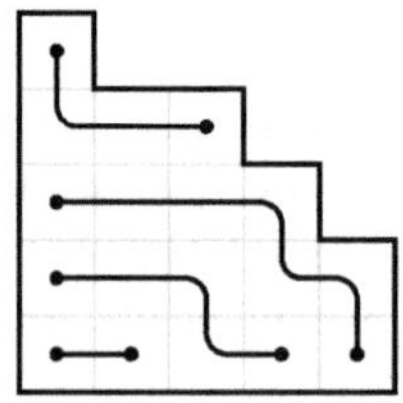

.

This "tail-switching" involution changes the sign of S by -1 since it either introduces or removes exactly one extra down step in the rim hook v_j.

The fixed points under this involution must have row i of C containing exactly i. This means that a special rim hook of length i begins in row i of S for all i. There is exactly one special rim hook tabloid S with this property—the special rim hook tabloid which features a completely flat rim hook in each row. This one fixed point of sign $+1$ only occurs when $\mu = \lambda$, verifying equation (2.4).

2.16 Given $\lambda \vdash n$, we will count the number of ways can we form the monomial $x_1^{\lambda_1} \cdots x_k^{\lambda_k}$ by multiplying out $h_\mu = h_{\mu_1} \cdots h_{\mu_\ell}$ by organizing our work into a table where rows are indexed by $x_1, \ldots, x_k$ and columns are indexed by $h_{\mu_1}, \ldots, h_{\mu_\ell}$. Place an m in the x_i row and h_{μ_j} column entry of the table if the monomial selected from h_{μ_j} to contribute to a final product contains x_i^m.

For example, when $\lambda = (3^2, 2, 1^2)$ and $\mu = (3^2, 2^2)$, one possible table is

$$\begin{array}{c} \\ x_1 \\ x_2 \\ x_3 \\ x_4 \\ x_5 \end{array} \begin{array}{c} \begin{array}{cccc} h_3 & h_3 & h_2 & h_2 \end{array} \\ \begin{bmatrix} 3 & 0 & 0 & 0 \\ 0 & 0 & 1 & 2 \\ 0 & 2 & 0 & 0 \\ 0 & 0 & 1 & 0 \\ 0 & 1 & 0 & 0 \end{bmatrix} \end{array}.$$

This table corresponds to the terms in each parenthesis in

$$h_{(3^2,2^2)}(x_1, x_2, \ldots) = (x_1^3 + \cdots + x_1^2 x_2 + \cdots + x_1 x_2 x_3 + \cdots)^2 (x_1^2 + \cdots + x_1 x_2 + \cdots)^2$$

which are selected to form the monomial $x_1^3 x_2^3 x_3^2 x_4^1 x_5^1$.

The number of ways to form a such a table is the coefficient of m_λ in h_μ. Each table is an element in $NM_{\lambda,\mu}$ as desired.

2.17 The assertion is true when $n = 1$ because $1! e_1 = p_1$. We proceed by induction.

Removing the ith column and last row of the $n \times n$ determinant leaves a determinant of the form

$$\begin{vmatrix} A & 0 \\ B & C \end{vmatrix}$$

where A is an $(i-1)\times(i-1)$ matrix of the same form as the original $n\times n$ matrix, 0 is the $(i-1)\times(n-i)$ zero matrix, B is an $(n-i)\times(i-1)$ matrix, and C is an $(n-i)\times(n-i)$ lower triangular matrix diagonal entries equal to $i, i+1, \ldots, n-1$. By the induction hypothesis, the determinant of this matrix is $(i-1)!e_{i-1}i\,(i-1)\cdots(n-1) = (n-1)!e_{i-1}$.

Expanding the determinant of the original $n\times n$ matrix along the last row,

$$\sum_{i=1}^{n}(-1)^{n-i}p_{n-(i-1)}(n-1)!e_{i-1} = (-1)^{n-1}(n-1)!\sum_{i=0}^{n-1}(-1)^{i}p_{n-i}e_{i},$$

which, by Theorem 2.9, is equal to $(-1)^{n-1}(n-1)!(-1)^{n-1}ne_n = n!e_n$.

Applying ω to both sides of the identity gives

$$n!h_n = \begin{vmatrix} p_1 & 1 & 0 & \cdots & 0 & 0 \\ -p_2 & p_1 & 2 & \cdots & 0 & 0 \\ \vdots & \vdots & \vdots & & \vdots & \vdots \\ (-1)^{n-2}p_{n-1} & (-1)^{n-3}p_{n-2} & (-1)^{n-4}p_{n-3} & \cdots & -p_1 & n-1 \\ (-1)^{n-1}p_n & (-1)^{n-2}p_{n-1} & (-1)^{n-3}p_{n-2} & \cdots & -p_2 & p_1 \end{vmatrix}.$$

Notes

The theory of symmetric functions has a long history and with many applications to the representation theory of finite groups, special functions, and combinatorics. There are two books on the theory of symmetric functions that we would recommend.

The first is Macdonald's *Symmetric functions and Hall polynomials* [82], which contains a wealth of information not presented here, including several generalizations of the symmetric functions such as the Hall–Littlewood symmetric functions which involve an extra parameter q and Macdonald polynomials which involve two extra parameters q and t. There are many combinatorial applications of both Hall–Littlewood symmetric functions and Macdonald polynomials which are beyond the scope of this book. See, for example, Haglund's book [55].

A second account of the theory of symmetric functions is found in Stanley's *Enumerative Combinatorics, Volume 2* [108]. The latter text contains notes on the history of symmetric functions with numerous references.

There are many approaches of developing the theory of symmetric functions. Our approach has been to give direct combinatorial proofs of identities wherever possible. Moreover, we have made sure that our proofs work over the ring of symmetric functions in infinitely many variables; this will be needed for some of our applications.

The exercises allow us to add 11 edges to the directed graph giving the transition matrices between bases for the ring of symmetric functions featured on page 56. Specifically,

1. Exercise 2.6 says the λ,μ entry of the e-to-p matrix is $(-1)^{n-\ell(\lambda)}\frac{OB_{\lambda,\mu}}{z_\lambda}$.
2. Exercise 2.6 says the λ,μ entry of the m-to-p matrix is $(-1)^{\ell(\lambda)+\ell(\mu)}\frac{w(B_{\mu,\lambda})}{z_\lambda}$.
3. Exercise 2.12 says the λ,μ entry of the e-to-s matrix is $K_{\lambda',\mu}$.
4. Exercise 2.14 says the λ,μ entry of the p-to-s matrix is χ^λ_μ.
5. Exercise 2.15 says the λ,μ entry of the m-to-s matrix is $K^{-1}_{\mu,\lambda}$.
6. Exercise 2.15 says the λ,μ entry of the s-to-e matrix is $K^{-1}_{\lambda,\mu'}$.
7. Exercise 2.16 says the λ,μ entry of the h-to-m matrix is $NM_{\lambda,\mu}$.

The ω transformation implies that the λ,μ entry of the x-to-y transition matrix is the λ,μ entry of the $\omega(x)$-to-$\omega(y)$ transition matrix for all bases x,y. Applying ω to items 1, 3, 6 on the above list as well as Theorem 2.22 gives

8. The λ,μ entry of the h-to-p matrix is $\frac{OB_{\lambda,\mu}}{z_\lambda}$.
9. The λ,μ entry of the h-to-s matrix is $K_{\lambda,\mu}$.
10. The λ,μ entry of the s-to-h matrix is $K^{-1}_{\lambda,\mu}$.
11. The λ,μ entry of the p-to-h matrix is $(-1)^{\ell(\lambda)+\ell(\mu)}w(B_{\lambda,\mu})$.

Theorem 2.28 when applied to the dual bases $\{s_\lambda : \lambda \vdash n\}$ and $\{s_\lambda : \lambda \vdash n\}$ and the dual bases $\{p_\lambda : \lambda \vdash n\}$ and $\{p_\lambda/z_\lambda : \lambda \vdash n\}$ when applied to item 4 on the above list gives

12. The λ,μ entry of the s-to-p transition matrix is $\chi^\mu_\lambda/z_\lambda$.

All of the above information about transition matrices is recorded as a directed graph in Appendix A.

We have not provided a combinatorial interpretation of the entries are the m-to-e and m-to-h transition matrices. Combinatorial interpretations for the entries of these matrices are described in [9], but they do not have straightforward descriptions and are difficult to use in applications, and so we choose to omit them.

In [115] and [116], White gave somewhat lengthy but purely combinatorial proofs that the λ,μ entry of the s-to-p transition matrix is $\chi^\mu_\lambda/z_\lambda$. We shall give a different approach to this result in Chapter 5 where we give a second proof of the so-called Muraghan–Nakyama rule.

The combinatorial interpretations of the entries of the transition matrices for symmetric functions can be found in [33, 73] while the idea of using labeled abaci to prove results about transition matrices is due to Loehr [80, 81].

Chapter 3
Counting with the Elementary and Homogeneous Symmetric Functions

Let φ be a ring homomorphism on the ring of symmetric functions Λ. If we know the values of $\varphi(e_n)$ for all n, then, as described in the beginning of Section 2.5, we know $\varphi(f)$ for any symmetric function f because f can be expressed as sums and products of elementary symmetric functions.

In this chapter we will define certain ring homomorphisms φ on the elementary symmetric functions. By understanding what φ does to the homogeneous symmetric functions, we will be able to find generating functions for permutation statistics.

3.1 Counting Descents

A first example of how to use ring homomorphisms to find generating functions will involve the distribution of descents in the permutations in the symmetric group S_n. Let $E_0(x) = 1$ and $E_n(x) = \sum_{\sigma \in S_n} x^{\mathrm{des}(\sigma)}$. The first of these polynomials are

$$\begin{aligned}
E_0(x) &= 1,\\
E_1(x) &= 1,\\
E_2(x) &= 1 + x,\\
E_3(x) &= 1 + 4x + x^2,\\
E_4(x) &= 1 + 11x + 11x^2 + x^3,\\
E_5(x) &= 1 + 26x + 66x^2 + 26x^3 + x^4,\\
E_6(x) &= 1 + 57x + 302x^2 + 302x^3 + 57x^4 + x^5.
\end{aligned}$$

We will find a simple, closed expression for the generating function $\sum_{n=0}^{\infty} E_n(x) z^n / n!$.

A. Mendes, J. Remmel, *Counting with Symmetric Functions*, Developments in Mathematics 43, DOI 10.1007/978-3-319-23618-6_3

Theorem 3.1. *Define a ring homomorphism* $\varphi : \Lambda \to \mathbb{Q}[x]$ *by* $\varphi(e_0) = 1$ *and*

$$\varphi(e_n) = \frac{(-1)^{n-1}}{n!}(x-1)^{n-1}$$

for $n \geq 1$*. Then* $\varphi(h_n) = E_n(x)/n!$.

Proof. Theorem 2.18 says that the expansion of h_n in terms of elementary symmetric functions can be expressed in terms of brick tabloids. We have

$$\begin{aligned} n!\varphi(h_n) &= n! \sum_{\lambda \vdash n} (-1)^{n-\ell(\lambda)} |B_{\lambda,(n)}| \varphi(e_\lambda) \\ &= n! \sum_{\lambda \vdash n} (-1)^{n-\ell(\lambda)} |B_{\lambda,(n)}| \varphi(e_{\lambda_1}) \varphi(e_{\lambda_{\lambda_2}}) \cdots \\ &= n! \sum_{\lambda \vdash n} (-1)^{n-\ell(\lambda)} |B_{\lambda,(n)}| \frac{(-1)^{\lambda_1 - 1}}{\lambda_1!}(x-1)^{\lambda_1 - 1} \frac{(-1)^{\lambda_2 - 1}}{\lambda_2!}(x-1)^{\lambda_2 - 1} \cdots \\ &= \sum_{\lambda \vdash n} \binom{n}{\lambda} |B_{\lambda,(n)}| (x-1)^{n-\ell(\lambda)}, \end{aligned} \tag{3.1}$$

where $\binom{n}{\lambda}$ denotes the multinomial coefficient $n!/(\lambda_1!\lambda_2!\cdots)$.

The plan is to create a collection of signed, weighed objects from (3.1) and then define a sign reversing involution which leaves fixed points corresponding to $E_n(x)$. To begin, construct combinatorial objects in the following manner:

1. Select a permutation $\lambda \vdash n$ and a brick tabloid in $B_{\lambda,(n)}$.
2. If the lengths of the bricks in the brick tabloid are $b_1, \ldots, b_\ell$, then select ℓ disjoint subsets of size $b_1, \ldots, b_\ell$ from $\{1, \ldots, n\}$. Write these subsets in decreasing order within the bricks of the brick tabloid.
3. Place a "1" in the last cell of each brick and place a choice of "x" or "-1" in every other cell.

Let $\mathcal{T}$ be the set of objects created in this manner. For $T \in \mathcal{T}$, define $w(T)$ to be the product of the -1s and xs appearing in T. One possible $T \in \mathcal{T}$ with weight $(-1)^2x^3$ when $n = 12$ is

1	x	x	1	-1	1	1	1	1	-1	x	1
6	7	3	1	5	2	11	12	10	9	8	4

.

In our three-step process of creating $T \in \mathcal{T}$, step 1 accounts for the sum and the $|B_{\lambda,(n)}|$ term in (3.1), step 2 accounts for the multinomial coefficient, and step 3 accounts for the $(x-1)^{n-\ell(\lambda)}$ term. This means

$$n!\varphi(h_n) = \sum_{T \in \mathcal{T}} w(T).$$

We now define an involution on $\mathcal{T}$. Scan $T \in \mathcal{T}$ from left to right looking for the first occurrence of either a -1 or two consecutive bricks with a decrease appearing in the integer labeling between them.

If neither is found in T, leave T fixed. If a -1 is found first, then break the brick with the -1 into two smaller bricks after the occurrence of the -1 and change the -1 to a 1. If we find two consecutive bricks with a decrease appearing in the integer labeling between them, combine the two bricks into one larger brick and change the 1 in the middle to a -1. The image of the $T \in \mathcal{T}$ displayed earlier in this proof under this operation is

1	x	x	1	1	1	1	1	1	-1	x	1
6	7	3	1	5	2	11	12	10	9	8	4

This process changes from an occurrence of a -1 into a descent and vice versa. This is a sign reversing and weight preserving involution. Fixed points under this involution must look like this:

1	x	x	1	x	x	1	x	x	x	x	1
3	11	6	1	7	4	2	12	10	9	8	5

Fixed points must have no -1s and no decreases between bricks. These fixed points naturally correspond to permutations in S_n with an x for each descent. This shows

$$n!\varphi(h_n) = \sum_{\sigma \in S_N} x^{\mathrm{des}(\sigma)} = E_n(x),$$

as desired. □

Corollary 3.2. *We have*

$$\sum_{n=0}^{\infty} \frac{E_n(x)}{n!} z^n = \frac{x-1}{x - e^{(x-1)z}}.$$

Proof. Applying the φ in Theorem 3.1 to the identity in Theorem 2.5 gives

$$\begin{aligned}
\sum_{n=0}^{\infty} \frac{E_n(x)}{n!} z^n &= \varphi\left(\sum_{n=0}^{\infty} h_n z^n\right) \\
&= \frac{1}{\varphi\left(\sum_{n=0}^{\infty} e_n(-z)^n\right)} \\
&= \frac{1}{1 + \sum_{n=1}^{\infty} \frac{(-1)^{n-1}}{n!}(x-1)^{n-1}(-z)^n} \\
&= \frac{x-1}{x - e^{(x-1)z}},
\end{aligned}$$

as required. □

In this chapter we will change the ring homomorphism φ in Theorem 3.1 in various ways, thereby finding many generating functions from the relationship between the elementary and homogeneous symmetric functions. In preparation, let us take a closer look at φ:

$$\varphi(e_n) = \frac{(-1)^{n-1}}{n!}(x-1)^{n-1}.$$

(Whenever we define a ring homomorphisms on the ring of symmetric functions by describing what happens to e_n, we will assume that $n \geq 1$ since $\varphi(e_0) = \varphi(1)$ must always equal 1 in order to be a homomorphism.) There are three main components to this definition: the $(-1)^{n-1}$ sign, the division by $n!$, and the $(x-1)^{n-1}$ term. Each of the three components plays different roles in the proof of Theorem 3.1. Let us examine each term carefully.

Since the coefficient of e_λ in h_n is $(-1)^{n-\ell}|B_{\lambda,(n)}|$, the $(-1)^{n-1}$ sign in the definition of φ allows us to cancel the $(-1)^{n-\ell}$ sign when expanding the homogeneous symmetric functions in terms of the elementary symmetric functions. This $(-1)^{n-1}$ will be a mainstay in our upcoming variations on the homomorphism φ.

The division by $n!$ gave rise to the multinomial coefficient $\binom{n}{\lambda}$ in equation (3.1). This multinomial coefficient enabled us to fill our bricks with decreasing sequences of integers. By changing the $n!$ in the definition of φ we will be able to change how we label the bricks with integers.

The $(x-1)^{n-1}$ term in the definition of φ allowed us to weigh a brick of length n by placing a choice of x or -1 in each nonterminal cell in a brick. By changing this $(x-1)^{n-1}$ term we will be able to change how we weight each brick. We can find many different generating functions by getting creative with this weighting term.

We begin showing how to modify φ to find new generating functions by showing how to count the number of permutations in S_n where descents must be arranged in certain ways. A permutation $\sigma = \sigma_1 \cdots \sigma_n \in S_n$ has a 2-descent if there is an index i for which $\sigma_{i-1} > \sigma_i > \sigma_{i+1}$. To count the number of permutations in S_n which do not have a 2-descent, we will restrict how the x and -1 weights appear in each brick.

In particular, let $R_{n-1,i}$ be the number of ways to rearrange i copies of x and $n-1-i$ copies of -1 such that no two xs appear consecutively. Changing the $(x-1)^{n-1}$ term in the definition of φ to

$$f(n) = \sum_{i\geq 0} R_{n-1,i}x^i(-1)^{n-1-i} \tag{3.2}$$

will allow us to weight each cell in a brick with either an x or a -1 such that no two consecutive xs appear within a brick. This choice will force us to never have a 2-descent, as seen in the proof of Theorem 3.3.

Theorem 3.3. *The generating function for permutations in S_n which do not have a 2-descent is*

$$\sum_{n=0}^{\infty} \frac{z^n}{n!} \sum_{\substack{\sigma \in S_n \text{ does not} \\ \text{have a 2-descent}}} x^{des(\sigma)} = \frac{e^{z/2}}{\cos\left(\frac{z\sqrt{4x-1}}{2}\right) - \frac{1}{\sqrt{4x-1}}\sin\left(\frac{z\sqrt{4x-1}}{2}\right)}.$$

Proof. Define a ring homomorphism φ by $\varphi(e_0) = 1$ and

$$\varphi(e_n) = \frac{(-1)^{n-1}}{n!} f(n)$$

for $n \geq 1$ where $f(n)$ is the function defined in (3.2). Following the same logic as what led to (3.1), we find

$$n!\varphi(h_n) = \sum_{\lambda=(\lambda_1,\ldots,\lambda_\ell)\vdash n} \binom{n}{\lambda} |B_{\lambda,(n)}| f(\lambda_1) f(\lambda_2) \cdots. \tag{3.3}$$

Begin to create combinatorial objects from (3.3) by using the summation, the $|B_{\lambda,(n)}|$ term, and the multinomial coefficient to follow steps 1 and 2 in the proof of Theorem 3.1. This creates a brick tabloid of shape (n) with the numbers $1,\ldots,n$ written in the cells such that each brick contains a decreasing sequence. To account for the $f(\lambda_1)\cdots f(\lambda_\ell)$ term in (3.3), place a 1 at the end of each brick and an x or -1 in every other cell such that no two xs appear in consecutive cells.

Let $\mathcal{T}$ be the set of objects created in this manner and let $w(T)$ be the product of the -1s and xs appearing in $T \in \mathcal{T}$. An example of a $T \in \mathcal{T}$ with weight $(-1)^5x^4$ is

-1	x	-1	-1	x	1	x	1	-1	-1	x	1
12	11	9	8	7	2	5	1	10	6	4	3

.

The signed, weighed sum over all possible $T \in \mathcal{T}$ is equal to $n!\varphi(h_n)$.

Perform the same sign reversing, weight preserving involution in the proof of Theorem 3.1: scan from left to right looking for the first -1 or consecutive bricks with a decrease between them. Break or combine bricks accordingly. The fixed points under this involution look like

x	1	1	x	1	x	1	1	x	1	1	1
12	5	7	8	2	11	4	6	10	1	3	9

as there can be no -1s or decreases between bricks. These fixed points correspond to permutations in S_n which do not have a 2-descent with a weight giving the number of descents. This shows

$$\varphi(h_n) = \frac{1}{n!} \sum_{\substack{\sigma \in S_n \text{ does not} \\ \text{have a 2-descent}}} x^{\mathrm{des}(\sigma)}.$$

To find a generating function, we apply φ to both sides of the identity

$$\sum_{n=0}^{\infty} h_n z^n = \frac{1}{\sum_{n=0}^{\infty} e_n(-z)^n}$$

to find

$$\sum_{n=0}^{\infty} \frac{z^n}{n!} \sum_{\substack{\sigma \in S_n \text{ does not} \\ \text{have a 2-descent}}} x^{\text{des}(\sigma)} = \frac{1}{1+\sum_{n=1}^{\infty} \frac{(-1)^{n-1}}{n!} f(n)(-z)^n}. \tag{3.4}$$

Our only remaining task is to use the definition of $f(n)$ to transform the right-hand side of (3.4) into the function displayed in the statement of the theorem. This is not a particularly difficult task but does require showing that a number of identities are true. Since this task is not strictly necessary—after all, (3.4) already gives us the desired generating function—and since showing these manipulations right now might distract from our development, we have chosen to describe how to turn (3.4) into the statement of the theorem in Exercises 3.5 and 3.6. □

If we take $x = 1$ in the generating function in Theorem 3.3, we find

$$\sum_{n=0}^{\infty} \frac{z^n}{n!} |\{\sigma \in S_n \text{ does not have a 2-descent}\}| = \frac{e^{z/2}}{\cos\left(\frac{z\sqrt{3}}{2}\right) - \frac{1}{\sqrt{3}} \sin\left(\frac{z\sqrt{3}}{2}\right)}. \tag{3.5}$$

The coefficients $|\{\sigma \in S_n \text{ does not have a 2-descent}\}|/n!$ tell us the probability that a permutation in S_n will not have a 2-descent.

This generating function in (3.5) is ripe for using the methods described in the second part of Section 1.3. The singularities of this function are when the denominator is 0, which happens when $z = 2\pi\sqrt{3}/9 + 2k\pi\sqrt{3}/3$ for integers k. The singularity closest to 0, namely $2\pi\sqrt{3}/9$, is the radius of convergence. Using L'Hôpitals rule,

$$\lim_{z \to 2\pi\sqrt{3}/9} \left(z - \frac{2\pi\sqrt{3}}{9}\right) \frac{e^{z/2}}{\cos\left(\frac{z\sqrt{3}}{2}\right) - \frac{1}{\sqrt{3}} \sin\left(\frac{z\sqrt{3}}{2}\right)} = -e^{\pi\sqrt{3}/9},$$

and so multiplying by $(z - 2\pi\sqrt{3}/9)$ removes this singularity. This means that our generating function behaves like that of

$$\frac{-e^{\pi\sqrt{3}/9}}{(z - 2\pi\sqrt{3}/9)} = e^{\pi\sqrt{3}/9} \sum_{n=0}^{\infty} \left(\frac{9}{2\pi\sqrt{3}}\right)^{n+1} z^n$$

for values of z close to $2\pi\sqrt{3}/9$. The singularity second closest to 0 has magnitude $|-4\pi\sqrt{3}/9|$, a number with reciprocal less than 0.42. Putting everything together,

$$\left| \frac{|\{\sigma \in S_n \text{ does not have a 2-descent}\}|}{n!} - e^{\pi\sqrt{3}/9} \left(\frac{9}{2\pi\sqrt{3}}\right)^{n+1} \right| < (0.42)^n$$

for large enough n. We now have a wonderful approximation for the coefficients of our generating function.

We can extract even more information from Theorem 3.3 to answer questions like this: What is the expected number of descents in a permutation without a 2-descent? Differentiating the function in Theorem 3.3 with respect to x and evaluating the result at $x = 1$ gives

$$\sum_{n=0}^{\infty}\frac{z^n}{n!}\sum_{\substack{\sigma\in S_n \text{ does not}\\ \text{have a 2-descent}}}\mathrm{des}(\sigma)=\frac{e^{z/2}\left(\sqrt{3}(2-3z)\sin\left(\frac{\sqrt{3}z}{2}\right)-3z\cos\left(\frac{\sqrt{3}z}{2}\right)\right)}{\sqrt{3}\left(\cos\left(\frac{\sqrt{3}z}{2}\right)-\frac{1}{\sqrt{3}}\sin\left(\frac{\sqrt{3}z}{2}\right)\right)^2}.$$

Again we have a function amenable to the techniques found in the second part of Section 1.3. Doing similar but more intricate calculations as shown above, we can find that

$$\left(\left(\frac{4\pi-3\sqrt{3}}{6\pi}\right)(n+1)-\frac{9+2\pi\sqrt{3}}{27}\right)e^{\pi\sqrt{3}/9}\left(\frac{9}{2\pi\sqrt{3}}\right)^{n+1}$$

is within $(0.42)^n$ of $\frac{1}{n!}\sum_{\sigma\in S_n \text{ does not have a 2-descent}}\mathrm{des}(\sigma)$ for large enough n. Combining this last approximation with the approximation for the number of permutations without a 2-descent, we can find an approximation for the expected number of descents in a permutation without a 2-descent:

$$\frac{\sum_{\sigma\in S_n \text{ does not have a 2-descent}}\mathrm{des}(\sigma)}{|\{\sigma\in S_n \text{ does not have a 2-descent}\}|}\approx\left(\frac{4\pi-3\sqrt{3}}{6\pi}\right)(n+1)-\frac{9+2\pi\sqrt{3}}{27}.$$

The ring homomorphism

$$\varphi(e_n)=\frac{(-1)^{n-1}}{n!}f(n),$$

where $f(n)$ is given by (3.2), was used in the proof of Theorem 3.3. This $f(n)$ was designed so that no two consecutive xs would appear in a brick, which in turn allowed us to find the generating function for the number of permutations without a 2-descent. Suppose we change this $f(n)$ to

$$\sum_{i\geq 0}R_{n-1,i,j}x^i(-1)^{n-1-i} \tag{3.6}$$

where $R_{n-1,i,j}$ is the number of ways to rearrange i copies of x and $n-1-i$ copies of -1 such that no j xs appear consecutively. This new $f(n)$ would give the generating function for the number of permutations without j consecutive descents, otherwise known as j-descents. Indeed, only slight modifications to the proof of Theorem 3.3 are needed to show that

$$\sum_{n=0}^{\infty}\frac{z^n}{n!}\sum_{\substack{\sigma\in S_n \text{ does not}\\ \text{have a } j\text{-descent}}}\mathrm{des}(\sigma)=\frac{1}{1+\sum_{n=1}^{\infty}\frac{(-1)^{n-1}}{n!}(-z)^n\sum_{i\geq 0}R_{n-1,i,j}x^i(-1)^{n-1-i}}. \tag{3.7}$$

When $j=2$, we were able to manipulate the generating function in (3.7) to get an explicit formula involving sines and cosines as displayed in Theorem 3.3, but such a simplification is not usually possible. However, if we specialize (3.7) by taking $x=1$, we can find a nice form for the resulting generating function:

Theorem 3.4. *For $j \geq 1$,*

$$\sum_{n=0}^{\infty} |\{\sigma \in S_n \text{ does not have a } j\text{-descent}\}| \frac{z^n}{n!} = \frac{j+1}{(1-\zeta^j)e^{\zeta z} + \cdots + (1-\zeta)e^{\zeta^j z}},$$

where $\zeta = e^{2\pi i/(j+1)}$ is a primitive $(j+1)$th root of unity.

Proof. We begin by showing the identity

$$\sum_{i \geq 0} (-1)^i R_{n-1,i,j} = \begin{cases} (-1)^{n-1} & \text{if } j+1 \text{ divides } n-1, \\ (-1)^n & \text{if } j+1 \text{ divides } n, \\ 0 & \text{otherwise}, \end{cases} \tag{3.8}$$

with a sign reversing involution I.

Define the sign of $r = r_1 \dots r_{n-1} \in R_{n-1,i,j}$ to be $(-1)^i$. If $r_1 = x$, then let $I(r)$ be the rearrangement r with r_1 changed to (-1). If $r_1 = (-1)$ and changing this value to x does not create j copies x which appear consecutively in r, then let $I(r)$ be the rearrangement r with r_1 changed to x.

If $r_1 = (-1)$ and $x r_2 r_3 \cdots r_{n-1}$ has j consecutive xs, then either

$$r = (-1)\, \underbrace{x\, x\, \cdots\, x}_{j-1} \qquad \text{or} \qquad r = (-1)\, \underbrace{x\, x\, \cdots\, x}_{j-1}\, (-1)\, r_{j+2}\, \cdots\, r_{n-1}.$$

In the first case, define $I(r) = r$, and in the second case, inductively define $I(r)$ such that $I(r) = r_1 \cdots r_{j+1} I(r_{j+2} \cdots r_{n-1})$.

There are only two possible fixed points r under this sign reversing involution I:

$$r = \underbrace{(-1)\, x\, x\, \cdots\, x\, (-1)}_{j+1} \underbrace{(-1)\, x\, x\, \cdots\, x\, (-1)}_{j+1} \cdots \underbrace{(-1)\, x\, x\, \cdots\, x\, (-1)}_{j+1}$$

and

$$r = \underbrace{(-1)\, x\, x\, \cdots\, x\, (-1)}_{j+1} \underbrace{(-1)\, x\, x\, \cdots\, x\, (-1)}_{j+1} \cdots \underbrace{(-1)\, x\, x\, \cdots\, x}_{j}$$

In the first case, $j+1$ divides $n-1$ and the sign is $(-1)^{\frac{n-1}{j+1}(j-1)} = (-1)^{n-1}$. In the second case, $j+1$ divides n and the sign is $(-1)^{\frac{n}{j+1}(j-1)} = (-1)^n$. We have now verified (3.8).

Taking $x = 1$ in (3.7) and using (3.8), we can find that

$$\sum_{n=0}^{\infty} |\{\sigma \in S_n \text{ does not have a } j\text{-descent}\}| \frac{z^n}{n!} = \frac{1}{\sum_{n=0}^{\infty} \frac{z^{(j+1)n}}{((j+1)n)!} - \frac{z^{(j+1)n+1}}{((j+1)n+1)!}}.$$

Using the idea in Exercise 1.18, the right side of the above equation is

$$\left(\frac{e^{\zeta^0 z} + \cdots + e^{\zeta^j z}}{j+1} - \int \frac{e^{\zeta^0 z} + \cdots + e^{\zeta^j z}}{j+1} dz \right)^{-1},$$

which in turn may be simplified to look like the statement of the theorem. □

As indicated in the last two theorems, changing the definition of the function $f(n)$ in the ring homomorphism $\varphi(e_n) = (-1)^{n-1} f(n)/n!$ can produce generating functions for permutations where the appearances of descents are restricted in some way.

We end this section with one last example of this phenomenon, with more examples given in Exercises 3.7 and 3.8.

A permutation $\sigma = \sigma_1\sigma_2\sigma_3 \cdots \sigma_n \in S_n$ is alternating provided

$$\sigma_1 > \sigma_2 < \sigma_3 > \sigma_4 < \cdots.$$

Descents occur exactly at odd indices in an alternating permutation.

To find the generating function for the number of alternating permutations in S_{2n}, we would like to apply the involution in Theorems 3.1 and 3.3 to find fixed points which are brick tabloids with bricks of exactly length 2. Defining

$$f(n) = \begin{cases} (-1)^{n/2-1} & \text{if } n \text{ is even,} \\ 0 & \text{if } n \text{ is odd} \end{cases} \tag{3.9}$$

will do this for us, as shown below.

Theorem 3.5. *We have* $\sum_{n=0}^{\infty} |\{\sigma \in S_n \text{ is alternating}\}| \frac{z^n}{n!} = \sec z + \tan z.$

Proof. Let φ be the ring homomorphism on the ring of symmetric functions defined by $\varphi(e_n) = (-1)^{n-1} f(n)/n!$ where $f(n)$ is given in (3.9). This definition of φ will imply that $\varphi(h_k) = 0$ when k is odd, so we focus our attention on $\varphi(h_{2n})$. The same steps as what led to (3.1) shows that

$$(2n)!\varphi(h_{2n}) = \sum_{\lambda \vdash 2n} \binom{2n}{\lambda} |B_{\lambda,(2n)}| f(\lambda_1) f(\lambda_2) \cdots.$$

Using this sum, construct combinatorial objects by selecting a brick tabloid of shape $(2n)$ such that every brick is an even length. Write the integers $1, \ldots, 2n$ in the cells of the bricks such that each brick contains a decreasing sequence. In addition to forcing each bricks to be of an even length, the function $f(n)$ instructs us to place a 1 in the final cell of each brick and a -1 in all other even cells. For example, one such combinatorial object is

	-1		-1		1		1		-1		1
12	11	9	8	7	2	5	1	10	6	4	3

.

Perform the usual involution, first described in the proof of Theorem 3.1: scan the bricks from left to right looking for either a -1 or a decrease between consecutive bricks and break or combine bricks accordingly. Fixed points cannot have a -1 or a decrease between bricks, and so they must look like the object below.

	1		1		1		1		1		1
12	9	11	2	8	7	10	5	6	1	4	3

These fixed points correspond to alternating permutations, thereby proving that $(2n)!\varphi(h_{2n})$ is the number of alternating permutations in S_{2n}. Applying φ to Theorem 2.5 gives a generating function for the number of alternating permutations of an even number:

$$\sum_{n=0}^{\infty} |\{\sigma \in S_{2n} \text{ is alternating}\}| \frac{z^{2n}}{(2n)!} = \varphi\left(\sum_{n=0}^{\infty} h_n z^n\right)$$
$$= \frac{1}{\sum_{n=0}^{\infty} \varphi(e_n)(-z)^n}$$
$$= \frac{1}{\sum_{n=0}^{\infty} \frac{(-1)^n}{(2n)!} z^{2n}}.$$

This last line is equal to $\sec z$.

To understand the alternating permutations of an odd number, we apply the ring homomorphism to the power symmetric functions. Using Theorem 2.22,

$$(2n-1)!\varphi(p_{2n}) = (2n-1)! \sum_{\lambda \vdash 2n} (-1)^{n-\ell(\lambda)} w(B_{\lambda,(n)})\varphi(e_{\lambda_1})\varphi(e_{\lambda_2})\cdots$$
$$= \sum_{\lambda \vdash 2n} \sum_{\substack{T \in B_{\lambda,(2n)} \text{ has even} \\ \text{length bricks } b_1,\ldots,b_\ell}} \frac{(2n-1)!}{b_1!\cdots b_k!}(b_k)(-1)^{\frac{b_1}{2}+\cdots+\frac{b_k}{2}-k}.$$
$$= \sum_{\lambda \vdash 2n} \sum_{\substack{T \in B_{\lambda,(2n)} \text{ has even} \\ \text{length bricks } b_1,\ldots,b_\ell}} \binom{2n-1}{b_1,\ldots,b_{\ell-1},b_\ell-1}(-1)^{\frac{b_1}{2}+\cdots+\frac{b_\ell}{2}-\ell}.$$

From this sum we can create the same objects as we did for $(2n)!\varphi(h_{2n})$ with the exception that we use the integers $1,\ldots,2n-1$ instead of $1,\ldots,2n$, leaving the final cell without an integer. These combinatorial objects look like this:

	-1		-1		1		1		-1		1
11	10	9	8	7	2	5	1	6	4	3	

The brick breaking or combining involution leaves fixed points corresponding to alternating permutations, and so $(2n-1)!\varphi(p_{2n})$ is equal to the number of alternating permutations in S_{2n-1}. Applying φ to both sides of Corollary 2.10 shows that the generating function $\sum_{n=1}^{\infty} |\{\sigma \in S_{2n-1} \text{ is alternating}\}| \frac{z^{2n-1}}{(2n-1)!}$ is equal to

$$\begin{aligned}\varphi\left(\frac{1}{z}\sum_{n=0}^{\infty} p_n z^n\right) &= \frac{\sum_{n=1}^{\infty}(-1)^{n-1} n\varphi(e_n) z^{n-1}}{\sum_{n=0}^{\infty}\varphi(e_n)(-z)^n} \\ &= \left(\sum_{n=1}^{\infty}\frac{(-1)^{n-1}z^{2n-1}}{(2n-1)!}\right) \Big/ \left(\sum_{n=0}^{\infty}\frac{(-1)^n z^{2n}}{(2n)!}\right).\end{aligned}$$

This last line is equal to $\tan z$.

Adding together the generating functions for the even alternating permutations with the generating function for the odd alternating permutations gives $\sec z + \tan z$, as desired. □

The singularity of $\sec z + \tan z$ closest to 0 is at $\pi/2$ and the singularity second closest to 0 is at $-3\pi/2$. Since $\lim_{z\to\pi/2}(z-\pi/2)(\sec z + \tan z) = -2$, the methods in the second part of Section 1.3 tell us that

$$\left|\frac{|\{\sigma \in S_n \text{ is alternating}\}|}{n!} - 2\left(\frac{2}{\pi}\right)^{n+1}\right| < \left(\frac{2}{3\pi} + 0.0001\right)^n$$

for large enough n. Therefore the probability that a random permutation in S_n is alternating is approximately $2^{n+2}/\pi^{n+1}$.

3.2 Changing Brick Labels

In Section 3.1, the function $f(n)$ in the definition of the ring homomorphism $\varphi(e_n) = (-1)^{n-1} f(n)/n!$ was changed, enabling us to count permutations with descents appearing in prescribed ways. In this section we will change the $n!$ in the definition of φ, enabling us to keep track of more permutation statistics and count objects other than permutations.

As a first example, consider the ring homomorphism defined by

$$\varphi(e_n) = \frac{(-1)^{n-1}}{[n]_q!} q^{\binom{n}{2}} (x-1)^{n-1}. \tag{3.10}$$

Without the extra powers of q, this is the same ring homomorphism defined in Theorem 3.1. Our goal is to show that the extra powers of q keep track of inversions. The next lemma will help in the process.

A descending run in a permutation is a consecutive decreasing subsequence. For example, 9 8 2 6 5 4 3 1 7 has descending runs 982, 65431, and 7.

Lemma 3.6. *Let $b_1,\ldots,b_\ell$ be nonnegative integers which sum to n. Then*

$$\begin{bmatrix} n \\ b_1,\ldots,b_\ell \end{bmatrix}_q q^{\binom{b_1}{2}+\cdots+\binom{b_\ell}{2}} = \sum_{\substack{\sigma\in S_n \text{ has descending} \\ \text{runs of lengths } b_1,\ldots,b_\ell}} q^{inv(\sigma)}.$$

Proof. Exercise 1.2, which is a straightforward generalization of Theorem 1.4, says

$$\begin{bmatrix} n \\ \lambda \end{bmatrix}_q = \begin{bmatrix} n \\ b_1,\ldots,b_\ell \end{bmatrix}_q = \sum_{r\in R(1^{b_1},\ldots,\ell^{b_\ell})} q^{\mathrm{inv}(r)},$$

so the q-multinomial coefficient in (3.11) instructs us to select a $r\in R(1^{b_1},\ldots,\ell^{b_\ell})$. With r we will associate a permutation $\sigma^{-1}\in S_n$ by numbering from right to left all the 1s in r with $1,\ldots,b_1$, then numbering all the 2s in r with $b_1+1,\ldots,b_1+b_2$, and so on. For example, say $b_1=4, b_2=5$, and $b_3=3$. Let $r=2\,2\,3\,1\,2\,3\,1\,2\,2\,1\,3\,1$ be the element selected in $R(1^4,2^5,3^3)$. The table below records r, σ^{-1}, and σ:

		1	2	3	4	5	6	7	8	9	10	11	12
r	=	2	2	3	1	2	3	1	2	2	1	3	1
σ^{-1}	=	9	8	12	4	7	11	3	6	5	2	10	1
σ	=	12	10	7	4	9	8	5	2	1	11	6	3

We have designed σ to have decreasing sequences of lengths $b_1,\ldots,b_\ell$. Furthermore, by construction,

$$\mathrm{inv}(\sigma) = \mathrm{inv}\left(\sigma^{-1}\right) = \mathrm{inv}(r) + \binom{b_1}{2} + \cdots + \binom{b_\ell}{2}$$

because $\mathrm{inv}(r)$ gives the inversions between descending runs and $\binom{b_i}{2}$ gives the inversions within the ith descending run in σ. □

The proof of Lemma 3.6 can be easily modified to show that $\begin{bmatrix} n \\ b_1,\ldots,b_\ell \end{bmatrix}$ is equal to $\sum q^{\mathrm{inv}(\sigma)}$ where the sum runs over the permutations $\sigma\in S_n$ with increasing runs of lengths $b_1,\ldots,b_\ell$.

Theorem 3.7. *We have*

$$\sum_{n=0}^{\infty} \frac{z^n}{[n]_q!} \sum_{\sigma\in S_n} x^{des(\sigma)} q^{inv(\sigma)} = \frac{x-1}{x-e_q^{(x-1)z}},$$

where $e_q^z = \sum_{n=0}^{\infty} q^{\binom{n}{2}} z^n/[n]_q!$ is a q-analogue of the exponential function.

Proof. We will retrace the proof of Theorem 3.1, keeping track of what happens to powers of q along the way. Applying the homomorphism φ in (3.10) to $[n]_q!h_n$ gives

$$[n]_q!\varphi(h_n) = \sum_{\lambda\vdash n} \begin{bmatrix} n \\ \lambda \end{bmatrix}_q q^{\binom{\lambda_1}{2}+\binom{\lambda_2}{2}+\cdots} |B_{\lambda,(n)}|(x-1)^{\lambda_1-1}(x-1)^{\lambda_2-1}\cdots. \quad (3.11)$$

Begin to create combinatorial objects by selecting a brick tabloid $T \in B_{\lambda,(n)}$ for some $\lambda \vdash n$. Fill each nonterminal cell in each brick in T with a choice of either x or -1 and place a $+1$ in the terminal cell of each brick. These choices use the summand, the $|B_{\lambda,(n)}|$ term, and the $(x-1)$ terms in (3.11).

Suppose the bricks in T have lengths $b_1,\ldots,b_\ell$ when read from left to right. Lemma 3.6 uses the powers of q in (3.1) to fill the cells of T with a permutation $\sigma \in S_n$ such that the integers in σ decrease within bricks. In each cell, write a power of q counting the number of integers to the right which are larger.

For example, one such combinatorial object created in the above manner is

x	x	-1	1	x	-1	-1	x	1	-1	-1	1
q^{11}	q^9	q^6	q^3	q^6	q^5	q^3	q^1	q^0	q^2	q^1	q^0
12	10	7	4	9	8	5	2	1	11	6	3

.

If we define the weight of such a combinatorial object to be the product of all xs, (-1)s, and powers of q, then $[n]_q!\varphi(h_n)$ is the weighted sum over all possible combinatorial objects.

The usual brick breaking and combining involution does not change the integers in the underlying permutation and therefore does not affect the powers of q. This involution leaves fixed points corresponding to $\sum_{\sigma\in S_n} x^{\mathrm{des}(\sigma)} q^{\mathrm{inv}(\sigma)}$. The generating function in the statement of the theorem follows from applying $\varphi(e_n)$ to Theorem 2.5. □

Every choice for $f(n)$ in section 3.1 when defining $\varphi(e_n) = (-1)^{n-1} f(n)/n!$ can be used to define $\varphi(e_n) = (-1)^{n-1} f(n)/[n]_q!$. This means all of our previous results can include a power of q to keep track of inversions. For example, one possible such q-analogue is

$$\sum_{n=0}^{\infty} \frac{z^n}{n!} \sum_{\sigma \in S_n \text{ is alternating}} q^{\mathrm{inv}(\sigma)} = \frac{1+\sin_q z}{\cos_q z},$$

where $\cos_q z = \sum_{n=0}^{\infty} (-1)^n q^{\binom{2n}{2}} \frac{z^{2n}}{[2n]_q!}$ and $\sin_q z = \sum_{n=0}^{\infty} (-1)^n q^{\binom{2n+1}{2}} \frac{z^{2n+1}}{[2n+1]_q!}$.

A second way to change the labels on a brick tabloid can be used to understand the distribution of common descents in two permutations. For $\sigma = \sigma_1,\ldots,\sigma_n$ and $\tau = \tau_1,\ldots,\tau_n \in S_n$, let $\mathrm{comdes}(\sigma,\tau)$ be the number of indices i for which $\sigma_i > \sigma_{i+1}$ and $\tau_i > \tau_{i+1}$. In order to find a generating function for $\sum_{\sigma,\tau\in S_n} x^{\mathrm{comdes}(\sigma,\tau)}$, consider the ring homomorphism defined by

$$\varphi(e_n) = \frac{(-1)^{n-1}}{(n!)^2}(x-1)^{n-1}. \tag{3.12}$$

Changing the "$1/n!$" to a "$1/(n!)^2$" will permit us to include two permutations in our brick tabloids instead of one.

Theorem 3.8. *We have* $\sum_{n=0}^{\infty} \frac{z^n}{(n!)^2} \sum_{\sigma,\tau \in S_n} x^{comdes(\sigma,\tau)} = \frac{x-1}{x - \sum_{n=0}^{\infty} \frac{(x-1)^n z^n}{(n!)^2}}$.

Proof. Using Theorem 2.18 to apply the homomorphism in (3.12) to $(n!)^2 h_n$ gives

$$(n!)^2 \varphi(h_n) = \sum_{\lambda \vdash n} \binom{n}{\lambda}^2 |B_{\lambda,(n)}|(x-1)^{\lambda_1 - 1}(x-1)^{\lambda_2 - 1} \cdots .$$

This sum tells us to create combinatorial objects similar to those in the proof of Theorem 3.1 but which contain two permutations instead of one:

x	1	1	-1	x	-1	-1	1	x	-1	x	1
10	2	4	12	9	7	6	1	11	8	5	3
11	5	3	12	10	8	4	2	9	7	6	1

.

Scan these objects from left to right looking for the first common descent or -1 and break or combine the bricks accordingly. Fixed points under this involution have no -1s and have powers of x registering the number of common descents, showing that $(n!)^2 \varphi(h_n) = \sum_{\sigma,\tau \in S_n} x^{\text{comdes}(\sigma,\tau)}$. The generating function in the statement of the theorem follows from applying φ to Theorem 2.5. □

We could easily find a similar generating function for common descents in m-tuples of permutations and we can change the "$(x-1)^{n-1}$" in (3.12) to any choice of "$f(n)$" in order to restrict the appearances of common descents in some way. In order to keep track of the inversions for each permutation, these results can be q-analogued by considering homomorphisms of the form $(-1)^{n-1} f(n)/([n]_{q_1}! \, [n]_{q_2}!)$.

A third way to change the brick labels will help us understand words. A word with letters in $\{0,\ldots,k-1\}$ is a finite sequence $w = w_1 \cdots w_n$ with each $w_i \in \{0,\ldots,k-1\}$. The set of all words of length n with letters in $\{0,\ldots,k-1\}$ is denoted by $\{0,\ldots,k-1\}_n^*$. The ring homomorphism defined by

$$\varphi(e_n) = (-1)^{n-1} q^{\binom{n}{2}} \begin{bmatrix} k \\ n \end{bmatrix}_q (x-1)^{n-1}, \tag{3.13}$$

where k is a positive integer, will enable us to prove the result below.

Theorem 3.9. *If $sum(w)$ denotes the sum of the integers in $w \in \{0,\ldots,k-1\}_n^*$, then*

$$\sum_{n=0}^{\infty} z^n \sum_{w \in \{0,\ldots,k-1\}_n^*} x^{des(w)} q^{sum(w)} = \frac{x-1}{x-(z-zx;q)_k},$$

where $(a;q)_k$ denotes the product $(1-aq^0)(1-aq^1)\cdots(1-aq^{k-1})$.

Proof. Each $r \in R(0^n, 1^{k-n})$ is associated with nonincreasing word $w \in \{0,\ldots,k-1\}_n^*$ by interpreting the number of 0s before the ith 1 in r to be the number of $(i-1)$s in w. To make w strictly decreasing, add $n-i$ to the ith integer in w. For instance, $r = 0\,0\,1\,1\,0\,0\,0\,1\,0\,0\,1$ corresponds to $w = 3\,3\,2\,2\,2\,0\,0$. Adding $6,5,4,3,2,1$, and 0 to w gives the strictly decreasing word $9\,8\,6\,5\,4\,3\,1\,0$.

Theorem 1.4 says that $q^{\binom{n}{2}} \left[{k \atop n}\right]_q = q^{\binom{n}{2}} \sum_{r \in R(0^n, 1^{k-n})} q^{\mathrm{inv}(r)}$. The number of inversions in each rearrangement r is equal to the sum of the integers in the corresponding word w. Since adding $n-i$ to the ith integer increases this sum by $\binom{n}{2}$, we have

$$q^{\binom{n}{2}} \begin{bmatrix} k \\ n \end{bmatrix}_q = \sum_{\substack{w \in \{0,\dots,k-1\}_n^* \text{ is} \\ \text{strictly decreasing}}} q^{\mathrm{sum}(w)}. \tag{3.14}$$

If φ is as defined in (3.13), then

$$\varphi(h_n) = \sum_{\lambda \vdash n} q^{\binom{\lambda_1}{2}} \begin{bmatrix} k \\ \lambda_1 \end{bmatrix}_q q^{\binom{\lambda_1}{2}} \begin{bmatrix} k \\ \lambda_1 \end{bmatrix}_q \cdots |B_{\lambda,(n)}| (x-1)^{\lambda_1 - 1} (x-1)^{\lambda_2 - 1} \cdots .$$

This sum tells us to create combinatorial objects like this (using $k = 8$):

x	x	1	1	1	-1	1	x	-1	x	x	1
q^7	q^5	q^3	q^3	q^7	q^7	q^2	q^5	q^4	q^3	q^2	q^1
7	5	3	3	7	7	2	5	4	3	2	1

Instead of a permutation, (3.14) allows us to fill each brick with a strictly decreasing word in $\{0,\dots,k-1\}_n^*$ and places a power of q to match integer in each cell.

Scan the bricks from left to right looking for either the first -1 or a weak decrease between consecutive bricks. Break or combine the bricks accordingly. This involution shows $\varphi(h_n) = \sum_{w \in \{0,\dots,k-1\}_n^*} x^{\mathrm{des}(w)} q^{\mathrm{sum}(w)}$. Applying φ to Theorem 2.5 gives

$$\sum_{n=0}^{\infty} z^n \sum_{w \in \{0,\dots,k-1\}_n^*} x^{\mathrm{des}(w)} q^{\mathrm{sum}(w)} = \frac{x-1}{x - \sum_{n=0}^{\infty} q^{\binom{n}{2}} \left[{k \atop n}\right]_q (x-1)^n z^n}.$$

The q-binomial theorem, displayed in our Exercise 1.6, says that generating function be written to look like the statement of the theorem. □

Every choice for $f(n)$ in section 3.1 when defining $\varphi(e_n) = (-1)^{n-1} f(n)/n!$ can be used to define a ring homomorphism $\varphi(e_n) = (-1)^{n-1} q^{\binom{n}{2}} \left[{k \atop n}\right] f(n)$. This allows us to restrict the appearances of descents in words.

To illustrate this point, suppose we wanted to count the words in $\{0,\dots,k-1\}_n^*$ which never have j consecutive descents, paralleling our result in Theorem 3.4. If we define a ring homomorphism by $\varphi(e_n) = (-1)^{n-1} \binom{k}{n} f(n)$ where $f(n)$ is given in (3.6), then $\varphi(h_n)$ is a sum of combinatorial objects built from brick tabloids with the exception that bricks contain decreasing sequences in $\{0,\dots,k-1\}_n^*$ instead of decreasing runs in a permutation. Replacing "$1/n!$" with "$\binom{k}{n}$" in (3.7) gives

$$\sum_{n=0}^{\infty} z^n \sum_{\substack{w \in \{0,\ldots,k-1\}^*_n \text{ does not} \\ \text{have } j \text{ consec. descents}}} x^{\text{des}(w)} = \frac{1}{1+\sum_{n=1}(-z)^n \binom{k}{n} \sum_{i\geq 1} R_{n-1,i,j}(-x)^i},$$

where $R_{n-1,i,j}$ is the number of ways to rearrange i copies of x and $n-i-1$ copies of -1 such that no j xs appear consecutively. Specializing by taking $x=1$ and simplifying using the approach found in the proof of Theorem 3.4, we can find that

$$\sum_{n=0}^{\infty} |\{w \in \{0,\ldots,k-1\}^*_n \text{ does not have } j \text{ consecutive descents}\}| z^n$$
$$= \frac{j+1}{(1-\zeta^j)(1+\zeta z)^k + \cdots + (1-\zeta)(1+\zeta^j z)^k},$$

where $\zeta = e^{2\pi i/(j+1)}$ is a primitive $(j+1)^{th}$ root of unity.

Although the result in Theorem 3.9 is about words, it can be translated into a statement about permutations in S_n; in particular, it gives information about the joint distribution of the descents, descents of the inverse, and major index statistics.

Theorem 3.10. *We have*

$$\sum_{n=0}^{\infty} \frac{z^n}{(u;q)_{n+1}} \sum_{\sigma \in S_n} x^{des(\sigma^{-1})} u^{des(\sigma)} q^{maj(\sigma)} = \sum_{k=0}^{\infty} \frac{(x-1)u^k}{x-(z-zx;q)_{k+1}}.$$

Proof. Select a term in $\sum_{w\in\{0,\ldots,k-1\}^*_n} x^{\text{wdes}(w)} q^{\text{sum}(w)}$ and let r be the reverse of the word w. With r we associate a permutation $\sigma^{-1} \in S_n$ by numbering from left to right the $(k-1)$s in r, then from left to right numbering the $(k-2)$s in r, and so on. This forces the $(k-1)$s in r to correspond to the first block of numbers in σ, the $(k-2)$s in r to correspond to the second block of numbers in σ, and so on. These blocks sort the exponents on q in nonincreasing order.

Next we associate with r a nonnegative integer sequence $a = a_1 \cdots a_n$ such that a_i is the difference between consecutive exponents on q in σ for $i = 1,\ldots,n-1$ and a_n is the final q exponent. For example, if $k=8$ and $w = 7\,5\,3\,3\,7\,7\,2\,5\,4\,3\,2\,1$, then here are r, σ^{-1}, σ, the q exponents, and a:

		1	2	3	4	5	6	7	8	9	10	11	12
r	=	1	2	3	4	5	2	7	7	3	3	5	7
σ^{-1}	=	12	10	7	6	4	11	1	2	8	9	5	3
σ	=	7	8	12	5	11	4	3	9	10	2	6	1
q exponents	=	7	7	7	5	5	4	3	3	3	2	2	1
a	=	0	0	2	0	1	1	0	0	1	0	1	1

This permutation σ and sequence a have the following properties:

1. The word w can be reconstructed from σ and a, so the process of associating w with the pair (σ,a) is invertible.
2. The number of descents in σ^{-1} is the number of descents in w.
3. Since the ith exponent is $a_i + \cdots + a_n$, the q exponents sum to $1a_1 + \cdots + na_n$.

4. As a special case of the last fact, the maximum q exponent is $a_1 + \cdots + a_n$.
5. We have $a_i = 0$ if and only if $\sigma_i < \sigma_{i+1}$.

Therefore the sum $\sum_{w \in \{0,\ldots,k-1\}_n^*} x^{\mathrm{des}(w)} q^{\mathrm{sum}(w)}$ is equal to

$$\sum_{j=0}^{k-1} \left(\sum_{\sigma \in S_n} x^{\mathrm{des}(\sigma^{-1})} \sum_{\substack{\text{sequences } a_1 \cdots a_n \\ \text{with } a_i = 0 \text{ iff } \sigma_i < \sigma_{i+1}}} q^{1a_1 + \cdots + na_n} u^{a_1 + \cdots + a_n} \right) \Bigg|_{u^j}$$

$$= \frac{1}{1-u} \sum_{\sigma \in S_n} x^{\mathrm{des}(\sigma^{-1})} \sum_{\substack{\text{sequences } a_1 \cdots a_n \\ \text{with } \sigma_i < \sigma_{i+1} \text{ iff } a_i = 0}} (uq^1)^{a_1} \cdots (uq^n)^{a_n} \Bigg|_{u^{k-1}}.$$

Let $\chi(\sigma_i < \sigma_{i+1})$ equal 1 if $\sigma_i < \sigma_{i+1}$ is true and 0 if false. With this notation, the above sum is equal to

$$\frac{1}{1-u} \sum_{\sigma \in S_n} x^{\mathrm{des}(\sigma^{-1})} \sum_{a_1 \geq \chi(\sigma_1 > \sigma_2)} (uq)^{a_1} \sum_{a_1 \geq \chi(\sigma_2 > \sigma_3)} (uq^2)^{a_2} \cdots \sum_{a_n \geq 0} (uq^n)^{a_n} \Bigg|_{u^{k-1}}$$

$$= \sum_{\sigma \in S_n} x^{\mathrm{des}(\sigma^{-1})} \frac{(uq)^{\chi(\sigma_1 > \sigma_2)} (uq^2)^{\chi(\sigma_2 > \sigma_3)} \cdots 1}{(1-u)(1-uq)(1-uq^2) \cdots (1-uq^n)} \Bigg|_{u^{k-1}}$$

$$= \frac{1}{(u;q)_{n+1}} \sum_{\sigma \in S_n} x^{\mathrm{des}(\sigma^{-1})} u^{\mathrm{des}(\sigma)} q^{\mathrm{maj}(\sigma)} \Bigg|_{u^{k-1}}.$$

The result displayed in the statement of the theorem now follows by applying the result in Theorem 3.9 and summing over all k. □

Taking $q = 1$ in Theorem 3.9, we find

$$\sum_{n=0}^{\infty} z^n \sum_{w \in \{0,1,\ldots,k-1\}_n^*} x^{\mathrm{des}(w)} = \frac{x-1}{x - (1 + z(x-1))^k}.$$

In Theorem 3.11, we formally extend this generating function to words with letters taken from an infinite alphabet. This infinite version of Theorem 3.9 is not remarkable because of the generating function itself but because the ring homomorphism used in the proof is defined by sending elementary symmetric functions to a specialization of the elementary symmetric functions. Examples where we define similar ring homomorphisms by sending the elementary symmetric functions to specializations of the homogeneous symmetric functions and the power symmetric functions can be found in Exercise 3.15.

Theorem 3.11. *We have*

$$\sum_{n=0}^{\infty} z^n \sum_{w_1 \cdots w_n \in \{1,2,\ldots\}_n^*} x^{des(w)} y_{w_1} \cdots y_{w_n} = \frac{x-1}{x - \prod_{i \geq 1} (1 + z(x-1) y_i)},$$

where $\{1,2,\ldots\}_n^*$ *denotes the set of words of length* n *with letters in* $\{1,2,\ldots\}$.

Proof. Define a ring homomorphism φ on Λ_n by

$$\varphi(e_n) = (-1)^{n-1}(x-1)^{n-1}e_n(y_1,y_2,\ldots)$$

for $n \geq 1$. It follows that

$$\varphi(h_n) = \sum_{\lambda \vdash n} \left|B_{\lambda,(n)}\right| (x-1)^{n-\ell(\lambda)} e_\lambda(y_1,y_2,\ldots).$$

From this sum we create combinatorial objects by selecting a brick tabloid of shape (n) for some $\lambda \vdash n$ and make a choice of x or -1 for each nonterminal cell of each brick. Use the elementary symmetric function $e_\lambda(y_1,y_2,\ldots)$ to place a square-free monomial in $y_1,y_2,\ldots$ in each brick by placing a variable y_i in each cell such that subscripts strictly decrease from left to right. One such example of a combinatorial object created in this way is

x	x	1	1	1	-1	1	x	-1	x	x	1
y_6	y_5	y_1	y_3	y_{88}	y_7	y_2	y_9	y_4	y_3	y_2	y_1

.

The weight of such an object is the product of all x, -1, and y variables.

Apply the usual involution on this collection of combinatorial objects by looking for decreases in the subscripts of the y variables between two bricks and breaking or combining accordingly. Fixed points correspond to words $w \in \{1,2,\ldots\}_n^*$ with a power of x for each strict decrease in w and a subscript on the variable y for each letter in w.

This proves

$$\varphi(h_n) = \sum_{w_1\cdots w_n \in \{1,2,\ldots\}_n^*} x^{\mathrm{des}(w)} y_{w_1}\cdots y_{w_n}.$$

The generating function in the statement of the theorem follows from applying φ to Theorem 2.5 and using the definition of the generating function $E(z)$ on page 35 to simplify. □

Our next two results, Theorems 3.12 and 3.13, illustrate how sums in the definition of the ring homomorphism φ can permit different choices of bricks in a brick tabloid.

Again considering words over the infinite alphabet $w = w_1\cdots w_n \in \{1,2,\ldots\}_n^*$, we can keep track of descents between consecutive even integers in w and consecutive odd integers in w separately. Let even_des(w) be the number of indices i for which $w_i > w_{i+1}$ and both w_i and w_{i+1} are even. Similarly, let odd_des(w) be the number of indices i for which $w_i > w_{i+1}$ and both w_i and w_{i+1} are odd.

Theorem 3.12. *The generating function*

$$\sum_{n=0}^{\infty} z^n \sum_{w_1\cdots w_n \in \{1,2,\ldots\}_n^*} u^{even\_des(w)} v^{even\_des(w)} y_{w_1}\cdots y_{w_n}$$

is equal to

$$\frac{(u-1)(v-1)}{uv-1-(v-1)\left(\prod_{i=1}^{\infty}(1+z(u-1)y_{2i})\right)-(u-1)\left(\prod_{i=1}^{\infty}(1+z(v-1)y_{2i-1})\right)}.$$

Proof. Let $f(n)$ be the function

$$\left(e_n(y_2,y_4,y_6,\dots)(u-1)^{n-1}+e_n(y_1,y_3,y_5,\dots)(v-1)^{n-1}\right)$$

and define a ring homomorphism φ by $\varphi(e_n)=(-1)^{n-1}f(n)$. Then we have

$$\varphi(h_n)=\sum_{\lambda\vdash n}|B_{\lambda,(n)}|f(\lambda_1)f(\lambda_2)\cdots.$$

Start building combinatorial objects by selecting $T\in B_{\lambda,(n)}$ for some $\lambda\vdash n$. Each brick of length k is assigned a term of the form $f(k)$.

The sum in the definition of $f(k)$ gives us a choice for each brick: use the $e_k(y_2,y_4,\dots)(u-1)^{k-1}$ term or use the $e_k(y_1,y_3,\dots)(v-1)^{k-1}$ term. If we select the former, place a sequence of y_is in the cells of the brick such that subscripts decrease, just like in the proof of Theorem 3.11, with the further specification that only even subscripts appear in the brick. Additionally, place a choice of either u or -1 in each nonterminal cell of the brick. Let us call these bricks "even bricks". If we select the $e_k(y_1,y_3,\dots)(v-1)^{k-1}$ term for a brick of length k, do the same operation but with odd subscripts in the brick and vs instead of us. Let us call these bricks "odd bricks."

The involution we define on such combinatorial objects is our usual brick breaking and combining involution with the modification that we only consider breaking or combining two even bricks or two odd bricks. That is, scan from left to right looking for the first -1, two consecutive even bricks which can be combined to create a larger even brick or two consecutive odd bricks which can be combined to create a larger odd brick. Break or combine accordingly.

Fixed points under this involution look like this:

u	u	1	v	v	v	1	u	u	u	u	1
y_6	y_4	y_2	y_{15}	y_{11}	y_7	y_5	y_{12}	y_{10}	y_6	y_4	y_2

These fixed points correspond to words $w\in\{1,2,\dots\}_n^*$ with a u weight corresponding to even_des(w), a v weight corresponding to odd_des(w), and a y_i each time i appears in w. It follows that $\varphi(h_n)=\sum_{w_1\cdots w_n\in\{1,2,\dots\}_n^*}u^{\text{even\_des}(w)}v^{\text{even\_des}(w)}y_{w_1}\cdots y_{w_n}$. To find the generating function in the statement of the theorem, apply φ to both sides of Theorem 2.5 and perform routine simplifications similar to those found in the proof of Theorem 3.11. □

We can specialize the generating function in Theorem 3.12 to the finite alphabet $\{1,\ldots,k\}$ by setting $y_i = 0$ for $i > k$. For example, by setting $y_i = 1$ for $i \le 3$, setting $y_i = 0$ for $i > 3$, and setting $u = v = 0$, we find

$$\sum_{n=0}^{\infty} z^n \{w \in \{1,2,3\}_n^* \text{ has even\_des}(w) = \text{odd\_des}(w) = 0\} = \frac{1}{1-3z+z^2}.$$

This happens to equal the generating function of even Fibonacci numbers $\sum_{n=0}^{\infty} F_{2n} z^n$ where $F_0 = F_1 = 1$ and $F_n = F_{n-1} + F_{n-2}$.

For a word $w = w_1 \cdots w_n \in \{0,\ldots,k-1\}_n^*$, define $\text{block}_j(w)$ to be the number of maximal consecutive strings of js in w. For example, if $w = 11022211001$, then $\text{block}_0(w) = 2$, $\text{block}_1(w) = 3$, and $\text{block}_2(w) = 1$.

Theorem 3.13. *We have*

$$\sum_{n=0}^{\infty} z^n \sum_{w \in \{0,\ldots,k-1\}_n^*} x_0^{\mathit{block}_0(w)} \cdots x_{k-1}^{\mathit{block}_{k-1}(w)} = \left(1 - \sum_{i=0}^{k-1} \frac{x_i z}{1-(1-x_i)z}\right)^{-1}.$$

Proof. Define $\varphi(e_n) = (-1)^{n-1} f(n)$ where

$$f(n) = x_0(1-x_0)^{n-1} + \cdots + x_{k-1}(1-x_{k-1})^{n-1}.$$

Applying φ to h_n gives $\sum_{\lambda \vdash n} |B_{\lambda,(n)}| f(\lambda_1) f(\lambda_2) \cdots$ from which we create combinatorial objects by selecting a brick tabloid $T \in B_{\lambda,(n)}$ for some $\lambda \vdash n$. For each brick in T, select an $i \in \{0,\ldots,k-1\}$, place a choice of 1 or $-x_i$ in each nonterminal cell in the brick, and place a x_i in the terminal cell. One such combinatorial object is

$$\boxed{1 \;\; -x_2 \;\; -x_2 \;\; x_2} \boxed{1 \;\; 1 \;\; x_1} \boxed{x_1} \boxed{1 \;\; 1 \;\; -x_0 \;\; x_0}.$$

Scan from left to right looking for either a $-$ sign or two bricks which contain the same variable x_i. Break or combine bricks accordingly, reversing the $-$ in the middle. This involution leaves fixed points corresponding to terms in the sum $\sum_{w \in \{0,\ldots,k-1\}_n^*} x_0^{\text{block}_0(w)} \cdots x_{k-1}^{\text{block}_{k-1}(w)}$. That is, we interpret a brick in a fixed point which consists of k 1's followed by x_i as a block of $(k+1)$ 1's. We can find our generating function from applying φ to Theorem 2.5:

$$\begin{aligned}
\sum_{n=0}^{\infty} z^n \sum_{w \in \{0,\ldots,k-1\}_n^*} x_0^{\text{block}_0(w)} \cdots x_{k-1}^{\text{block}_{k-1}(w)} &= \left(1 - \sum_{n=1}^{\infty} \sum_{i=0}^{k-1} \frac{x_i}{1-x_i} (z(1-x_i))^n\right)^{-1} \\
&= \left(1 - \sum_{i=0}^{k-1} \frac{x_i z}{1-(1-x_i)z}\right)^{-1},
\end{aligned}$$

which is the generating function in the statement of the theorem.

What is the expected value of $\text{block}_0(w)$ for $w \in \{0,1\}_n^*$? We calculate

$$\frac{\partial}{\partial x_0}\left(1-\frac{x_0 z}{1-(1-x_0)z}-\frac{x_1 z}{1-(1-x_1)z}\right)^{-1}\Bigg|_{x_0=1,x_1=1}=\frac{z(1-z)}{(1-2z)^2}.$$

The coefficient of z^n for $n \geq 1$ in the above series is $(n+1)2^{n-2}$. There are 2^n words in $\{0,\ldots,k-1\}_n^*$, so the expected number of 0 blocks is $(n+1)2^{n-2}/2^n = (n+1)/4$.

We end this chapter with one final application of how the relationship between the elementary and homogeneous symmetric functions can give us information about permutation statistics.

Theorem 3.14. *We have*

$$\sum_{n=0}^{\infty}\frac{z^n}{[n]_q!(x;u)_{n+1}}\sum_{\sigma\in S_n}x^{des(\sigma)}q^{inv(\sigma)}u^{maj(\sigma)}=\sum_{k=0}^{\infty}\frac{x^k}{e_q^{-u^0 z}\cdots e_q^{-u^k z}}.$$

Proof. This result will be proved using the ring homomorphism defined by

$$\varphi_k(e_n)=\frac{1}{[n]_q!}\sum_{\substack{i_0,\ldots,i_k\geq 0\\ i_0+\cdots+i_k=n}}\begin{bmatrix}n\\ i_0,\ldots,i_k\end{bmatrix}_q q^{\binom{i_0}{2}+\cdots+\binom{i_k}{2}}u^{0i_0+\cdots+ki_k},$$

where k is a nonnegative integer. Applying φ to $[n]_q!h_n$, we have

$$[n]_q!\varphi_k(h_n)=\sum_{\lambda\vdash n}\begin{bmatrix}n\\ \lambda\end{bmatrix}_q(-1)^{n-\ell(\lambda)}|B_{\lambda,(n)}|\prod_{\text{parts }\lambda_j\text{ in }\lambda}\varphi_k(e_{\lambda_j}). \tag{3.15}$$

From this sum we build combinatorial objects by first selecting a brick tabloid in $B_{\lambda,(n)}$ for some $\lambda \vdash n$. The $(-1)^{n-\ell(\lambda)}$ term tells us to place a $+1$ in the terminal cell of each brick and a -1 in all other cells. The note after the proof of Lemma 3.6 allows us to use the $\begin{bmatrix}n\\ \lambda\end{bmatrix}_q$ term in (3.15) to fill the cells of the brick tabloid with a permutation σ such that σ has increasing runs within each brick, keeping a power of q counting the inversions in σ.

For each brick of length n, the definition of $\varphi_k(e_n)$ tells us to select nonnegative integers $i_0,\ldots,i_k$ which sum to n. Record these choices of $i_0,\ldots,i_k$ by placing i_j copies of u^{i_j} into the cells such that the exponents on u are weakly increasing within each brick.

Each brick of length n currently contains an increasing sequence of integers. By Lemma 3.6, the $\begin{bmatrix}n\\ i_0,\ldots,i_k\end{bmatrix}_q q^{\binom{i_0}{2}+\cdots+\binom{i_k}{2}}$ term in the definition of $\varphi_k(e_n)$ allows us to rearrange this increasing sequence as to have descending runs of lengths $i_0,\ldots,i_k$. This means we can order the integers within each brick such that there is a decrease whenever two consecutive cells contain the same power of u. The powers of q allow us to register a q in each cell according to the inversions caused by the integer in that cell.

For example, here is one combinatorial object created in the indicated manner:

-1	-1	1	-1	-1	-1	-1	-1	-1	1	-1	1
u^1	u^1	u^3	u^0	u^0	u^0	u^0	u^2	u^3	u^3	u^1	u^1
q^9	q^1	q^4	q^8	q^7	q^2	q^0	q^0	q^3	q^1	q^1	q^0
10	2	6	12	11	4	1	3	9	7	8	5

By defining the weight of such a combinatorial object as the product of all -1s, powers of u, and powers of q, the weighted sum over all possible objects is equal to $[n]_q!\varphi_k(h_n)$.

Scan the cells from left to right looking for either a -1 or two consecutive bricks which can be combined to create another combinatorial object in our collection. Either break or combine the bricks as needed. This involution leaves fixed points with only bricks of length 1 such that the powers of u must weakly decrease and there must be an increase in the permutation whenever two consecutive bricks have the same power of u. One such fixed point is below:

1	1	1	1	1	1	1	1	1	1	1	1
u^3	u^3	u^3	u^2	u^2	u^1	u^1	u^1	u^1	u^1	u^0	u^0
q^3	q^4	q^5	q^1	q^1	q^1	q^1	q^1	q^2	q^2	q^0	q^0
4	6	8	2	3	5	7	9	11	12	1	10

We can count the number of fixed points using an approach similar to that found in the proof of Theorem 3.10. Given a fixed point, let a_i be the difference in u exponents on bricks i and $i+1$ for $i = 1,\ldots,n-1$ and let a_n be the final u exponent. For example, the sequence $a_1,\ldots,a_n$ corresponding to the fixed point displayed above is 0 0 1 0 1 0 0 0 0 1 0 0.

From the definition of a, the u exponent on the ith brick is $a_i+\cdots+a_n$ and therefore the exponents on u in the fixed point total $a_1+2a_2+\cdots+a_n$. This also means that the largest u exponent is $a_1+\cdots+a_n$. Furthermore, if σ is the permutation in the bottom row of the fixed point, then $a_i = 0$ if and only if $\sigma_i > \sigma_{i+1}$.

Therefore the weighted sum over all fixed points is equal to

$$\sum_{j=0}^{k}\left(\sum_{\sigma\in S_n} q^{\mathrm{inv}(\sigma)} \sum_{\substack{\text{sequences } a_1\cdots a_n \\ \text{with } a_i=0 \text{ iff } \sigma_i>\sigma_{i+1}}} u^{1a_1+\cdots+na_n}x^{a_1+\cdots+a_n}\right)\Bigg|_{x^j}.$$

Multiplying by $1/(1-x)$ allows us to extract the coefficient of x^k from the inside term instead of summing over all coefficients of x^j. We can sum over each variable a_i with $\sigma_i > \sigma_{i+1}$ individually and, since we are extracting the coefficient of x^k, we can include extra infinite sums of the form $\sum_{a_i\geq 0}(xu^i)^{a_i}$ when $\sigma_i \not> \sigma_{i+1}$. What this means is that the above expression is equal to

$$\frac{1}{1-x}\sum_{\sigma\in S_n} q^{\mathrm{inv}(\sigma)} \sum_{a_1\geq\chi(\sigma_1>\sigma_2)} (xu^1)^{a_1} \sum_{a_2\geq\chi(\sigma_2>\sigma_3)} (xu^2)^{a_2}\cdots\Bigg|_{x^k}$$
$$= \sum_{\sigma\in S_n} q^{\mathrm{inv}(\sigma)} \frac{x^{\mathrm{des}(\sigma)}u^{\mathrm{maj}(\sigma)}}{(1-x)(1-xu)\cdots(1-xu^n)}\Bigg|_{x^k},$$

where we summed each of the geometric series.

We have shown that $[n]_q!\varphi_k(h_n) = \sum_{\sigma\in S_n} q^{\mathrm{inv}(\sigma)}x^{\mathrm{des}(\sigma)}u^{\mathrm{maj}(\sigma)}/(x;u)_{n+1}\big|_{x^k}$. Applying φ_k to Theorem 2.5 and summing over all $k\geq 0$ gives

$$\sum_{n=0}^{\infty} \frac{z^n}{[n]_q!(x;u)_{n+1}} \sum_{\sigma\in S_n} q^{\mathrm{inv}(\sigma)}x^{\mathrm{des}(\sigma)}u^{\mathrm{maj}(\sigma)} = \sum_{k=0}^{\infty} x^k\varphi_k\left(\sum_{n=0}^{\infty} h_n z^n\right)$$
$$= \sum_{k=0}^{\infty} \frac{x^k}{\sum_{n=0}^{\infty}\varphi_k(e_n)(-z)^n}.$$

The generating function in the statement of the theorem follows by noticing that $\varphi(e_n)$ is the coefficient of z^n in $e_q^{u^0 z}\cdots e_q^{u^k z}$. □

In this section we changed the "$1/n!$" in $\varphi(e_n) = (-1)^{n-1}f(n)/n!$ in various ways, allowing us to keep track of inversions, the major index statistic, descents in the inverse permutation, common descents, and analogous statistics for words. By combining these different changes to "$1/n!$" and by modifying the function $f(n)$, we can find all sorts of generating functions for permutations and words.

Exercises

3.1. For $\sigma = \sigma_1\cdots\sigma_n \in S_n$, the permutation statistic "rise," denoted $\mathrm{ris}(\sigma)$, is the number of indices i for which $\sigma_i < \sigma_{i+1}$. Suppose that for each n, T_n is a subset of S_n. Further, suppose we know

$$f(z,x) = \sum_{n=0}^{\infty} \frac{z^n}{n!} \sum_{\sigma\in T_n} x^{\mathrm{des}(\sigma)}.$$

How can $f(z,x)$ be used to find a generating function for $\sum_{n=0}^{\infty}\frac{z^n}{n!}\sum_{\sigma\in T_n} x^{\mathrm{ris}(\sigma)}$?

The point of this exercise is to illustrate that if we know the generating function for $\sum_{\sigma\in T_n} x^{\mathrm{stat}(\sigma)}$ for some permutation statistic stat, then we can use it to find the generating function for the complement statistic defined by $n-\mathrm{stat}(\sigma)$ for $\sigma\in T_n$.

3.2. Let stat be a permutation statistic, let k be a nonnegative integer, let T_n be a subset of S_n for all n, and let a_n be the number of permutations in T_n with $\mathrm{stat}(\sigma) = k$. Suppose we know

$$f(z,x) = \sum_{n=0}^{\infty} \frac{z^n}{n!} \sum_{\sigma\in T_n} x^{\mathrm{stat}(\sigma)}. \tag{3.16}$$

How can $f(z,x)$ be used to find a generating function for $\sum_{n=0}^{\infty} a_n\frac{z^n}{n!}$?

3.3. Let stat be a permutation statistic, let m be a nonnegative integer, and let T_n be a subset of S_n for all n. Suppose we know the generating function $f(z,x)$ in (3.16). How can $f(z,x)$ be used to find a generating function for

$$\sum_{n=0}^{\infty} \frac{z^n}{n!} \sum_{\sigma \in T_n} \text{stat}(\sigma)^m?$$

This allows us to find the mth moment of stat(σ), which is $\sum_{\sigma \in T_n} \text{stat}(\sigma)^m/|T_n|$. If a permutation $\sigma \in T_n$ is randomly chosen such that each element in T_n has an equal probability of being selected, then the first moment is the expected value of stat and the second moment can be used to find the variance, which is equal to

$$\frac{1}{|T_n|} \sum_{\sigma \in T_n} \text{stat}(\sigma)^2 - \left(\frac{1}{|T_n|} \sum_{\sigma \in T_n} \text{stat}(\sigma) \right)^2.$$

3.4. The normal distribution with expected value μ and variance v is $\frac{1}{\sqrt{2\pi v}} e^{-\frac{(x-\mu)^2}{2v}}$. Use Exercise 3.3 to find the expected value and variance for descents in S_n, thereby finding the normal distribution which best approximates the distribution of descents in S_n.

3.5. Prove $2^{n-2i}\binom{n-i}{i} = \sum_{j \geq 0} \binom{n+1}{2j+1}\binom{j}{i}$ by a double counting argument.

3.6. With the help of the identity in Exercise 3.5, show that the function $f(n)$ in (3.2) is equal to

$$f(n) = (-1)^{n-1} \sum_{i \geq 0} \binom{n-i}{i} (-x)^i = \frac{\left(1+\sqrt{1-4x}\right)^{n+1} - \left(1-\sqrt{1-4x}\right)^{n+1}}{(-2)^{n+1}\sqrt{1-4x}}.$$

Then manipulate the generating function in (3.4) into the function in Theorem 3.3.

3.7. Define

$$f(n) = (-1)^{n-1} \sum_{i \geq 0} \binom{n-1-i}{i} x^{2i}.$$

Show that $f(n)$ gives the number of rearrangements of xs and (-1)s of length $n-1$ such that every maximal consecutive subsequence of xs has an even length. Use the identity displayed in Exercise 3.6 to find an explicit formula for $f(n)$.

Let φ be the ring homomorphism defined by $\varphi(e_n) = (-1)^{n-1} f(n)/n!$ where $f(n)$ is given above. Show that $n!\varphi(h_n) = \sum x^{\text{des}(\sigma)}$ where the sum runs over all permutations $\sigma \in S_n$ with maximal descending runs of only odd lengths. Use this to find an explicit generating function for

$$\sum_{n=0}^{\infty} \frac{z^n}{n!} \sum_{\substack{\sigma \in S_n \text{ has descending} \\ \text{runs of only odd lengths}}} x^{\text{des}(\sigma)}.$$

3.8. Find a generating function for $\sum x^{\mathrm{des}(\sigma)}$ where the sum runs over all permutations in $\sigma \in S_n$ which do not have a maximal descending run of length 1. Approximately what is the probability that a random permutation in S_n will not have a maximal descending run of length 1?

3.9. A Dyck path of length $2n$ is a path in the plane which starts at $(0,0)$, ends at $(2n,0)$, uses steps of the form $(1,1)$ or $(1,-1)$, and never travels below the x-axis. A labeled Dyck path is a Dyck path where each step between $y=k-1$ and $y=k$ is labeled with a number in $\{1,\dots,k\}$. For example, one labeled Dyck path is

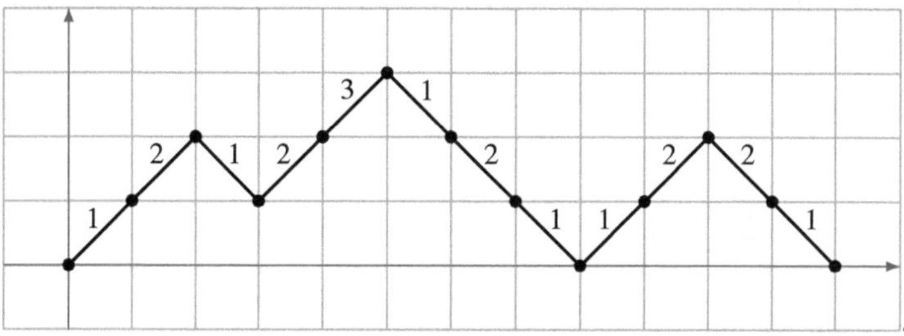

Use a bijection to show that the number of labeled Dyck paths of length $2n$ is equal to the number of alternating permutations in S_{2n}.

3.10. Let $A_k(z) = \sum_{n=0}^{\infty} a_{2n,k} z^{2n}$ where $a_{2n,k}$ is the number of labeled Dyck paths of length $2n$ which start at $(0,k)$, end at $(2n,k)$, and never travel below the line $y=k$ (see Exercise 3.9). Show that $A_{k-1}(z) = 1/(1-k^2z^2A_k(z))$ and deduce that

$$\sum_{n=0}^{\infty} |\{\sigma \in S_{2n} : \sigma \text{ is alternating}\}| z^{2n} = \cfrac{1}{1-\cfrac{1^2z^2}{1-\cfrac{2^2z^2}{1-\cfrac{3^2z^2}{1-\cdots}}}}.$$

3.11. Modifying the methods introduced in exercises 3.9 and 3.10, show that

$$\sum_{n=0}^{\infty} |\{\sigma \in S_{2n+1} : \sigma \text{ is alternating}\}| z^{2n+1} = \cfrac{z}{1-\cfrac{1\cdot 2\cdot z^2}{1-\cfrac{2\cdot 3\cdot z^2}{1-\cfrac{3\cdot 4\cdot z^2}{1-\cdots}}}}.$$

Use this result together with the continued fraction in Exercise 3.10 to find the remainder when the number of alternating permutations in S_n is divided by 4.

3.12. The hyperoctahedral group B_n, also called the set of signed permutations, is the set of permutations σ of $\{-n,\dots,-1,1,\dots,n\}$ such that $\sigma(-i) = -\sigma(i)$ for all i. Elements in B_n are denoted as permutations of $\{1,\dots,n\}$ in one-line notation with a line over i if $\sigma(i)$ is negative. For example, $\overline{2}\ 1 \in B_2$ denotes the permutation σ which satisfies $\sigma(1) = -2$, $\sigma(-1) = 2$, $\sigma(2) = 1$, and $\sigma(-2) = -1$.

Define a total ordering $\prec$ on the integers that all nonnegative numbers are larger than all positive numbers and nonnegative numbers and positive numbers are ordered among themselves in the usual way. Thus, for example,

$$0 \prec 1 \prec 2 \prec 3 \prec 4 \prec -4 \prec -3 \prec -2 \prec -1.$$

For $\sigma = \sigma_1 \cdots \sigma_n \in B_n$, set $\sigma_{n+1} = +\infty$ and define $\operatorname{des}_B(\sigma)$ to be the number of indices i for which $\sigma_{i+1} \prec \sigma_i$ and $\operatorname{neg}(\sigma)$ to be the number of indices i for which σ_i is negative.

By defining $\varphi(e_n)$ to equal $\dfrac{(-1)^{n-1}}{n!}\left((x-1)^{n-1} + v^n x(1-x)^{n-1}\right)$, prove that

$$\sum_{n=0}^{\infty} \frac{z^n}{n!} \sum_{\sigma \in B_n} v^{\operatorname{neg}(\sigma)} x^{\operatorname{des}_B(\sigma)} = \frac{x-1}{xe^{v(1-x)z} - e^{(x-1)z}}.$$

Use this result to find the generating function for $\sum_{n=0}^{\infty} \frac{z^n}{n!} \sum_{\sigma \in D_n} v^{\operatorname{neg}(\sigma)} x^{\operatorname{des}_B(\sigma)}$ where the demihyperoctahedral group D_n is the subgroup of B_n containing those permutations σ with an even number of indices i for which $\sigma(i)$ is negative.

3.13. For $w = w_1 \cdots w_n \in \{0, \ldots, k-1\}_n^*$, define the number of weak descents, denoted $\operatorname{wdes}(w)$, to be the number of indices i for which $w_i \geq w_{i+1}$. Use the ring homomorphism φ defined by

$$\varphi(e_n) = (-1)^{n-1} \begin{bmatrix} n+k-1 \\ k-1 \end{bmatrix}_q (x-1)^{n-1} \tag{3.17}$$

to find a generating function for $\sum_{n=0}^{\infty} z^n \sum_{w \in \{0,\ldots,k-1\}_n^*} x^{\operatorname{wdes}(w)} q^{\operatorname{sum}(w)}$. Then find simple expressions for $\sum_{n=0}^{\infty} z^n \sum_{w \in \{0,\ldots,k-1\}_n^*} x^{\operatorname{wdes}(w)}$ and the generating function for the number of words which do not have j consecutive weak descents.

3.14. Use Exercise 3.13 to show that

$$\sum_{n=0}^{\infty} \frac{(-1)^n z^n q^n}{(u;q)_{n+1}} \sum_{\sigma \in S_n} x^{\operatorname{des}(\sigma^{-1})} u^{\operatorname{des}(\sigma)} q^{\operatorname{maj}(\sigma)} = \sum_{k=1}^{\infty} \frac{(x-1)u^{-k-1}}{x - \sum_{n=0}^{\infty} \begin{bmatrix} n+k-1 \\ k-1 \end{bmatrix}_{\frac{1}{q}} (x-1)^n z^n}.$$

3.15. If a ring homomorphism φ is defined on Λ_n by setting

$$\varphi(e_n) = (-1)^{n-1}(x-1)^{n-1} h_n(y_1, y_2, \ldots)$$

for $n \geq 1$, then what enumeration result arises from applying φ to h_n? What is the corresponding result if φ is changed to

$$\varphi'(e_n) = (-1)^{n-1}(x-1)^{n-1} p_n(y_1, y_2, \ldots)?$$

Exercises 3.16, 3.17, and 3.18 illustrate how information about certain orthogonal polynomials can be found by defining homomorphisms on Λ.

3.16. The Chebyshev polynomial of the first kind $T_n(x)$ is the coefficient of z^n in the series expansion of $(1-xz)/(1-2xz+z^2)$ and the Chebyshev polynomial of the second kind $U_n(x)$ is the coefficient of z^n in the series expansion of $1/(1-2xz+z^2)$. Let φ be the homomorphism defined by

$$\varphi(e_n) = \begin{cases} 1 & \text{if } n=0 \text{ or } n=2, \\ 2x & \text{if } n=1, \text{ and} \\ 0 & \text{otherwise.} \end{cases}$$

Show that $\varphi(p_n) = 2T_n(x)$ for $n \geq 1$ and $\varphi(h_n) = U_n(x)$ for $n \geq 0$. Then use previously established relationships between e_n, h_n, and p_n to prove these identities:

$$T_n(x) = \frac{1}{2}\sum_{i=0}^{\lfloor n/2 \rfloor}\left(\binom{n-i-1}{i} + 2\binom{n-i-1}{i-1}\right)(-1)^i(2x)^{n-2i}, \tag{3.18}$$

$$U_n(x) = \sum_{i=0}^{\lfloor n/2 \rfloor}\binom{n-i}{i}(-1)^i(2x)^{n-2i}, \tag{3.19}$$

$$U_n(x) = \frac{2}{n}\sum_{i=0}^{n-1} U_i(x)T_{n-i}(x), \tag{3.20}$$

$$U_n(x) = 2xU_{n-1}(x) - U_{n-2}(x), \tag{3.21}$$

$$U_n(x) = \sum_{\lambda \vdash n}\frac{2^{\ell(\lambda)}}{z_\lambda}T_{\lambda_1}(x)\cdots T_{\lambda_\ell}(x), \tag{3.22}$$

$$0 = \sum_{\lambda \vdash n}\frac{(-2)^{\ell(\lambda)}}{z_\lambda}T_{\lambda_1}(x)\cdots T_{\lambda_\ell}(x) \tag{3.23}$$

are true for $n \geq 3$. Additionally, show that

$$T_n(x) = \begin{vmatrix} x & 1 & 0 & 0 & 0 & 0 \\ 1 & 2x & 1 & 0 & 0 & 0 \\ 0 & 1 & 2x & 1 & 0 & 0 \\ & & & \ddots & & \\ 0 & 0 & 0 & 1 & 2x & 1 \\ 0 & 0 & 0 & 0 & 1 & 2x \end{vmatrix} \quad \text{and} \quad U_n(x) = \begin{vmatrix} 2x & 1 & 0 & 0 & 0 & 0 \\ 1 & 2x & 1 & 0 & 0 & 0 \\ 0 & 1 & 2x & 1 & 0 & 0 \\ & & & \ddots & & \\ 0 & 0 & 0 & 1 & 2x & 1 \\ 0 & 0 & 0 & 0 & 1 & 2x \end{vmatrix}$$

where these are $n \times n$ determinants.

3.17. The Legendre polynomial $P_n(x)$ is the coefficient of z^n in the power series expansion of $1/\sqrt{1-2xz+z^2}$. Define a homomorphism φ by $\varphi(p_n) = T_n(x)$ where $T_n(x)$ is defined in Exercise 3.16. Use Exercise 2.3 to show that $\varphi(h_n) = P_n(x)$. This means we can apply φ to previously established relationships between h_n and p_n to find numerous identities involving $P_n(x)$ and $T_n(x)$.

3.18. The Hermite polynomial $H_n(x)$ is the coefficient of $z^n/n!$ in e^{2xz-z^2}. Define a homomorphism φ by $\varphi(p_0)=1$, $\varphi(p_1)=2x$, $\varphi(p_2)=-2$, and $\varphi(p_n)=0$ for $n\geq 3$. Show $\varphi(n!h_n)=H_n(x)$ and

$$H_n(x)=\sum_{i=0}^{\lfloor n/2\rfloor}\frac{n!}{(n-2i)!i!}(-1)^i(2x)^{n-2i}.$$

3.19. Define a homomorphism φ by $\varphi(h_n)=p(n)$ where $p(n)$ is the number of integer partitions of n. Using the fact that the logarithm of a product is a sum of logarithms and the series expansion of $\ln 1/(1-z)$, show that $\varphi(p_n)=\sigma(n)$ where $\sigma(n)$ is the sum of the divisors of n. Conclude that $n\sigma(n)=\sum_{k=1}^{n}\sigma(k)p(n-k)$.

Solutions

3.1 Subtracting the first term from the series and then taking "z" as "zx" and "x" as "x^{-1}" in Corollary 3.2, we find

$$\sum_{n=1}^{\infty}\frac{z^n}{n!}\sum_{\sigma\in T_n}x^{n-\operatorname{des}(\sigma)}=f(zx,1/x).$$

Since each index from $1,\ldots,n-1$ is either a descent or a rise, $\operatorname{des}(\sigma)+\operatorname{ris}(\sigma)=n-1$ for all permutations $\sigma\in S_n$, implying $n-\operatorname{des}(\sigma)=\operatorname{ris}(\sigma)+1$. Dividing the above equation by x and then adding the first term back into the series gives

$$\sum_{n=0}^{\infty}\frac{z^n}{n!}\sum_{\sigma\in T_n}x^{\operatorname{ris}(\sigma)}=\left(f(zx,1/x)-1\right)/x+1.$$

3.2 Taking partial derivatives,

$$\begin{aligned}\frac{1}{k!}\frac{\partial}{\partial x}f(z,x)&=\sum_{n=0}^{\infty}\frac{z^n}{n!}\sum_{\sigma\in T_n}\frac{\operatorname{stat}(\sigma)\left(\operatorname{stat}(\sigma)-1\right)\cdots\left(\operatorname{stat}(\sigma)-k+1\right)}{k!}x^{\operatorname{stat}(\sigma)-k}\\&=\sum_{n=0}^{\infty}\frac{z^n}{n!}\sum_{\sigma\in T_n}\binom{\operatorname{stat}(\sigma)}{k}x^{\operatorname{stat}(\sigma)-k}.\end{aligned}$$

Evaluating this expression at $x=0$ gives the constant terms with respect to x, thereby counting all permutations σ with $\operatorname{stat}(\sigma)-k=0$. In this case $\binom{\operatorname{stat}(\sigma)}{k}=1$, which means $\sum_{\sigma\in T_n}\binom{\operatorname{stat}(\sigma)}{k}x^{\operatorname{stat}(\sigma)-k}\Big|_{x=0}=a_n$. We have found

$$\frac{1}{k!}\frac{\partial}{\partial x}f(z,x)\bigg|_{x=0}=\sum_{n=0}^{\infty}a_n\frac{z^n}{n!}.$$

3.3 Let A_x be the operator $x\frac{\partial}{\partial x}$. Then

$$A_x f(z,x) = \sum_{n=0}^{\infty} \frac{z^n}{n!} \sum_{\sigma \in T_n} \mathrm{stat}(\sigma) x^{\mathrm{stat}(\sigma)}.$$

Iterating m times and evaluating at $x = 1$ (or taking the limit as $x \to 1$ as necessary),

$$A_x^m f(z,x)|_{x=1} = \sum_{n=0}^{\infty} \frac{z^n}{n!} \sum_{\sigma \in T_n} \mathrm{stat}(\sigma)^m,$$

which gives a way to find the desired generating function.

3.4 Using the operator A_x defined in the solution to Exercise 3.3, in the example of descents we take $f(z,x) = (x-1)/(x - e^{(x-1)z})$ and see that

$$A_x^1 f(z,x)\big|_{x=1} = \frac{z^2}{2(1-z)^2} = \frac{1}{2}\sum_{n=0}^{\infty}(n+1)z^{n+2}$$

where we used L'Hôpital's rule to take the limit as $x \to 1$ and Newton's binomial theorem to simplify. The expected number of descents in S_n is $(n-1)/2$.

As for the variance, we calculate

$$\begin{aligned} A_x^2 f(z,x)\big|_{x=1} &= \frac{3z^2 - z^3 + z^4}{6(1-z)^3} \\ &= \frac{1}{6}\sum_{n=0}^{\infty} 3\binom{n+2}{2} z^{n+2} - \binom{n+2}{2} z^{n+3} + \binom{n+2}{2} z^{n+4}. \end{aligned}$$

The coefficient of z^n in the above expression is the second moment. Therefore the variance is

$$\frac{1}{2}\binom{n}{2} - \frac{1}{6}\binom{n-1}{2} + \frac{1}{6}\binom{n-2}{2} - \left(\frac{n+1}{2}\right)^2 = \frac{n+1}{12}.$$

The normal distribution which best approximates the number of descents in S_n is therefore $\sqrt{6}e^{-\frac{3(2x-n+1)^2}{2(n+1)}}/\sqrt{(n+1)\pi}$. Below we plot this normal distribution when $n = 25$ along with bars showing the exact probabilities that a permutation in S_{25} has x descents:

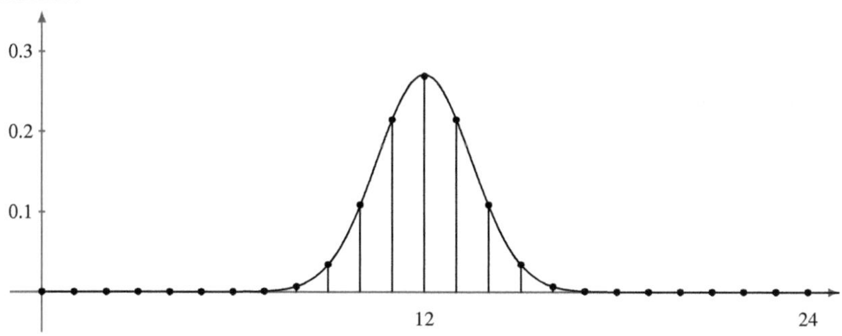

3.5 Create an ordered pair of subsets (S,T) by first selecting a subset S of size i from $\{1,\ldots,n-i\}$ and then selecting a subset T of any size from $\{1,\ldots,n\}\setminus S$. If there are odd number of elements in $S\cup T$, add $n+1$ to T. There are $\binom{n-i}{i}$ ways to select S and an independent 2^{n-2i} ways to select T, so there are $2^{n-2i}\binom{n-i}{i}$ such ordered pairs.

There is a second way to create such an ordered pair (S,T). First choose a subset X of an odd size larger than $2i$ from $\{1,\ldots,n+1\}$. If X has $2j+1$ elements, choose S to be a subset of size i from the smallest j elements in X and let $T=X\setminus S$. Then S is a subset of size i selected from $\{1,\ldots,n-i\}$, T is a subset of $\{1,\ldots,n+1\}$, and $S\cup T$ is odd, as desired. The number of ways to follow this procedure is $\sum_{j\geq 0}\binom{2n+1}{2j+1}\binom{j}{i}$ since the summand and $\binom{n+1}{2j+1}$ select X while $\binom{j}{i}$ selects S.

3.6 Take a sequence, like

$$x\ \ -1\ \ x\ \ -1\ \ -1\ \ x\ \ -1\ \ -1\ \ x\ \ -1\ \ x\ \ -1\ \ -1\ \ x\ \ -1\ \ -1\ \ -1\ \ x$$

which contains i copies of x and $n-1-i$ copies of -1 such that no two xs appear consecutively and interpret the first x and subsequent pairs $-1\ x$ as a bar and the remaining -1s as a star. This changes the above sequence into

$$|\,|\star|\star|\,|\star|\star\star|.$$

There are $\binom{n-i}{i}$ rearrangements i bars and $n-2i$ stars, so the number of desired sequences is also equal to $\binom{n-i}{i}$. Therefore $f(n)=(-1)^{n-1}\sum_{i\geq 1}\binom{n-i}{i}(-x)^i$.

Expanding both terms with the help of the binomial theorem,

$$\begin{aligned}(1+\sqrt{1-4x})^{n+1}-(1-\sqrt{1-4x})^{n+1}&=\sum_{k\geq 0}\binom{n+1}{k}\left(1-(-1)^k\right)\left(\sqrt{1-4x}\right)^k\\&=2\sum_{j\geq 0}\binom{n+1}{2j+1}\left(\sqrt{1-4x}\right)^{2j+1}.\end{aligned}$$

Therefore the expression on the right-hand side of statement of this exercise is

$$\begin{aligned}\frac{(-1)^{n+1}}{2^n}\sum_{j\geq 0}\binom{n+1}{2j+1}(1-4x)^j&=\frac{(-1)^{n+1}}{2^n}\sum_{i\geq 0}\sum_{j\geq 0}\binom{n+1}{2j+1}\binom{j}{i}(-4x)^i\\&=(-1)^{n-1}\sum_{i\geq 0}\binom{n-i}{i}(-x)^i,\end{aligned}$$

where the last line used Exercise 3.5. This is $f(n)$, as desired.

Let $a=\sqrt{1-4x}$ and substitute $f(n)$ into (3.4) to find

$$\left(\sum_{n=0}^{\infty}\frac{(-1)^{n-1}}{n!}\left(\frac{(1+a)^{n+1}-(1-a)^{n+1}}{(-2)^{n+1}a}\right)(-z)^n\right)^{-1}$$

$$=\left(\frac{1}{2a}(1+a)\sum_{n=0}^{\infty}\frac{\left(\frac{-z(1+a)}{2}\right)^n}{n!}-\frac{1}{2a}(1-a)\sum_{n=0}^{\infty}\frac{\left(\frac{-z(1-a)}{2}\right)^n}{n!}\right)^{-1}$$

$$=\left(\frac{1}{2a}(1+a)e^{-z/2}e^{-za/2}-\frac{1}{2a}(1-a)e^{-z/2}e^{za/2}\right)^{-1}.$$

This last line is equal to

$$\frac{e^{z/2}}{\cosh(za/2)-\frac{1}{a}\sinh(za/2)},$$

which, after using $a=i\sqrt{4x-1}$ and the identities $\cosh(iz)=\cos(z)$ and $\sinh(iz)=i\sin(z)$, can be manipulated into the desired expression.

3.7 Take a rearrangement of $2i$ copies of x and $n-1-2i$ copies of -1 such that every maximal consecutive subsequence of xs has an even length, such as

$$x\ \ x\ \ x\ \ x\ \ -1\ \ -1\ \ -1\ \ x\ \ x\ \ x\ \ x\ \ x\ \ x\ \ -1\ \ x\ \ x\ \ -1\ \ x\ \ x,$$

and interpret "xx" as a star and "-1" as a bar. This changes the above sequence into $\star\star\,|\,|\,|\star\star\star\,|\,\star\,|\,\star$. There are $\binom{n-1-i}{i}$ rearrangements of i stars and $n-2i-1$ bars, so the number of desired sequences is also equal to $\binom{n-1-2i}{i}$. Therefore

$$f(n)=(-1)^{n-1}\sum_{i\geq 0}\binom{n-i}{i}(-x)^i=-\frac{\left(1+\sqrt{1+4x^2}\right)^n-\left(1-\sqrt{1+4x^2}\right)^n}{(-2)^n\sqrt{1+4x^2}}\tag{3.24}$$

where the second equality follows from taking "n" as "$n-1$" and "$-x$" as "x^2" in the identity displayed in Exercise 3.6.

Using the same steps as what led to (3.1),

$$n!\varphi(h_n)=\sum_{\lambda\vdash n}\binom{n}{\lambda}|B_{\lambda,(n)}|f(\lambda_1)f(\lambda_2)\cdots.$$

From this sum we create combinatorial objects by selecting a brick tabloid in $B_{\lambda,(n)}$ for some $\lambda\vdash n$, writing the integers $1,\ldots,n$ in the cells so that each brick contains a decreasing sequence, placing a sequence of xs and -1s such that every maximal consecutive subsequence of xs has an even length in the first $k-1$ cells in each brick of length k, and placing a $+1$ in the last cell of each brick.

Use the usual brick breaking or combining involution first described in the proof of Theorem 3.1 on this set of combinatorial objects. Fixed points which look like

x	x	1	x	x	x	x	1	1	x	x	1
10	7	3	12	11	9	6	1	5	8	4	2

,

have bricks of only odd lengths. Therefore $n!\varphi(h_n) = \sum x^{\mathrm{des}(\sigma)}$ where the sum runs over all permutations $\sigma \in S_n$ with descending runs of only odd lengths.

Applying φ to both sides of Theorem 2.5,

$$\sum_{n=0}^{\infty} \frac{z^n}{n!} \sum_{\substack{\sigma \in S_n \text{ has descending} \\ \text{runs of only odd lengths}}} x^{\mathrm{des}(\sigma)} = \varphi\left(\sum_{n=0}^{\infty} h_n z^n\right) = \frac{1}{1 - \sum_{n=1}^{\infty} z^n f(n)/n!}.$$

Using (3.24) to simplify, the above generating function can be shown to equal

$$\frac{\sqrt{1+4x^2}}{\sqrt{1+4x^2} + e^{-z/2}\left(e^{-z\sqrt{1+4x^2}/2} - e^{z\sqrt{1+4x^2}/2}\right)}.$$

3.8 Take a rearrangement of i copies of -1 and $n-i-1$ copies of x which end in x and have every -1 immediately preceded by an x, such as

$$x \ \ -1 \ \ x \ \ x \ \ -1 \ \ x \ \ -1 \ \ x \ \ x \ \ -1 \ \ x \ \ -1 \ \ x \ \ -1 \ \ x \ \ x \ \ -1 \ \ x \ \ x,$$

and interpret each "$x(-1)$" as a star and all other copies of "x" as a bar. This changes the above sequence into $\star \mid \star\star \mid \star\star\star \mid \star \mid \mid$. There are $\binom{n-i-2}{i}$ rearrangements of i stars and $n-2i-1$ bars ending with a bar, so there are $\binom{n-i-2}{i}$ such sequences.

Define a function $f(n)$ by

$$f(n) = \sum_{i \geq 0} \binom{n-i-2}{i} x^{n-i-1}(-1)^i = x^{n-1} \sum_{i \geq 0} \binom{n-i-2}{i} \left(-\frac{1}{x}\right)^i.$$

Taking "n" as "$n-2$" and "x" as "$1/x$" in the identity displayed in Exercise 3.6 allows us to rewrite this function as

$$f(n) = x \frac{\left(x + \sqrt{x^2-4x}\right)^{n-1} - \left(x - \sqrt{x^2-4x}\right)^{n-1}}{2^{n-1}\sqrt{x^2-4x}}. \tag{3.25}$$

Define a ring homomorphism φ such that $\varphi(e_n) = (-1)^{n-1} f(n)/n!$ where $f(n)$ is given in (3.25). Using the same steps as what led to (3.1),

$$n!\varphi(h_n) = \sum_{\lambda \vdash n} \binom{n}{\lambda} |B_{\lambda,(n)}| f(\lambda_1) f(\lambda_2) \cdots.$$

From this sum we create combinatorial objects by selecting a brick tabloid in $B_{\lambda,(n)}$ for some $\lambda \vdash n$, writing the integers $1, \ldots, n$ in the cells so that each brick contains a decreasing sequence, placing a sequence of xs and -1s such that every -1 is immediately preceded by an x in the first $k-1$ cells in each brick of length k, and placing a $+1$ in the last cell of each brick.

The usual brick breaking or combining involution first described in the proof of Theorem 3.1 leaves fixed points which have no bricks of length 1. This shows that $n!\varphi(h_n)$ is equal to $\sum x^{\mathrm{des}(\sigma)}$ where the sum runs over all permutations in $\sigma \in S_n$ which do not have descending runs of length 1.

Applying φ to both sides of Theorem 2.5,

$$\sum_{n=0}^{\infty} \frac{z^n}{n!} \sum_{\substack{\sigma \in S_n \text{ has no} \\ \text{descending runs of length 1}}} x^{\text{des}(\sigma)} = \varphi\left(\sum_{n=0}^{\infty} h_n z^n\right) = \frac{1}{1 - \sum_{n=1}^{\infty} z^n f(n)/n!}.$$

Using 3.25 and performing very similar manipulations as shown in the solution to Exercise 3.6, the above generating function can be shown to equal

$$\frac{e^{-xz/2}}{\cos\left(\frac{z\sqrt{4x-x^2}}{2}\right) - \frac{x}{\sqrt{4x-x^2}} \sin\left(\frac{z\sqrt{4x-x^2}}{2}\right)}.$$

Taking $x = 1$ in this function,

$$\sum_{n=0}^{\infty} \frac{|\{\sigma \in S_n \text{ has no descending runs of length 1}\}|}{n!} z^n = \frac{e^{-z/2}}{\cos\left(\frac{z\sqrt{3}}{2}\right) - \frac{1}{\sqrt{3}} \sin\left(\frac{z\sqrt{3}}{2}\right)}.$$

With the exception of the negative sign in the exponent of $e^{-z/2}$, this is the same function as found in (3.5). Doing the same calculations as in the discussion which follows (3.5), we find that the approximate probability that a permutation in S_n has no descending runs of length 1 is $e^{-\pi\sqrt{3}/9}\left(9/(2\pi\sqrt{3})\right)^{n+1}$.

3.9 Starting with an alternating permutation $\sigma = \sigma_1\sigma_2 \cdots \sigma_{2n-1}\sigma_{2n}$, we will describe how to create a labeled Dyck path. Begin by drawing a Dyck path with $(1,-1)$ steps ending at x coordinates $\sigma_1, \sigma_3, \ldots, \sigma_{2n-1}$ and $(1,1)$ steps ending at x coordinates $\sigma_2, \sigma_4, \ldots, \sigma_{2n}$. Color the $(1,1)$ steps blue.

Suppose the most left $(1,-1)$ step ends at x coordinate σ_{2i-1}. Label this $(1,-1)$ step with the number of blue $(1,1)$ steps found between x coordinates $\sigma_{2i} - 1$ and σ_{2i-1}. Recolor the $(1,1)$ step ending at x coordinate σ_{2i} black.

Continue inductively: Find the next most left $(1,-1)$ step, say it ends x coordinate σ_{2j-1}. Label this step with the number of blue $(1,1)$ steps found between x coordinates $\sigma_{2j} - 1$ and σ_{2j-1}, and recolor the $(1,1)$ step ending at x coordinate σ_{2j} black. After completing this process, all of the $(1,-1)$ steps are labeled and the $(1,1)$ steps remain unlabeled.

Suppose the most right $(1,1)$ step ends at $x = \sigma_{2i}$. Remove the pair $\sigma_{2i-1}\sigma_{2i}$ from σ, leaving a sequence $\sigma_1\sigma_2\ \sigma_3\sigma_4\ \cdots\ \sigma_{2a-1}\sigma_{2a}$ with alternating descents. There are $a+1$ positions before and after each pair in this sequence:

$$\underbrace{\qquad}_{1}\ \sigma_1\sigma_2\ \underbrace{\qquad}_{2}\ \sigma_3\sigma_4\ \underbrace{\qquad}_{3}\ \cdots\ \underbrace{\qquad}_{a}\ \sigma_{2a-1}\sigma_{2a}\ \underbrace{\qquad}_{a+1}$$

Identify those positions for which the pair $\sigma_{2i-1}\sigma_{2i}$ could be reinserted as to maintain alternating descents. If $\sigma_{2i-1}\sigma_{2i}$ was originally in the ℓth such position, label the $(1,1)$ step ending at x coordinate σ_{2i} with ℓ.

Continue inductively: Find the next most right unlabeled $(1,1)$ step, say it ends at x coordinate σ_{2j}. Remove $\sigma_{2j-1}\sigma_{2j}$ from σ. If $\sigma_{2j-1}\sigma_{2j}$ was originally in the ℓth

position it may have been placed in the remaining portion of σ, label the $(1,1)$ step with ℓ. After completing this process, all steps are labeled. The resulting labeled Dyck path is the image of σ.

For example, the image of $\sigma = 3\ 2\ 7\ 5\ 12\ 9\ 10\ 1\ 11\ 4\ 8\ 6$ after applying the above operations is the labeled Dyck path

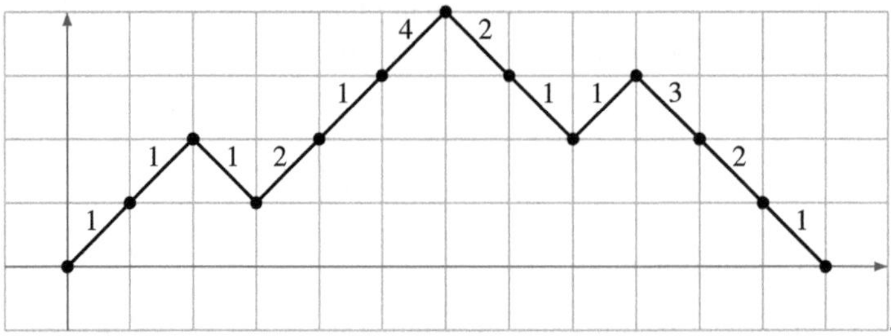

This process is a bijection. Indeed, the map described above is reversible. Given a labeled Dyck path, we can first identify the pairs $\sigma_1\sigma_2, \cdots, \sigma_{2n-1}\sigma_{2n}$ in the permutation σ with reading the labels on the steps of the form $(1,-1)$ from left to right. The order in which to insert these pairs as to form an alternating permutation can be deduced by reading the labels on the steps of the form $(1,1)$ from left to right. For example, working backward from the labeled Dyck path in the statement of the exercise produces the permutation 7 1 8 4 3 2 6 5 11 9 12 10.

3.10 Suppose that the first time after $(0,k-1)$ that a labeled Dyck path counted by $a_{2n,k-1}$ returns to the line $y = k-1$ is at $(2i, k-1)$. The underlying Dyck path must look like this:

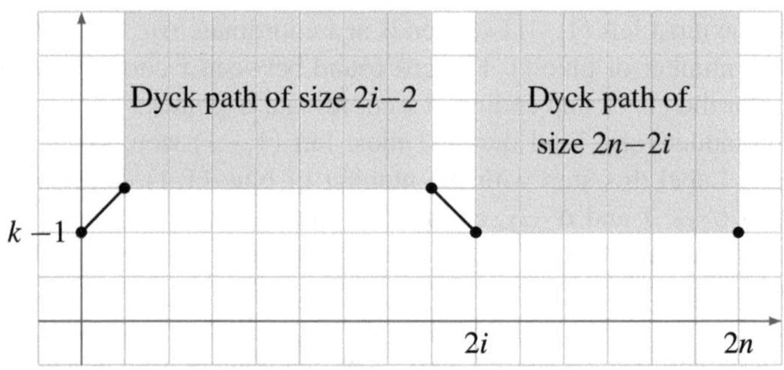

Since there are k ways to label the first $(1,1)$ step and k ways to label the $(1,-1)$ step ending at $x = 2i$, the number of labeled Dyck paths is $k^2 a_{2i-2,k} a_{2n-2i,k-1}$. Summing over all possible i gives $a_{n,k-1} = k^2 \sum_{i=1}^{n} a_{2i-2,k} a_{2n-2i,k-1}$ for $n \geq 1$. Therefore

$$\begin{aligned} A_{k-1}(z) - 1 &= k^2 \sum_{n=1}^{\infty} \sum_{i=1}^{n} a_{2i-2,k} a_{2n-2i,k-1} z^{2n} \\ &= k^2 z^2 A_k(z) A_{k-1}(z). \end{aligned}$$

Solving for $A_{k-1}(z)$ gives $A_{k-1}(z) = 1/(1 - k^2 z^2 A_k(z))$.

Exercise 3.9 says $A_0(z) = \sum_{n=0}^{\infty} |\{\sigma \in S_{2n} : \sigma \text{ is alternating}\}| z^{2n}$ and so the continued fraction follows from repeatedly applying $A_{k-1}(z) = 1/(1 - k^2 z^2 A_k(z))$, starting with $k = 1$.

3.11 The number of alternating permutations of $2n+1$ is equal to the number of alternating permutations of $2n+2$ which end with the integer 1. Looking back at the bijection in the solution to Exercise 3.9, the labeled Dyck paths of length $2n+2$ which correspond to alternating permutations ending in 1 are those labeled Dyck paths such that

1. except for the most left $(1,1)$ step, the labels on each $(1,1)$ step between $y = k-1$ and $y = k$ must have a label in $\{1,\ldots,k-1\}$, and
2. the Dyck path touches the x-axis at only $(0,0)$ and $(2n+2,0)$.

In pictures, we must have a labeled Dyck path which looks like

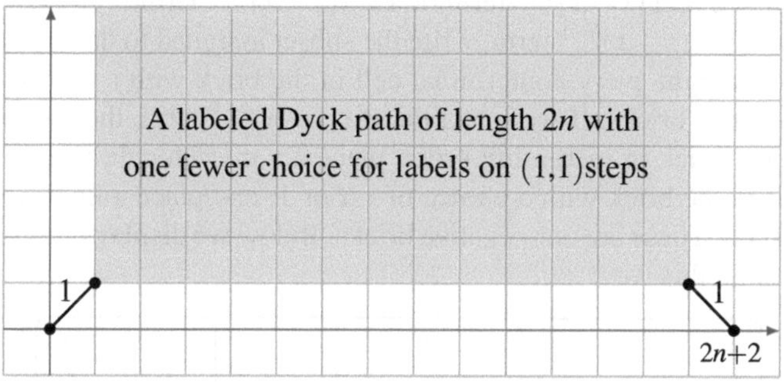

Let $A_k(z) = \sum_{n=0}^{\infty} a_{2n,k} z^{2n}$ be the generating function for the number of labeled Dyck paths which start at $(0,k)$, end at $(2n,k)$, and have one fewer choice for the labels on the $(1,1)$ steps. A very similar calculation to that found in the solution to Exercise 3.10 shows that $A_{k-1}(z) = 1/(1-(k-1)\cdot k \cdot z^2 A_k(z))$. This gives

$$\sum_{n=0}^{\infty} |\{\sigma \in S_{2n+1} : \sigma \text{ is alternating}\}| z^{2n+1} = zA_1(z) = \frac{z}{1 - 1\cdot 2\cdot z^2 A_2(z)}.$$

The continued fraction in the statement of the exercise follows from repeatedly applying $A_{k-1}(z) = 1/(1-(k-1)kz^2 A_k(z))$.

The two continued fractions in this exercise and Exercise 3.10 allow us to find the number of alternating permutations in S_n modulo k. For example, by replacing each 4 with 0 in the continued fraction expressions, a generating function with coefficient z^n congruent to the number of alternating permutations in S_n modulo 4 is

$$\frac{1}{1-1^2z^2} + \frac{z}{1 - \dfrac{1\cdot 2\cdot z^2}{1 - 2\cdot 3\cdot z^2}}.$$

Going further and replacing both the -2 and the -6 with 2, the coefficients of the above generating function are congruent to the coefficients in

$$\frac{1}{1-1^2z^2}+\frac{z}{1+\frac{2\cdot z^2}{1+2\cdot z^2}}=\sum_{n=0}^{\infty} z^{2n}+\frac{z+2z^3}{1+4z^2}.$$

Therefore the remainder after the number of alternating permutations in S_n is divided by 4 is 1 if n is even, 1 if $n=1$, 2 if $n=3$, and 0 otherwise.

3.12 Let $f(n)=\left((x-1)^{n-1}+v^n x(1-x)^{n-1}\right)$ so that we can write $\varphi(e_n)$ using the more compact notation $(-1)^{n-1}f(n)/n!$. Applying φ to $n!h_n$ gives equation (3.3). From this expression we create combinatorial objects with types of bricks: positive bricks and negative bricks.

With the sum, the $|B_{\lambda,(n)}|$, and multinomial coefficients in (3.3), select a brick tabloid $T\in B_{\lambda,(n)}$ for some $\lambda\vdash n$ and associate a disjoint subset of $\{1,\ldots,n\}$ of size k to each brick of length k in T. For each brick of length k, the function $f(k)$ gives us a choice of a $(x-1)^{k-1}$ term or a $v^k x(1-x)^{k-1}$ term.

If we select the $(x-1)^{n-1}$ term, write the subset assigned to the brick in decreasing order and weight every nonterminal cell in the brick with a choice of x or -1. These are positive bricks. If we select the $v^k x(1-x)^{k-1}$ term, then write the subset assigned to the brick in increasing order, place a v in each cell, weight every nonterminal cell in the brick with a choice of $-x$ or 1, and place a final weight of x in the terminal cell. These are our negative bricks. Below we display one combinatorial object created in this way:

x	x	x	1	$-x$	1	x	$-x$	x	x	x	1
				v	v	v	v	v			
8	5	3	1	2	4	7	11	12	10	9	6

The involution we would like to apply to these objects is a modification to our usual brick breaking and combining involution in that we will only combine two positive bricks together or two negative bricks together. That is, to apply our involution we scan a combinatorial object from left to right looking for the first cell containing

1. a -1,
2. two consecutive positive bricks straddling a decrease,
3. a $-x$, or
4. two consecutive negative bricks straddling an increase.

If we first find case 1 above, break the single positive brick into two bricks and reverse the sign on the -1 to $+1$. If we find case 2, then combine the two positive bricks, changing the sign on the 1 in the middle. If we first find case 3 above, break the single positive brick into two bricks and reverse the sign on the $-x$ to x. If we find case 4, then combine the two negative bricks, changing the sign on the x in the middle.

Fixed points must look like this:

x	x	x	1	1	1	1	x	1	x	x	1
				v	v	v	v	v			
8	5	3	1	2	4	11	7	12	10	9	6

.

These fixed points correspond to elements $\sigma \in B_n$ with a power of v counting the number of integers i for which $\sigma(i)$ is negative and a power of x counting the number of times we have a descent in B_n. The generating function follows from applying φ to both sides of Theorem 2.5 and simplifying.

The demihyperoctahedral group D_n contains those elements $\sigma \in B_n$ with $\mathrm{neg}(\sigma)$ is even. Thus if we let $f(x,v,z) = (x-1)/(xe^{v(1-x)z} - e^{(x-1)z})$ the generating function we just found for B_n, then

$$\sum_{n=0}^{\infty} \frac{z^n}{n!} \sum_{\sigma \in D_n} v^{\mathrm{neg}(\sigma)} x^{\mathrm{des}_B(\sigma)} = \frac{1}{2}\left(f(x,v,z) + f(x,-v,z)\right)$$
$$= \frac{1}{2}\left(\frac{x-1}{xe^{v(1-x)z} - e^{(x-1)z}} + \frac{x-1}{xe^{-v(1-x)z} - e^{(x-1)z}}\right).$$

3.13 Theorem 1.4 says $\left[{n+k-1 \atop k-1}\right]_q = \sum_{r \in R(0^n,1^{k-1})} q^{\mathrm{inv}(r)}$, so the q-multinomial coefficient tells us to select $r \in R(0^n, 1^{k-1})$ for each brick of length n. With r, interpret the number of 0s before the ith 1 to be the number of $(i-1)$s in a nonincreasing word $w \in \{0,\ldots,k-1\}_n^*$. It follows that $\mathrm{inv}(r) = \mathrm{sum}(w)$.

Applying the homomorphism φ in (3.17) to h_n gives

$$\varphi(h_n) = \sum_{\lambda \vdash n} \left[{\lambda_1 + k - 1 \atop k-1}\right]_q \left[{\lambda_2 + k - 1 \atop k-1}\right]_q \cdots |B_{\lambda,(n)}| (x-1)^{\lambda_1 - 1}(x-1)^{\lambda_2 - 1} \cdots .$$

This sum tells us to create combinatorial objects like this (using $k = 8$):

x	x	1	1	1	-1	-1	x	1	-1	-1	1
q^7	q^1	q^1	q^3	q^4	q^7	q^5	q^5	q^4	q^1	q^1	q^1
7	1	1	3	4	7	5	5	4	1	1	1

.

The usual involution shows $\varphi(h_n) = \sum_{w \in \{0,\ldots,k-1\}_n^*} x^{\mathrm{wdes}(w)} q^{\mathrm{sum}(w)}$. Applying φ to Theorem 2.5 gives

$$\sum_{n=0}^{\infty} z^n \sum_{w \in \{0,\ldots,k-1\}_n^*} x^{\mathrm{wdes}(w)} q^{\mathrm{sum}(w)} = \frac{x-1}{x - \sum_{n=0}^{\infty} \left[{n+k-1 \atop k-1}\right]_q (x-1)^n z^n}.$$

Taking $q = 1$ and using Newton's binomial theorem (see exercises 1.19 and 1.20), this generating function specializes to $(x-1)/(x - (1-(x-1)z)^{-k})$.

Using the ring homomorphism defined by $\varphi(e_n) = (-1)^{n-1}\binom{n+k-1}{k-1} f(n)$, where $f(n)$ is given in (3.6), gives

$$\sum_{n=0}^{\infty} z^n \sum_{\substack{w \in \{0,\ldots,k-1\}_n^* \text{ does not} \\ \text{have } j \text{ consec. descents}}} x^{\operatorname{des}(w)} = \frac{1}{1+\sum_{n=1}^{\infty}(-z)^n\binom{n+k-1}{n}\sum_{i\geq 1} R_{n-1,i,j}(-x)^i}.$$

Specializing by taking $x = 1$ and simplifying using the approach found in the proof of Theorem 3.4, we find that the generating function for number of words without j consecutive weak descents is

$$\frac{j+1}{(1-\zeta^j)(1-\zeta z)^{-k}+\cdots+(1-\zeta)(1-\zeta^j z)^{-k}},$$

where $\zeta = e^{2\pi i/(j+1)}$ is a primitive $(j+1)^{th}$ root of unity.

3.14 Select a term in $\sum_{w\in\{0,\ldots,k-1\}_n^*} x^{\operatorname{wdes}(w)} q^{\operatorname{sum}(w)}$ and let r be the reverse of the word w. With r we associate a permutation $\sigma^{-1} \in S_n$ by numbering from right to left the $(k-1)$s in r, then from right to left numbering the $(k-2)$s in r, and so on. This forces the $(k-1)$s in r to correspond to the first block of numbers in σ, the $(k-2)$s in r to correspond to the second block of numbers in σ, and so on. These blocks sort the exponents on q in nonincreasing order.

Also associate with r a nonnegative integer sequence $a = a_1\cdots a_n$ such that a_i is the difference between consecutive exponents on q in σ for $i = 1,\ldots,n-1$ and a_n is the final q exponent. This permutation σ and sequence a have the same properties as the σ and a in the proof of Theorem 3.10 except that property 5 should instead state that $a_i = 0$ if and only if $\sigma_i > \sigma_{i+1}$. Therefore

$$\sum_{w\in\{0,\ldots,k-1\}_n^*} x^{\operatorname{wdes}(w)} q^{\operatorname{sum}(w)}$$
$$= \sum_{j=0}^{k-1}\left(\sum_{\sigma\in S_n} x^{\operatorname{des}(\sigma^{-1})} \sum_{\substack{\text{sequences } a_1\cdots a_n \\ \text{with } a_i=0 \text{ iff } \sigma_i>\sigma_{i+1}}} q^{1a_1+\cdots+na_n} u^{a_1+\cdots+a_n}\right)\Bigg|_{u^j}.$$

Using similar steps as found in the proof of Theorem 3.10, this simplifies to

$$\frac{(-1)^n}{q^n(\frac{1}{u};\frac{1}{q})_{n+1}} \sum_{\sigma\in S_n} \frac{x^{\operatorname{des}(\sigma^{-1})}}{u^{\operatorname{des}(\sigma)} q^{\operatorname{maj}(\sigma)}}\Bigg|_{u^{k+1}}.$$

The desired result follows by using the result in Exercise 3.13, summing over all k, and replacing "q" with "$1/q$" and "u" with "$1/u$".

3.15 It follows that

$$\varphi(h_n) = \sum_{\lambda\vdash n} \left|B_{\lambda,(n)}\right| (x-1)^{n-\ell(\lambda)} h_\lambda(y_1,y_2,\ldots).$$

Create combinatorial objects in the exact same way as in the proof of Theorem 3.11 with the exception that the subscripts of the y variables are allowed to weakly decrease within bricks.

Apply the usual involution on this collection of combinatorial objects by looking for weak decreases in the subscripts of the y variables between two bricks and breaking or combining accordingly. Fixed points correspond to words $w \in \{1,2,\dots\}_n^*$ with a power of x for each weak decrease (weak decreases are defined in Exercise 3.13) in w and a subscript on the variable y for each letter in w.

This proves

$$\varphi(h_n) = \sum_{w_1\cdots w_n \in \{1,2,\dots\}_n^*} x^{\mathrm{wdes}(w)} y_{w_1} \cdots y_{w_n}.$$

The generating function

$$\sum_{n=0}^{\infty} z^n \sum_{w_1\cdots w_n \in \{1,2,\dots\}_n^*} x^{\mathrm{wdes}(w)} y_{w_1} \cdots y_{w_n} = \frac{x-1}{x - \prod_{i\geq 1} 1/(1-z(x-1)y_i)}$$

follows from applying φ to Theorem 2.5 and using the definition of the generating function $H(z)$.

The adjustment to the above argument when φ is changed to φ' is that each cell in a brick must be filled with the same variable y_i instead of having the subscripts within a brick weakly decrease. If we combine or break bricks when there are the same subscripts straddling two bricks, then we find fixed points with powers of x counting the number of times consecutive cells have the same subscript. This gives

$$\sum_{n=0}^{\infty} z^n \sum_{w_1\cdots w_n \in \{1,2,\dots\}_n^*} x^{\text{the number of } i \text{ for which } w_i = w_{i+1}} y_{w_1} \cdots y_{w_n}$$
$$= \frac{x-1}{x-1-\sum_{n=1}^{\infty} z^n (x-1)^n p_n(y_1,y_2,\dots)}.$$

3.16 Applying φ to both sides of Corollary 2.10,

$$\sum_{n=1}^{\infty} \varphi(p_n) z^n = \frac{\varphi\left(\sum_{n=1}^{\infty} (-1)^{n-1} n e_n z^n\right)}{\varphi(E(-z))} = \frac{2xz - 2z^2}{1-2xt+t^2}.$$

Therefore $1 + \sum_{n=1}^{\infty} \varphi(p_n) z^n/2 = (1-xz)/(1-2xz+z^2)$, showing $\varphi(p_n) = 2T_n(x)$.

Applying φ to both sides of Theorem 2.5 gives

$$\sum_{n=0}^{\infty} \varphi(h_n) z^n = \varphi(H(z)) = \frac{1}{\varphi(E(-z))} - \frac{1}{1-2xt+t^2},$$

and so $\varphi(h_n) = U_n(x)$.

Using Theorem 2.22,

$$2T_n(x) = \varphi(p_n) = \sum_{\lambda \vdash n} (-1)^{n-\ell(\lambda)} w(B_{\lambda,(n)}) \varphi(e_{\lambda_1}) \varphi(e_{\lambda_2}) \cdots. \tag{3.26}$$

Since $\varphi(e_n) = 0$ unless $n \leq 2$, the bricks in our weighted brick tabloid must have length 1 or 2. There are $\binom{n-i-1}{i}$ ways to select a brick tabloid with i bricks of length 2 and $n-2i$ bricks of length 1 which end in a brick of length 1. There are $\binom{n-i-1}{i-1}$ ways to select a brick tabloid with i bricks of length 2 and $n-2i$ bricks of length 1 which end in a brick of length 2. In either case, the definition of $\varphi(e_1)$ tells us that such a brick tabloid has an associated factor of $(2x)^{n-2i}$. There are $n-i$ total bricks, so the $(-1)^{n-\ell(\lambda)}$ term gives us $(-1)^i$. Putting everything together, (3.26) is equal to

$$\sum_{i=0}^{\lfloor n/2 \rfloor} \binom{n-i-1}{i}(-1)^i(2x)^{n-2i} + 2\binom{n-i-1}{i-1}(-1)^i(2x)^{n-2i}$$

with the extra 2 in front of the second binomial coefficient coming from the weight on a weighted brick tabloid. This proves (3.18).

Using Theorem 2.18,

$$U_n(x) = \varphi(h_n) = \sum_{\lambda \vdash n} (-1)^{n-\ell(\lambda)} |B_{\lambda,(n)}| \varphi(e_{\lambda_1})\varphi(e_{\lambda_2})\cdots. \tag{3.27}$$

Just as in (3.26), the bricks in our brick tabloid must have length 1 or 2. There are $\binom{n-i}{i}$ brick tabloids with i bricks of length 2 and $n-i$ bricks of length 1. The bricks of length 1 each contribute a $2x$ term and since $(-1)^{n-\ell(\lambda)} = (-1)^i$, (3.27) is equal to (3.19), as desired.

Equations (3.20), (3.21), (3.22), and (3.23) immediately follow from applying φ to Theorem 2.8, the identity $\sum_{i=0}^{n}(-1)^i e_i h_{n-i}$ which is implicit in Theorem 2.5, Theorem 2.11, and Theorem 2.12, respectively. The two identities involving determinants come from applying φ to Theorem 2.14 and to applying the ω transformation and then φ to Theorem 2.13.

3.17 By Exercise 2.3, $\left(-1+\sum_{n=0}^{\infty} p_n z^n\right)/z = H'(z)/H(z)$. Applying φ to both sides,

$$\frac{x-z}{1-2xz+z^2} = \varphi\left(\frac{H'(z)}{H(z)}\right).$$

Integrating with respect to z gives $\ln\left(1-2xz+z^2\right)^{-1/2} = \ln \varphi(H(z))$ from which we conclude that $\varphi(H(z)) = 1/\sqrt{1-2xz+z^2}$, as desired.

3.18 By Exercise 2.3, $\left(-1+\sum_{n=0}^{\infty} p_n z^n\right)/z = H'(z)/H(z)$. Applying φ to both sides,

$$2x - 2z = \varphi\left(\frac{H'(z)}{H(z)}\right).$$

Integrating with respect to z gives $2xz - z^2 = \ln \varphi(H(z))$ from which we conclude that $\varphi(H(z)) = e^{2xz-z^2}$, as desired.

Using the h-to-p transition matrix, we have

$$H_n(x) = \varphi(n!h_n) = \sum_{\lambda \vdash n} \frac{n!}{z_\lambda} OB_{\lambda,(n)} \varphi(p_\lambda).$$

Since $\varphi(p_n) = 0$ if $n \geq 3$, the parts in a partition λ in above sum must be either 1 or 2. Let i be the number of parts of size 2 in such a partition, meaning that the number of 1s is $n-2i$. It follows that $z_\lambda = 2^i i!(n-2i)!$, $OB_{\lambda,(n)} = 1$, and $\varphi(p_\lambda) = (-2)^i(2x)^{n-2i}$. Using these values in the above expression proves the result.

3.19 By Exercise 2.3,

$$\sum_{n=1}^{\infty} p_n z^{n-1} = \frac{1}{z}\left(-1 + \sum_{n=0}^{\infty} p_n z^n\right) = \frac{H'(z)}{H(z)}.$$

Integrating both sides gives $\sum_{n=1}^{\infty} p_n z^{n-1}/n = \ln H(z)$. Using Theorem 1.8 and the series expansion of $\ln 1/(1-z^i)$, we have

$$\sum_{n=1}^{\infty} \frac{\varphi(p_n)}{n} z^n = \ln \varphi(H(z)) = \ln \prod_{i=1}^{\infty} \frac{1}{1-z^i} = \sum_{i=1}^{\infty} \ln \frac{1}{1-z^i} = \sum_{i=1}^{\infty} \sum_{k=1}^{\infty} \frac{(z^i)^n}{n}.$$

Comparing coefficients of z^n on the extremes of these equalities,

$$\varphi(p_n) = n \sum_{i \cdot k = n} \frac{1}{k} = \sum_{i \cdot k = n} i = \sigma(n).$$

The identity involving $p(n)$ and $\sigma(n)$ given in the statement of the exercise follows from applying φ to Theorem 2.8.

Notes

The method of applying ring homomorphisms to symmetric functions in order to find generating functions started with the work of Brenti [14, 15]. The generating function in Corollary 3.2 is well known (see page 244 of [25], page 68 of [102], and page 215 of [45]), but it was Brenti who defined a ring homomorphism φ by setting $\varphi(e_n) = (-1)^{n-1}(x-1)/n!$ to find it.

Brenti then observed that the same ring homomorphism applied to $n!p_\lambda/z_\lambda$ gave the sum of $x^{\text{exc}(\sigma)}$ over the set of all permutations of S_n whose cycle type induces the partition λ, denoted $\mathcal{C}_\lambda$. This allowed Brenti to prove that $\sum_{\sigma \in \mathcal{C}_\lambda} x^{\text{exc}(\sigma)}$ was unimodal.

However, Brenti did not use any results on the combinatorics of the transition matrices between bases of symmetric functions in his paper. The first paper to combine homomorphisms with such combinatorics was [6]; this is where Theorems 3.1 and 3.7 come from. Beck and Remmel also gave a completely combinatorial proof of the fact that $\frac{n!}{z_\lambda}\varphi(p_\lambda) = \sum_{\sigma in \mathcal{C}_\lambda} x^{\text{exc}(\sigma)}$. Exercise 3.12 is also due to Beck, see [7].

Since then, numerous papers written by a variety of authors have tinkered with the proof to understand the permutation enumeration for signed permutations and multiples of permutations [5, 8, 47, 59, 66, 74, 75, 76, 77, 78, 94, 96, 99, 100, 113, 114].

Alternating permutations have been called up-down permutations and zigzag sequences in the literature. There have been various proofs and extensions of alternating permutations throughout the years. Leonhard Euler correctly gave the expansion of $\sec z$ up to z^{16} (his coefficient of z^{18} is incorrect) and their connection to permutations [39]. This work probably spurred Sylvester to call the number of alternating permutations Euler numbers [110].

Désiré André's 1879 and 1881 papers are credited as containing the first proof that the generating function for the alternating permutations is $\sec z + \tan z$ [2, 3]. After Roger Entringer reproved André's result in 1966 [37], his work was recounted and reworked in a series of papers by Leonard Carlitz and Richard Scoville over the next decade [16, 17, 19, 22]. The proof technique in these works was to use recursions given by the definition of alternating permutations to find a differential equation for the generating function. The q-analogue of the alternating permutations was discovered by Richard Stanley by working with binomial posets [107].

The common descent statistic and was first studied by Leonard Carlitz, Richard Scoville, and Theresa Vaughan [21, 23, 17]. Later, this statistic was studied by Jean-Marc Fédou with Don Rawlings and Thomas Langley with the second author; the latter paper used our approach of manipulating the relationships between symmetric functions [40, 41, 76].

The result in Theorem 3.14 is due to Adriano Garsia and Ira Gessel [48, 51]. The proof we have provided was published in [89].

The continued fraction expansions for the alternating permutations found in exercises 3.10 and 3.11 and for set partitions given in Exercise 4.9 are due to Philippe Flajolet [42].

Chapter 4
Counting with Nonstandard Bases

In Chapter 3 we found generating functions for permutation statistics by defining ring homomorphisms on e_n and then applying them to h_n. In this chapter we describe another layer of versatility by defining ring homomorphisms on e_n and then applying them to a brand new basis for the ring of symmetric functions, $p_{\nu,\lambda}$.

4.1 The Basis $p_{\nu,\lambda}$

The motivation for defining a new basis comes from Theorem 2.22, which says that the coefficient of e_λ in p_n is $(-1)^{n-\ell(\lambda)}w(B_{\lambda,(n)})$. We have seen in the proof of Theorem 3.5 that the extra weight on a brick tabloid can be useful when finding generating functions. These weights will be even more useful when the weight can be changed to be something other than the length of the last brick, and so that is how we will define $p_{\nu,n}$.

Let ν be a function on the set of nonnegative integers. Recursively define a symmetric function $p_{\nu,n}$ such that

$$p_{\nu,n} = (-1)^{n-1}\nu(n)e_n + \sum_{k=1}^{n-1}(-1)^{k-1}e_k p_{\nu,n-k}$$

for all $n \geq 1$. This definition of $p_{\nu,n}$ allows us to write the generating function for $p_{\nu,n}$ in terms of the elementary symmetric functions. We have

$$\left(\sum_{n=0}^{\infty}(-1)^n e_n z^n\right)\left(\sum_{n=1}^{\infty}p_{\nu,n}z^n\right) = \sum_{n=1}^{\infty}\left(\sum_{k=0}^{n-1}p_{\nu,n-k}(-1)^k e_k\right)z^n$$
$$= \sum_{n=1}^{\infty}(-1)^{n-1}\nu(n)e_n z^n,$$

A. Mendes, J. Remmel, *Counting with Symmetric Functions*, Developments in Mathematics 43, DOI 10.1007/978-3-319-23618-6_4

and so

$$\sum_{n=1}^{\infty} p_{\nu,n} z^n = \frac{\sum_{n=1}^{\infty}(-1)^{n-1}\nu(n)e_n z^n}{\sum_{n=0}^{\infty}(-1)^n e_n z^n}. \tag{4.1}$$

If $\nu(n)=1$ for all $n \geq 1$, equation (4.1) tells us that

$$1+\sum_{n=1}^{\infty} p_{1,n}z^n = 1+\frac{\sum_{n=1}^{\infty}(-1)^{n-1}e_n z^n}{\sum_{n\geq 0}(-1)^n e_n z^n} = \frac{1}{\sum_{n=0}^{\infty}(-1)^n e_n z^n} = 1+\sum_{n=1}^{\infty} h_n z^n.$$

This means that $p_{1,n}$ is the homogeneous symmetric function h_n. Other special cases for ν give well-known symmetric functions. When taking $\nu(n)=n$ for $n \geq 1$, Corollary 2.10 says that $p_{n,n}$ is the power symmetric function p_n. When taking

$$\nu(n) = \begin{cases} 0 & \text{if } n \leq k, \\ (-1)^k & \text{otherwise} \end{cases}$$

for some nonnegative integer k, then Corollary 2.7 says that $p_{\nu,n}$ is the Schur function corresponding to the partition $(1^k, n)$.

For $\lambda = (\lambda_1, \lambda_2, \ldots) \vdash n$, let $p_{\nu,\lambda} = p_{\nu,\lambda_1} p_{\nu,\lambda_2} \cdots$. The reason we have defined $p_{\nu,\lambda}$ in this way is because its expansion in terms of elementary symmetric functions is a collection of weighted brick tabloids.

Suppose $T \in B_{\lambda,\mu}$ has bricks of length $b_1, \ldots, b_\ell$ ending each row. Define $w_\nu(T)$ to be the product $\nu(b_1)\cdots\nu(b_\ell)$ and let $w_\nu(B_{\lambda,\mu})$ be the sum of weights of all $T \in B_{\lambda,\mu}$. These are brick tabloids when $\nu(n)=1$ and these are weighted brick tabloids when $\nu(n)=n$.

Theorem 4.1. *The coefficient of e_λ in $p_{\nu,\mu}$ is $(-1)^{n-\ell(\lambda)} w_\nu(B_{\lambda,\mu})$.*

Proof. Let $c_{\lambda,\mu}$ be the coefficient of e_λ in $p_{\nu,\mu}$. These numbers satisfy the following three recursive identities:

1. $c_{(n),(n)} = (-1)^{n-1}\nu(n)$.
2. If $\lambda \vdash n$ has more than one part, then $c_{\lambda,(n)} = \sum_{i=1}^{n-1}(-1)^{n-i} c_{\lambda\backslash i,(n-i)}$.
3. If $\alpha+\beta$ denotes the partition created by the multiset union of α and β where $\alpha \vdash \mu_1$ and $\beta \vdash n-\mu_1$, then

$$c_{\lambda,\mu} = \sum_{\alpha+\beta=\lambda} c_{\alpha,(\mu_i)} c_{\beta,\mu\backslash\mu_i}.$$

The only difference between these three statements and the statements in the proof of Theorem 2.22 is the first item. Therefore showing that both $c_{\lambda,\mu}$ and $(-1)^{n-\ell(\lambda)} w_\nu(B_{\lambda,\mu})$ satisfy the completely deterministic recursions above is so similar to the proof of Theorem 2.22 that it is left to the reader. □

Corollary 4.2. *If $\nu(n) \neq 0$ for all $n \geq 1$, the set $\{p_{\nu,\lambda} : \lambda \vdash n\}$ is a basis for Λ_n.*

Proof. There are no brick tabloids of shape λ and type μ when μ precedes λ in the reverse lexicographic order, and there is exactly one brick tabloid of nonzero weight when $\lambda = \mu$. Thus the p_ν-to-e transition matrix is triangular with nonzero diagonal entries. Since the elementary symmetric functions are a basis, so is $\{p_{\nu,\lambda} : \lambda \vdash n\}$. □

4.2 Counting with the Elementary and $p_{\nu,n}$

As a first example of how $p_{\nu,n}$ may be used, we will find a generating function for the number of final descents in permutations. Let $\mathrm{fd}(\sigma)$ be the number of final descents in $\sigma \in S_n$, that is, $\mathrm{fd}(\sigma)$ is the length of the last maximal descending run.

We defined the ring homomorphism $\varphi(e_n) = (-1)^{n-1}(x-1)^{n-1}/n!$ in section 3.1 in order to find a generating function registering descents. The factor of the form $(x-1)^{n-1}$ allowed us to assign an x or a -1 into each nonterminal cell in a brick of length n. In order to change this assignment of an x or -1 in the final brick, we define

$$\nu(n) = \begin{cases} 0 & \text{if } n < j, \\ \dfrac{x^{n-1}}{(x-1)^{n-1}} & \text{if } n = j, \\ \dfrac{(x-1)^{n-j-1}(-1)^1 x^{j-1}}{(x-1)^{n-1}} & \text{if } n > j, \end{cases}$$

where j is a positive integer. We will see in the proof below that ν will give us control over how the xs and -1s appear in the final brick in a brick tabloid.

Theorem 4.3. *We have*

$$\sum_{n=0}^{\infty} \frac{z^n}{n!} \sum_{\sigma \in S_n} x^{des(\sigma)} y^{fd(\sigma)} = \left(\frac{x-1}{x - e^{(x-1)z}} \right) \left(1 + (1-y) \frac{e^{xyz} - e^{(x-1)z}}{(x-1-xy)} \right).$$

Proof. This proof is similar to that of Theorem 3.1 but with changes made to the last brick in a brick tabloid. Using Theorem 4.1, we apply the ring homomorphism defined by $\varphi(e_n) = (-1)^{n-1}(x-1)^{n-1}/n!$ to $n! p_{\nu,n}$ to find

$$n! \varphi(p_{\nu,n}) = \sum_{\lambda \vdash n} \binom{n}{\lambda} w_\nu(B_{\lambda,(n)})(x-1)^{\lambda_1 - 1}(x-1)^{\lambda_2 - 1} \cdots .$$

From this sum we create combinatorial objects by selecting a brick tabloid $T \in B_{\lambda,(n)}$, filling T with a permutation such that each brick contains a decreasing sequence of integers, placing a choice of x or -1 in each nonterminal cell of each brick, and placing a $+1$ in the terminal cell of each brick.

To account for the extra weight on the last brick given by ν, we first demand that the final brick must have a length of j or greater. If the final brick has length at least

j, the division by $(x-1)^{n-1}$ in the definition of $\nu(n)$ tells us to erase the choices of x or -1. Then if the last brick has length j, place an x in every nonterminal cell of the final brick. If the last brick has length greater than j, place a sequence of xs or -1s in the nonterminal cells of the final brick such that the sequence ends with a -1 followed by exactly $j-1$ copies of x.

One example of such a combinatorial object when $j=6$ is below:

-1	1	1	x	-1	-1	x	x	x	x	x	1
7	6	11	12	10	9	8	5	4	3	2	1

We have forced the x and -1 labels to end with the sequence $-1\,x\,x\,x\,x\,x\,1$. Applying our usual brick breaking or combining involution introduced in Chapter 3, we are left with fixed points with exactly j final descents. Therefore $n!\varphi(p_{\nu,n}) = \sum x^{\mathrm{des}(\sigma)}$ where the sum runs over permutations $\sigma \in S_n$ with $\mathrm{fd}(\sigma) = j$. Applying φ to both sides of (4.1), we have

$$\begin{aligned}\sum_{n=0}^{\infty} \frac{z^n}{n!} \sum_{\sigma \in S_n} x^{\mathrm{des}(\sigma)} y^{\mathrm{fd}(\sigma)} &= 1 + \sum_{j=1}^{\infty} y^j \sum_{n=1}^{\infty} \frac{z^n}{n!} \sum_{\sigma \in S_n \text{ has } \mathrm{fd}(\sigma)=j} x^{\mathrm{des}(\sigma)} \\ &= 1 + \sum_{j=1}^{\infty} y^j \varphi\left(\sum_{n=1}^{\infty} p_{\nu,n} z^n \right) \\ &= 1 + \sum_{j=1}^{\infty} y^j \frac{\sum_{n=1}^{\infty} (-1)^{n-1} \nu(n) \varphi(e_n) z^n}{\sum_{n=0} (-1)^n \varphi(e_n) z^n}.\end{aligned}$$

The generating function in the statement of the theorem follows from replacing $\varphi(e_n)$ and $\nu(n)$ with their definitions and then performing routine manipulations to simplify. □

The function ν used in the proof of Theorem 4.3 can be used together with the many ways of changing the brick labels provided in section 3.2. In a straightforward manner, we can refine Theorem 4.3 by inversions, count the length of the final number of common descents in pairs of permutations, count the final number of decreases in a word in $\{0,\ldots,k-1\}_n^*$, and refine theorem 3.10 by the number of final descents in the inverse permutation. Modifying the result in Exercise 4.1 can also show that all of these same theorems can be refined by a statistic registering the length of the final increasing run.

We can also change the properties of the final brick when we have defined ring homomorphisms of the form $\varphi(e_n) = (-1)^{n-1} f(n)/n!$ for some function f as found in section 3.1. We give an example of this ability in the proof of the next theorem.

Theorem 4.4. *We have*

$$\sum_{n=1}^{\infty} \frac{z^n}{n!} \sum_{\substack{\sigma \in S_n \text{ does not have} \\ \text{a 2-descent and } \sigma_{n-1} < \sigma_n}} x^{des(\sigma)} = \frac{2}{\sqrt{4x-1}\cot\left(\frac{z\sqrt{4x-1}}{2}\right) - 1}.$$

Proof. Let $f(n)$ be as defined in (3.2) and let φ be the ring homomorphism defined by $\varphi(e_n) = (-1)^{n-1}f(n)/n!$. This was the definition of φ used in Theorem 3.3 in order to find a generating function for the number of permutations which do not have a 2-descent.

For this proof we would like the last brick in a brick tabloid to always have length 1. To make this happen, we define a function $\nu(n)$ by

$$\nu(n) = \begin{cases} 1 & \text{if } n = 1 \\ \dfrac{-f(n-1)}{f(n)} & \text{if } n \geq 2. \end{cases}$$

Applying φ to $n!p_{\nu,n}$ allows us to create the same combinatorial objects as in the proof of Theorem 3.3 except for the last brick. The definition of ν erases any choices for x or -1 in the last brick and replaces them with a sequence of xs and -1s such that no two xs can appear consecutively, a 1 appears in the terminal cell, and a -1 appears in the second to last cell.

After applying our usual brick breaking or combining involution, the placement of a -1 in the second to last cell in combinatorial object forces any fixed points to have a final brick of length 1. The weighed sum of all fixed points, and therefore $\varphi(n!p_{\nu,n})$, is equal to $\sum x^{\mathrm{des}(\sigma)}$ where the sum runs over permutations in $\sigma \in S_n$ without a 2-descent such that $\sigma_{n-1} < \sigma_n$.

A generating function comes from applying φ to (4.1):

$$\sum_{n=1}^{\infty} \frac{z^n}{n!} \sum_{\substack{\sigma \in S_n \text{ does not have} \\ \text{a 2-descent and } \sigma_{n-1} < \sigma_n}} x^{\mathrm{des}(\sigma)} = \frac{\sum_{n=1}^{\infty}(-1)^{n-1}\nu(n)\varphi(e_n)z^n}{\sum_{n=0}(-1)^n\varphi(e_n)z^n}. \tag{4.2}$$

The bottom half of this expression is

$$\frac{1}{\sum_{n=0}(-1)^n\varphi(e_n)z^n} = \frac{1}{-\sum_{n=0}^{\infty} f(n)z^n/n!}.$$

This is the function found in (3.4) which simplifies to the function in the statement of Theorem 3.3. The numerator in (4.2) simplifies to $-\sum_{n=1}^{\infty} f(n-1)z^n/n!$, which is the derivative of $-\sum_{n=0}^{\infty} f(n)z^n/n!$. This is the reciprocal of the function in the statement of Theorem 3.3. Putting this together, we have that (4.2) is equal to

$$\frac{e^{z/2}}{\cos\left(\frac{z\sqrt{4x-1}}{2}\right) - \frac{1}{\sqrt{4x-1}}\sin\left(\frac{z\sqrt{4x-1}}{2}\right)} \int \frac{\cos\left(\frac{z\sqrt{4x-1}}{2}\right) - \frac{1}{\sqrt{4x-1}}\sin\left(\frac{z\sqrt{4x-1}}{2}\right)}{e^{z/2}}\,dz.$$

The integral in the above equation evaluates to $2e^{-z/2}\sin\left(z\sqrt{4x-1}/2\right)/\sqrt{4x-1}$, which in turn allows us to find the function in the statement of the theorem. □

A valley in $\sigma = \sigma_1 \cdots \sigma_n \in S_n$ is an index between 2 and $n-1$ such that $\sigma_{i-1} > \sigma_i$ and $\sigma_i < \sigma_{i+1}$. Let $\mathrm{val}(\sigma)$ be the number of valleys in σ. Aside from displaying how ν can control what happens in the final part of a permutation, Theorem 4.4 secretly encodes information about the distribution of valleys in S_n.

Corollary 4.5. *We have* $\sum_{n=0}^{\infty}\frac{z^n}{n!}\sum_{\sigma\in S_n} x^{val(\sigma)} = \frac{1}{\sqrt{x-1}\cot\left(z\sqrt{x-1}\right)-1}$.

Proof. To begin we will show that the number of permutations of n with k valleys is equal to 2^{n-2k-1} times the number of permutations $\sigma=\sigma_1\cdots\sigma_n\in S_n$ such that σ does not have a 2-descent, $\sigma_{n-1}<\sigma_n$, and $\mathrm{des}(\sigma)=k$.

Suppose $\sigma_i>\cdots>\sigma_{i+j}$ is a maximal descending run of length at least 2 in the permutation $\sigma=\sigma_1\cdots\sigma_n\in S_n$. If $i+j\neq n$, circle the integers $\sigma_{i+1},\cdots,\sigma_{i+j-1}$ and interlace these circled integers into the increasing run which ends at σ_i. If $i+j=n$, circle the integers $\sigma_{i+1},\ldots,\sigma_{i+j}$ and interlace these circled integers into the increasing run which ends at σ_i. Let $\textcircled{\sigma}$ be the result of performing this operation on each maximal descending run of length at least 2.

For example, if σ is the permutation

$$\sigma = 10\ 9\ 1\ 7\ 11\ 8\ 6\ 5\ 2\ 13\ 17\ 18\ 12\ 4\ 14\ 16\ 13\ 3,$$

then

$$\textcircled{\sigma} = \textcircled{9}\ 10\ 1\ \textcircled{5}\ \textcircled{6}\ 7\ \textcircled{8}\ 11\ 2\ \textcircled{12}\ 13\ 17\ 18\ \textcircled{3}\ 4\ \textcircled{13}\ 14\ 16.$$

We can reconstruct σ from $\textcircled{\sigma}$ by placing any circled integers into the next descending run to the right, so this process of changing σ to $\textcircled{\sigma}$ is reversible.

By construction, the permutation $\textcircled{\sigma}$ cannot have a 2-descent and must end with an increase. Furthermore, the permutation $\textcircled{\sigma}$ has exactly one descent each time a maximal decreasing run of length at least 2 is followed by an increasing run in σ, that is, $\textcircled{\sigma}$ has exactly one descent for each valley in σ.

The places in σ which begin and end a maximal descending run of length at least 2 cannot be circled. Furthermore, the last integer in $\textcircled{\sigma}$ cannot be circled. If σ has k valleys, this means that there can be at most $n-2k-1$ circled integers in $\textcircled{\sigma}$.

The bijection which turns σ into $\textcircled{\sigma}$ shows that the number of permutations in S_n with k valleys is equal to the number of permutations $\sigma\in S_n$ without 2-descents, $\sigma_{n-1}<\sigma_n$, and $\mathrm{des}(\sigma)=k$ with at most $n-2k-1$ circled integers.

Therefore

$$\begin{aligned}\sum_{n=1}^{\infty}\frac{z^n}{n!}\sum_{\sigma\in S_n} x^{\mathrm{val}(\sigma)} &= \sum_{n=1}^{\infty}\frac{z^n}{n!}2^{n-1}\sum_{\substack{\sigma\in S_n \text{ does not have}\\ \text{a 2-Descent and } \sigma_{n-1}<\sigma_n}}\left(\frac{x}{4}\right)^{\mathrm{des}(\sigma)}\\ &= \frac{1}{2}\,\frac{2}{\sqrt{4\frac{x}{4}-1}\cot\left(\frac{2z}{2}\sqrt{4\frac{x}{4}-1}\right)-1}\end{aligned}$$

where the last line follows from Theorem 4.4. This is our generating function. □

As the impetus for defining the nonstandard basis $p_{v,n}$ came from the use of the power symmetric functions when finding a generating function for the alternating permutations in Theorem 3.5, we will show how this basis can generalize the alternating permutations. Define $\sigma=\sigma_1\cdots\sigma_n\in S_n$ to be j-alternating provided $\sigma_i<\sigma_{i+1}$ if and only if j divides i. Descents occur exactly at odd indices in an alternating permutation, so an alternating permutation is 2-alternating.

Theorem 4.6. *Let m be an integer in $\{0,\dots,j-1\}$. Then*

$$\sum_{n=1}^{\infty} \frac{z^{nj-m}}{(nj-m)!} |\{\sigma \in S_{nj-m} \text{ is } j \text{ alternating}\}| = \frac{\sum_{n=1}^{\infty}(-1)^{n-1} z^{nj-m}/(nj-m)!}{\sum_{n=0}^{\infty}(-1)^{n} z^{nj}/(nj)!}.$$

Proof. Let φ be the homomorphism defined by $\varphi(e_n) = (-1)^{n-1} f(n)/n!$ where $f(n)$ is $(-1)^{n/j-1}$ if j divides n and 0 otherwise. Let ν be the function defined by $\nu(n) = n!/(n-m)!$. The usual arguments give that $(jn-m)!\varphi(p_{\nu,jn})$ is equal to

$$\sum_{\lambda \vdash jn} \sum_{\substack{T \in B_{\lambda,(jn)} \text{ has bricks with} \\ \text{lengths } b_1,\dots,b_\ell \text{ divisible by } j}} \frac{(jn-m)!}{b_1! \cdots b_\ell!} \frac{b_\ell!}{(b_\ell - m)!} (-1)^{\frac{b_1}{j}+\dots+\frac{b_\ell}{j}-\ell}.$$

The factorials in this sum simplify to $\binom{jn-m}{b_1,\dots,b_\ell - m}$. This means we should create combinatorial objects which look like this (when $n=3$, $j=4$, and $m=3$):

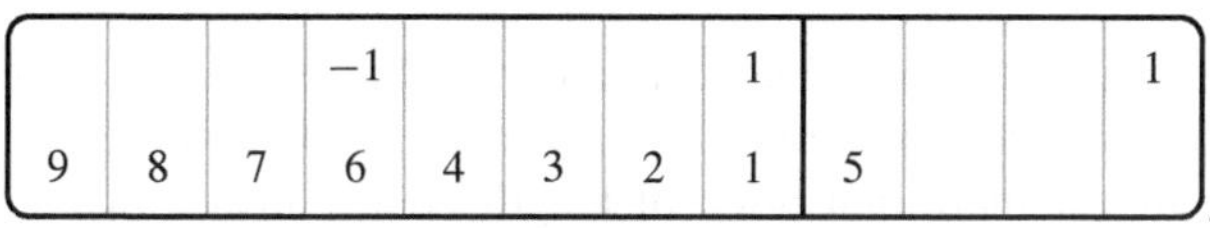

The -1 or 1 sign comes every j^{th} brick, and each brick is a multiple of j. The usual involution leaves fixed points corresponding to j alternating permutations. The generating function follows from applying φ to equation (4.1). □

Summing the cases of $m = 0,\dots,j-1$ in Theorem 4.6 can give nice expressions for the generating function for the j-alternating permutations in S_n. For instance, when $j=2$ we find the generating function for alternating permutations (starting at $n=1$) is $\sec z + \tan z - 1$ and the generating function for the 3-alternating permutations is

$$\frac{3 + 2\sqrt{3} e^{z/2} \sin\left(\frac{\sqrt{3}z}{2}\right)}{e^{-z} + 2e^{z/2}\cos\left(\frac{\sqrt{3}z}{2}\right)}.$$

For $\sigma = \sigma_1 \cdots \sigma_n \in S_n$, let $\text{step}(\sigma)$ be the number of indices i for which $\sigma_i = \sigma_{i+1} + 1$; these are indices which take one step down. We end this section by applying a different sort of involution on brick tabloids in order to find a generating function for the permutation statistic step.

Theorem 4.7. *We have* $\displaystyle\sum_{n=1}^{\infty} \frac{z^n}{(n-1)!} \sum_{\sigma \in S_n} x^{step(\sigma)} = \frac{z}{(1-z)^2 e^{z(1-x)}}.$

Proof. Define a ring homomorphism φ by $\varphi(e_n) = (-1)^{n-1} f(n)/n!$ where $f(n) = (-1)(1-x)^n$ and define a function ν by $\nu(n) = n \cdot n!/f(n)$. Then we have

$$(n-1)!\varphi(p_{\nu,n}) = \sum_{\lambda \vdash n} \sum_{\substack{T \in B_{\lambda,(n)} \text{ has} \\ \text{bricks } b_1,\dots,b_\ell}} \binom{n-1}{b_1,\dots,b_\ell - 1} \frac{1}{b_\ell} f(b_1)\cdots f(b_\ell)\nu(b_\ell). \quad (4.3)$$

From this we create combinatorial objects by selecting $T \in B_{\lambda,(n)}$ for some $\lambda \vdash n$ and using the multinomial coefficient in (4.3) to place decreasing sequences of integers in the bricks of T such that each integer in $\{1,\ldots,n\}$ appears once in T and the integer n appears in the final brick (the appearance of the multinomial coefficient $\binom{n-1}{b_1,\ldots,b_\ell-1}$ in (4.3) instead of $\binom{n-1}{b_1,\ldots,b_\ell}$ gives this condition).

The definition of $f(n)$ tells us to place a choice of either 1 or $-x$ in each cell and, with the extra power of -1 in $f(n)$, $f(n)$ tells us to reverse the $\pm$ sign on the terminal 1 or $-x$ in each brick. To account for the function ν, we notice that $\nu(b_\ell)/b_\ell = b_\ell!/f(b_\ell)$ and so we erase all 1 or $-x$ labels on the final brick coming from the definition of $f(n)$ and then permute the integers in the final brick.

For example, one combinatorial object created in this manner is

1	$-x$	-1	$-x$	$-x$	x	1	x				
11	10	2	8	5	1	4	3	9	6	12	7

.

The weighted sum over all possible combinatorial objects is equal to $(n-1)!\varphi(p_{\nu,n})$.

We will perform two involutions. First, scan the bricks from left to right looking for either the first nonterminal brick of length greater than 1 or the first brick of length 1 which sees a decrease in the integer labeling with the nonterminal brick to its right. If we find a nonterminal brick of length greater than 1, make the first cell of this brick into its own brick of length 1 and reverse the sign on the 1 or $-x$ in this new brick. If we find a brick of length 1 with a decrease in the integer labels, then reverse this operation.

This process is a weight preserving and sign reversing involution. Fixed points look like this

-1	x	-1	x	x	x	-1	x				
1	2	3	4	5	8	10	11	9	6	12	7

.

Fixed points under this first involution must consist of an increasing list of bricks of size 1, each containing a -1 or an x, followed by the final brick. Perform a second involution on these fixed points by first locating the largest integer i such that either

1. i appears immediately to the right of j in the final brick where j is the smallest integer in the final brick which is larger than i, or
2. i does not appear in the final brick and the label above i is a -1.

If the first case is found, remove i from the final brick and place it in the increasing sequence of bricks of length one with a label of -1. If the second case is found, undo this operation. For example, the image of the combinatorial object shown above under this second involution is

-1	x	-1	x	x	x	x					
1	2	3	4	5	8	11	9	6	12	10	7

.

Fixed points under the second involution cannot have either of the above two cases hold. One such object is

x	x	x	x	x							
2	4	5	8	11	9	3	6	1	12	7	10

.

If we take such a fixed point and place each i in a brick of length one after the appearance of $i+1$ in the final brick, then we find a permutation $\sigma = \sigma_1 \cdots \sigma_n \in S_n$ with one x for each time $\sigma_i = \sigma_{i+1}+1$. For instance, the fixed point above corresponds to the permutation 9 8 3 2 6 5 4 1 12 11 7 10. This means $(n-1)!\varphi(p_{\nu,n}) = \sum_{\sigma\in S_n} x^{\text{step}(\sigma)}$.

A generating function follows from applying φ to (4.1). We have

$$\sum_{n=1}^{\infty} \frac{z^n}{(n-1)!} \sum_{\sigma\in S_n} x^{\text{step}(\sigma)} = \frac{\sum_{n=1}^{\infty}(-1)^{n-1}\nu(n)\varphi(e_n)z^n}{\sum_{n=0}^{\infty}(-1)^n\varphi(e_n)z^n} = \frac{\sum_{n=1}^{\infty} nz^n}{\sum_{n=0}^{\infty} z^n(1-x)^n/n!},$$

which is equal to the generating function in the statement of the theorem. □

One of the curious facts that we can deduce from the generating function in Theorem 4.7 is the probability that a random permutation $\sigma \in S_n$ will have step(σ) even. To find this probability, we add the function in Theorem 4.7 to the same function with "x" replaced with "$-x$," divide by 2, and then take $x=1$ to find

$$\begin{aligned}\sum_{n=1}^{\infty} \frac{|\{\sigma\in S_n \text{ has step}(\sigma) \text{ even}\}|}{(n-1)!} z^n &= \frac{z(1+e^{-2z})}{2(1-z)^2}\\ &= \left(\frac{e^2+1}{2e^2}\right)\frac{1}{(1-z)^2} - \left(\frac{e^2-1}{2e^2}\right)\frac{1}{1-z} + g(z),\end{aligned}$$

where $g(z)$ is a function with no singularities. Using Newton's binomial theorem and accounting for the division by $(n-1)!$ instead of the usual $n!$ in Theorem 4.7, we find that a good approximation to $|\{\sigma \in S_n \text{ has step}(\sigma) \text{ even}\}|/n!$ is

$$\left(\left(\frac{e^2+1}{2e^2}\right)(n+1) - \frac{e^2-1}{2e^2}\right)/n = \frac{1}{2} + \frac{1}{2e^2} + \frac{1}{e^2 n}.$$

So, for large n, the probability that a random permutation $\sigma \in S_n$ has step(σ) even is approximately $1/2 + 1/(2e^2) \approx 0.57668$.

4.3 Recurrences

A linear homogeneous recurrence relation with constant coefficients is a sequence which is recursively defined by

$$a_n = c_1 a_{n-1} + c_2 a_{n-2} + \cdots + c_k a_{n-k} \tag{4.4}$$

for all $n > k$ where k is a positive integer and $c_1, \ldots, c_k$ are constants. We assume that $a_1, \ldots, a_k$ are known constants, from which the entire sequence can be found.

It is not difficult to find the generating function $A(z) = \sum_{n=1}^{\infty} a_n z^n$ for such sequences; indeed,

$$\begin{aligned} A(z) &= a_1 z + \cdots + a_k z^k + \sum_{n=k+1}^{\infty} a_n z^n \\ &= a_1 z + \cdots + a_k z^k + \sum_{n=k+1}^{\infty} (c_1 a_{n-1} + c_2 a_{n-2} + \cdots + c_k a_{n-k}) z^n \\ &= a_1 z + \cdots + a_k z^k + c_1 z \left(A(z) - a_1 z_1 - \cdots - a_{k-1} z^{k-1} \right) + \cdots + c_k z^k A(z). \end{aligned}$$

Solving for $A(z)$ gives

$$A(z) = \frac{a_1 z + (a_2 - c_1 a_1) z^2 + \cdots + (a_k - c_{k-1} a_1 - \cdots - c_1 a_{k-1}) z^k}{1 - c_1 z - \cdots - c_k z^k}. \tag{4.5}$$

In this section we will give a simple, combinatorial interpretation for the terms in this recursion using weighted brick tabloids. Such combinatorial interpretations can help us better understand the sequence a_n; see Exercises 4.3 and [12] for examples of how to use these combinatorial interpretations to prove identities.

As an immediate result of the combinatorial interpretation, we will be able to define a ring homomorphism φ and a function ν in order to find the generating function $A(z)$. This roundabout way of finding $A(z)$ uses more overhead than the above calculation, but we include it to exhibit the versatility of using brick tabloids and symmetric function identities in counting problems.

Let $B_{n,k}$ be the set of all brick tabloids $T \in B_{\lambda,(n)}$ for some $\lambda \vdash n$ such that there is only one brick in T or the last two bricks in T have lengths which sum to an integer larger than k. For a brick b of length ℓ in a brick tabloid $T \in B_{n,k}$, define

$$w(b) = \begin{cases} 0 & \text{if } \ell > k, \\ c_\ell & \text{if } \ell \le k \text{ and } b \text{ is not the last brick in } T, \text{ and} \\ a_\ell & \text{if } \ell \le k \text{ and } b \text{ is the last brick in } T, \end{cases}$$

and define $w(T)$ to be the product of the weights of the bricks in T. For example, one $T \in B_{12,3}$ with weight $c_1^2 c_2 c_3^2 a_2$ can be depicted by

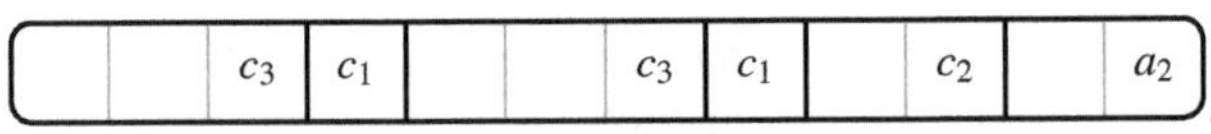

If $n \le k$, then $B_{n,k}$ contains exactly one brick tabloid T (which consists of exactly one brick of length n) and so $\sum_{T \in B_{n,k}} w(T) = a_n$. If $n > k$, then $T \in B_{n,k}$ contains at least two bricks. By summing over the length of the first brick,

$$\sum_{T \in B_{n,k}} w(T) = \sum_{i=1}^{k} \sum_{\substack{T \in B_{n,k} \text{ has first} \\ \text{brick of length } i}} w(T) = \sum_{i=1}^{k} c_i \sum_{T \in B_{n-i,k}} w(T).$$

Therefore $\sum_{T \in B_{n,k}} w(T)$ satisfies the recursion in (4.4). This means we have our combinatorial interpretation: $a_n = \sum_{T \in B_{n,k}} w(T)$.

To find the generating function $A(z) = \sum_{n=1}^{\infty} a_n z^n$, we can define a ring homomorphism φ by $\varphi(e_n) = (-1)^{n-1} c_n$ if $n \leq k$ and 0 otherwise. To go along with φ, we define ν by $\nu(n) = (a_n - c_{n-1}a_1 - \cdots - c_1 a_{n-1})/c_n$ if c_n is nonzero and 0 otherwise. Then

$$\varphi(p_{\nu,n}) = \sum_{\lambda \vdash n} \sum_{\substack{T \in B_{\lambda,(n)} \text{ has bricks} \\ \text{with lengths } b_1,\ldots,b_\ell}} c_{b_1} \cdots c_{b_\ell} \nu(b_\ell).$$

From this sum we select a brick tabloid $T \in B_{\lambda,(n)}$ for some $\lambda \vdash n$ and associate a weight of c_{b_i} to the brick of length b_i. The function $\nu(b_\ell)$ tells us to cancel the c_{b_ℓ} weight on the final brick in T and replace it with either a_{b_ℓ} or $-c_{b_\ell - i} a_i$ for some i.

On this collection of weighted brick tabloids we can perform the following sign reversing involution. If the last brick in such a combinatorial object T is $-c_{b_\ell - i} a_i$ for some i, then change T by breaking the last brick into two bricks, one of length $b_\ell - i$ and the other of length i. If the last two bricks in T sum to an integer less than or equal to k, reverse this operation and combine the last two bricks into one brick with weight $-c_{b_\ell - i} a_i$. For instance, this involution pairs

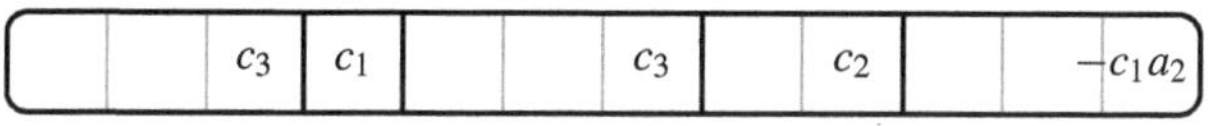

with

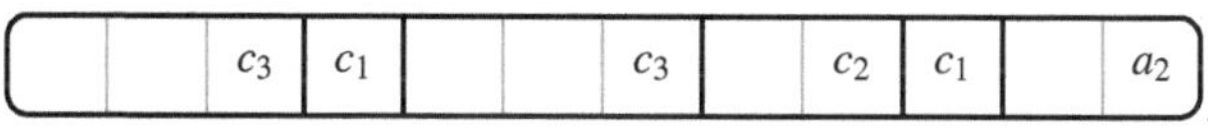

Fixed points can have no negative weights and the last two bricks cannot have lengths which sum to an integer smaller than k. These fixed points have a weighted sum equal to $a_n = \sum_{T \in B_{n,k}} w(T)$.

By applying φ to the identity in (4.1), we find

$$A(z) = \varphi\left(\sum_{n=1}^{\infty} p_{\nu,n} z^n\right) = \frac{\sum_{n=1}^{\infty} (-1)^{n-1} \nu(n) \varphi(e_n) z^n}{\sum_{n=0}^{\infty} (-1)^n \varphi(e_n) z^n},$$

which in turn is equal to the function in (4.5), as expected.

4.4 The Exponential Formula

In this section we apply the machinery we have developed to understand and refine the "exponential formula," which is a relationship between the generating functions for connected objects and collections of those connected objects.

A common problem in combinatorics is to count the number of ways that the integers $1, \ldots, n$ can be arranged in a structure which can be partitioned into disjoint, unordered components. Here are three main examples of this phenomenon:

1. Permutations of n are built from disjoint, unordered cycles.
2. A set partition of n is a collection of pairwise disjoint nonempty sets with union to $\{1,\ldots,n\}$. For instance, $\{\{1,3,6,8\},\{2,5,7,10,11\},\{4,12\},\{9\}\}$ is a set partition of 12. Set partitions of n are built from disjoint, unordered sets.
3. Labeled graphs on n nodes are built from disjoint, unordered connected components.

The exponential formula gives generating functions for these situations.

Let $\mathcal{P}_n$ be a set of "pictures" containing the integers $1,\ldots,n$. By a picture, we simply mean that the integers $1,\ldots,n$ are arranged in some way. These pictures will be our disjoint, connected components.

Let $\mathcal{S}_n$ be the collection of sets of the form $\{p_1,\ldots,p_k\}$ such that each p_i is a picture, the sizes of $p_1,\ldots,p_k$ sum to n, and the integer labels in the pictures have been replaced so that the total set contains the integers $1,\ldots,n$. For $s \in \mathcal{S}_n$, define the statistic $\mathrm{pic}(s)$ be the number of pictures in s.

As an example, suppose each element in $\mathcal{P}_n$ is a rooted labeled tree with n nodes (see Exercise 4.4 for the definition of rooted labeled tree). Here is an element in $\mathcal{P}_7$:

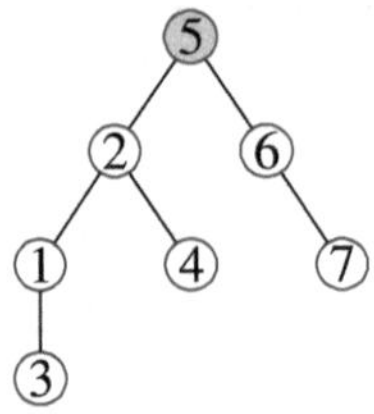

Objects in $\mathcal{S}_n$ are built from components found in the sets $\mathcal{P}_1,\mathcal{P}_2,\ldots$; in this example, we are creating rooted labeled forests on n nodes. Here is an element in $\mathcal{S}_{20}$:

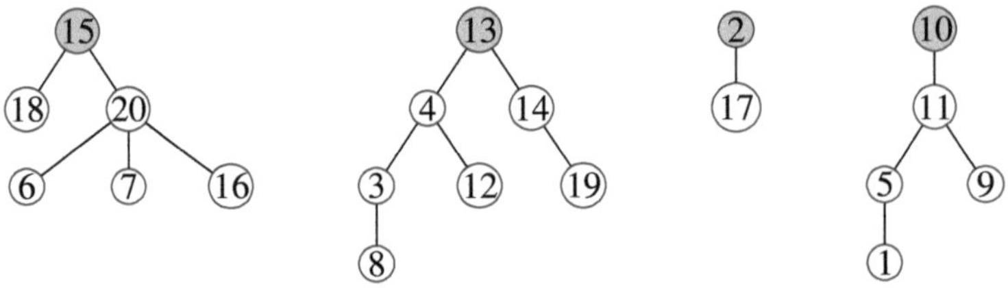

We have chosen to list the rooted trees in this labeled forest in decreasing order according to their smallest element, but since the rooted trees are unordered elements of a set, we could have displayed them in any order. For the above object s, $\mathrm{pic}(s) = 4$.

The next theorem is known as the exponential formula. It may be proved without the machinery we have developed, but proving it by defining a ring homomorphism on e_n provides two significant benefits. First, once understood in this way, we can use the ability to weight the last brick in a brick tabloid differently from the other bricks in order to refine the exponential formula a few different ways. Second, this style of combinatorial proof by sign reversing involution unifies many different ad hoc methods for finding generating functions.

Theorem 4.8. *If* $\mathcal{P}(z) = \sum_{n=1}^{\infty} |\mathcal{P}_n| z^n/n!$, *then* $\sum_{n=0}^{\infty} \frac{z^n}{n!} \sum_{s \in \mathcal{S}_n} x^{pic(s)} = e^{x\mathcal{P}(z)}$.

Proof. Define a ring homomorphism by $\varphi(e_n) = (-1)^{n-1} f(n)/n!$ where

$$f(n) = \sum_{m=1}^{n} \sum_{\substack{i_1,\ldots,i_m \geq 1 \\ i_1+\cdots+i_m=n}} \binom{n}{i_1,\ldots,i_m} \frac{|\mathcal{P}_{i_1}|\cdots|\mathcal{P}_{i_m}|}{m!} (-1)^{m-1} x^m. \tag{4.6}$$

Applying φ to $n!h_n$ gives (3.3). With this equation, begin to create combinatorial objects by using the summand and $|B_{\lambda,(n)}|$ terms to select a brick tabloid T. Use the multinomial coefficient $\binom{n}{\lambda}$ in (3.3) to associate with each brick in T of size k a subset of $\{1,\ldots,n\}$ of size k such that the subsets associated with the bricks in T are pairwise disjoint.

The function $f(n)$ in (4.6) tells us how to weight a brick of length n. With this function, select an integer m between 1 and n and select positive indices $i_1,\ldots,i_m$ which sum to n. Use the factor of $|\mathcal{P}_{i_1}|\cdots|\mathcal{P}_{i_m}|$ in (4.6) to select pictures $p_{i_1},\ldots,p_{i_m}$, each containing $i_1,\ldots,i_m$ integers. We have already been assigned a subset of positive integers of size n, say $\{j_1,\ldots,j_n\}$ where $j_1 < \cdots < j_n$. Use the binomial coefficient $\binom{n}{i_1,\ldots,i_m}$ to replace the numbers $1,\ldots,i_\ell$ on picture p_{i_ℓ} with elements of $\{j_1,\ldots,j_n\}$ such that each picture has distinct integer labels, the union of which is equal to $\{j_1,\ldots,j_n\}$.

Use the $\frac{1}{m!}$ term in (4.6) to sort these pictures containing the re-indexed labels in increasing order according to the smallest element; suppose the pictures are $p_{k_1},\ldots,p_{k_m}$ when listed in this way. Place p_{k_1} into the k_1^{th} cell of the brick when reading left to right, place p_{k_2} into the $(k_1+k_2)^{th}$ cell reading left to right, and so on. In each nonterminal cells which contain a picture, place one $-x$. Place an x in the terminal cell of the brick.

Performing the above operations uses all the terms in (3.3). One possible combinatorial object created in this manner when taking pictures to be rooted labeled trees is shown below.

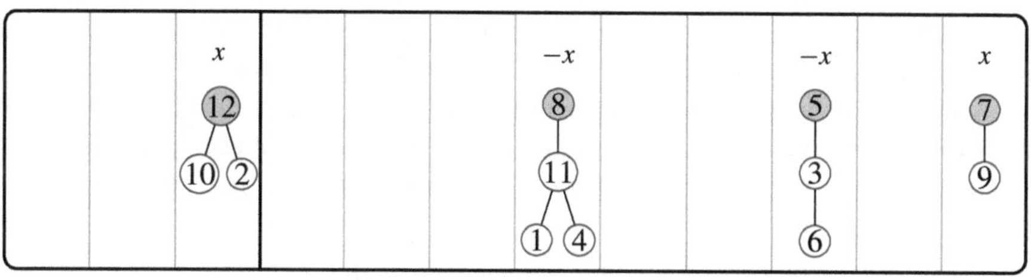

If we define the weight of such an object to be the product of the $-x$ and x terms, then it follows that the weighted sum over all possible combinatorial objects is equal to $n!\varphi(h_n)$.

To rid ourselves of any object with a negative sign, apply the following sign reversing weight preserving involution. Scan the bricks from left to right looking for a $-x$ or two consecutive bricks where the last picture in the first brick contains a smaller integer than the first picture in the second brick. If a $-x$ is scanned first, break the brick into two immediately after the $-x$ and reverse the sign on the $-x$. If two consecutive bricks with the correct order of pictures is found, then combine the two consecutive bricks and reverse the sign on x now appearing in the middle.

The image of the combinatorial object shown above under this process is shown below:

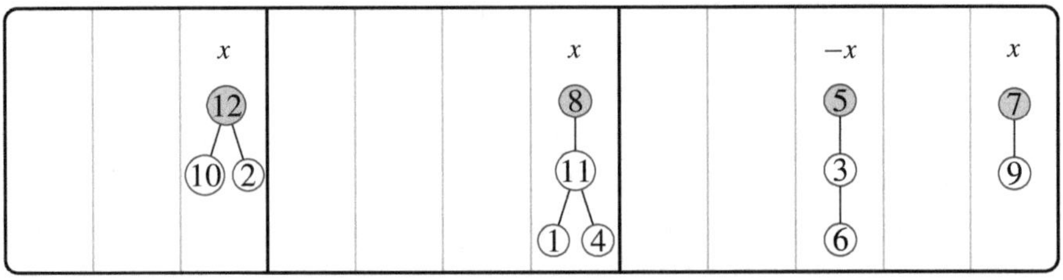

This is easily seen to be sign reversing and weight preserving. The fixed points cannot have any $-x$ labels and hence each brick must contain one and only one picture. Furthermore, in a fixed point, the pictures when read from left to right must be written in decreasing order according to their minimum elements. These fixed points correspond to $\sum_{s\in\mathcal{S}_n} x^{\mathrm{pic}(s)}$. Applying φ to Theorem 2.5 gives

$$\sum_{n=0}^{\infty}\frac{z^n}{n!}\sum_{s\in\mathcal{S}_n} x^{\mathrm{pic}(s)} = \left(1+\sum_{n=1}^{\infty}\frac{z^n}{n!}\sum_{m=1}^{n}\sum_{i_1+\cdots+i_m=n}(-x)^m\binom{n}{i_1,\ldots,i_m}\frac{|\mathcal{P}_{i_1}|\cdots|\mathcal{P}_{i_m}|}{m!}\right)^{-1}$$
$$= \left(\sum_{m=0}^{\infty}\frac{(-x)^m}{m!}\sum_{n=0}^{m}\sum_{i_1+\cdots+i_m=n}\frac{|\mathcal{P}_{i_1}|\cdots|\mathcal{P}_{i_m}|}{i_1!\cdots i_m!}z^{i_1+\cdots+i_m}\right)^{-1}$$

which may be seen to equal $e^{x\mathcal{P}(z)}$. □

Now that the exponential formula has been proved, let us show off some of its capabilities. As a first example, we will find generating functions for the cycle lengths in permutations in S_n.

Let $\mathcal{P}_n$ be the set of cycles of length n. Although we did not do this in Theorem 4.8, we have the ability to assign indeterminates to each picture in $\mathcal{P}_n$ if we so desire; the involution in the proof of Theorem 4.8 is not affected by the content of the pictures and hence is still weight preserving with respect to any indeterminates. This means that we can additionally assign a weight of q_n to go along with each picture in $\mathcal{P}_n$.

There are $(n-1)!$ cycles of length n and so $|\mathcal{P}_n| = q_n(n-1)!$. Therefore

$$\mathcal{P}(z) = \sum_{n=1}^{\infty}\frac{z^n}{n!}|\mathcal{P}_n| = q_1 z + q_2\frac{z^2}{2} + q_3\frac{z^3}{3} + \cdots.$$

Elements in $\mathcal{S}_n$ are permutations in S_n written in cyclic notation with powers of q_i counting the number of cycles of length i. Taking $x = 1$ in Theorem 4.8 (this power of x would count the number of cycles in a permutation, something we are already doing with the indeterminates q_i) gives

$$\sum_{n=0}^{\infty}\frac{z^n}{n!}\sum_{\sigma\in S_n} q_1^{\mathrm{cyc}_1(\sigma)}q_2^{\mathrm{cyc}_2(\sigma)}\cdots = e^{q_1 z+q_2\frac{z^2}{2}+q_3\frac{z^3}{3}+\cdots}, \tag{4.7}$$

where $\mathrm{cyc}_i(\sigma)$ is the number of cycles of length i in the permutation σ. The coefficient of $z^n/n!$ in (4.7) is known as the cycle index polynomial.

By taking special values of $q_1, q_2, \ldots$, equation (4.7) can produce some interesting generating functions. For example, if we let $\mathrm{cyc}(\sigma)$ be the total number of cycles in σ and let $q_1 = q_2 = \cdots = x$ in (4.7), we find

$$\sum_{n=0}^{\infty}\frac{z^n}{n!}\sum_{\sigma\in S_n} x^{\mathrm{cyc}(\sigma)} = e^{x(z+\frac{z^2}{2}+\frac{z^3}{3}+\cdots)} = e^{x\ln\frac{1}{1-z}} = \frac{1}{(1-z)^x}.$$

An expansion of the right-hand side of the above expression by Newton's binomial theorem shows that $\sum_{\sigma\in S_n} x^{\mathrm{cyc}(\sigma)} = (x+0)(x+1)\cdots(x+(n-1))$.

By taking certain values of q_n in (4.7) to equal zero, we can restrict the appearances of certain cycles. For instance, if j is a positive integer, then it may be shown that the permutations $\sigma \in S_n$ such that $\sigma^j = 1$ are those permutations with cycles of length dividing j. To find a generating function for those permutations in S_n with $\sigma^j = 1$, take $q_n = x$ if n divides j and $q_n = 0$ otherwise to find that

$$\sum_{n=0}^{\infty}\frac{z^n}{n!}\sum_{\sigma\in S_n \text{ with } \sigma^j=1} x^{\mathrm{cyc}(\sigma)} = \prod_{n \text{ divides } j} e^{xz^n/n}.$$

In the special case of $j = 2$, the generating function registering a power of x for each cycle in an involution (a permutation with cycles of length 1 or 2) is $e^{x(z+z^2/2)}$.

As another example, the generating function for the number of permutations in S_n with all cycles of length at least m is equal to

$$\begin{aligned}\sum_{n=0}^{\infty}\frac{z^n}{n!}\sum_{\substack{\sigma\in S_n \text{ has cycles}\\ \text{of length at least } m}} x^{\mathrm{cyc}(\sigma)} &= e^{xz^m/m+xz^{m+1}/(m+1)+\cdots}\\ &= e^{x\left(\ln\frac{1}{1-z}-\frac{z^1}{1}-\cdots-\frac{z^{m-1}}{m-1}\right)}\\ &= \frac{e^{x\left(-\frac{z^1}{1}-\cdots-\frac{z^{m-1}}{m-1}\right)}}{(1-z)^x}.\end{aligned}$$

Taking $x = 1$ in this last equation and then using the asymptotic techniques developed in the second part of section 1.3 shows that the approximate probability that a random permutation will have cycles of length at least m is $e^{-1-\frac{1}{2}-\cdots-\frac{1}{m-1}}$.

As a second application of Theorem 4.8 we will count the number of set partitions. Let $\mathcal{P}_n$ contain only one picture, the picture "$\{1,\ldots,n\}$". Then $\mathcal{P}(z) = e^z - 1$. It follows that elements in $\mathcal{S}_n$ are set partitions of n. The exponential formula gives

$$\sum_{n=0}^{\infty}\frac{z^n}{n!}\sum_{s \text{ is a set partition of } n} x^{\text{the number of sets in } s} = e^{x(e^z-1)}.$$

In particular, taking the coefficient of x^k in the expansion of the above expression, it follows that

$$\sum_{n=0}^{\infty} |\{\text{the number of set partitions of } n \text{ with } k \text{ parts}\}| \frac{z^n}{n!} = \frac{(e^z-1)^k}{k!}.$$

Additionally, we can restrict the sizes of the sets which appear in a set partition in the same way that we restricted the cycles in a permutation.

As a last example of the exponential formula, let $\mathcal{G}_n$ be the set of labeled graphs on n nodes. Since between any two nodes there is a choice of either placing an edge or not, there are $2^{\binom{n}{2}}$ total elements in $\mathcal{G}_n$. If we let $\mathcal{P}(z)$ the generating function for the number of connected components in $\mathcal{G}_n$, taking $x = 1$ in Theorem 4.8 gives

$$\sum_{n=0}^{\infty} 2^{\binom{n}{2}} \frac{z^n}{n!} = e^{\mathcal{P}(z)}.$$

The function $\mathcal{P}(z)$ can be found by taking logarithms. A second application of Theorem 4.8 gives

$$\sum_{n=0}^{\infty} \frac{z^n}{n!} \sum_{g \in \mathcal{G}_n} x^{\text{the number of connected components in } g} = e^{x \ln\left(\sum_{n=0}^{\infty} 2^{\binom{n}{2}} \frac{z^n}{n!}\right)} = \left(\sum_{n=0}^{\infty} 2^{\binom{n}{2}} \frac{z^n}{n!}\right)^x.$$

Next we will show how the ability to weight the last brick in a brick tabloid differently can refine Theorem 4.8 in a few different ways. Given an element $s \in \mathcal{S}_n$ built with components found in $\mathcal{P}_1, \ldots, \mathcal{P}_n$, let $\text{one}(s)$ be the size of the component in which label 1 may be found. For example, if s is the set partition

$$\{9\}, \{4,12\}, \{2,5,7,10,11\}, \{1,3,6,8\}$$

of 12, then $\text{one}(s) = 4$ since number 1 appears within a set of size 4.

Theorem 4.9. *If* $\mathcal{P}(z) = \sum_{n=1}^{\infty} |\mathcal{P}_n| z^n/n!$, *then*

$$\sum_{n=1}^{\infty} \frac{z^n}{n!} \sum_{s \in \mathcal{S}_n} x^{pic(s)} y^{one(s)} = \int x e^{x\mathcal{P}(z)} \frac{\partial}{\partial z} (\mathcal{P}(yz))\, dz.$$

Proof. Define a function ν in order to weight the last brick in a brick tabloid by

$$\nu(n) = \begin{cases} 0 & \text{if } n \neq j \\ \dfrac{nx|\mathcal{P}_n|}{f(n)} & \text{if } n = j. \end{cases}$$

for some positive integer j where $f(n)$ is given in (4.6). Applying the function φ in the proof of Theorem 4.8 to $(n-1)!p_{\nu,n}$ gives (4.3). From this equation we create the same combinatorial objects as found in the proof of Theorem 4.8 except for the following conditions on the final brick:

1. The last brick must be of length j,
2. The last brick must contain exactly one picture, and
3. The picture in the last brick contains the integer 1 (the appearance of the multinomial coefficient $\binom{n-1}{b_1,\ldots,b_\ell-1}$ in (4.3) instead of $\binom{n-1}{b_1,\ldots,b_\ell}$ gives this condition).

Apply the involution found in the proof of Theorem 4.8. Since the 1 appears in the last brick and that brick contains only one picture, this involution never combines the last two bricks. Fixed points correspond to elements in $\mathcal{S}_n$ with the 1 appearing in a picture of size j. Therefore we have

$$\begin{aligned}\frac{d}{dz}\left(\sum_{n=1}^{\infty}\frac{z^n}{n!}\sum_{s\in\mathcal{S}_n}x^{\mathrm{pic}(s)}y^{\mathrm{one}(s)}\right) &= \sum_{n=1}^{\infty}\frac{z^{n-1}}{(n-1)!}\sum_{s\in\mathcal{S}_n}x^{\mathrm{pic}(s)}y^{\mathrm{one}(s)}\\ &= \sum_{j=1}^{\infty}\frac{y^j}{z}\varphi\left(\sum_{n=1}^{\infty}p_{\nu,n}z^n\right)\\ &= \sum_{j=1}^{\infty}y^j\frac{\sum_{n=1}^{\infty}(-1)^{n-1}\nu(n)\varphi(e_n)z^{n-1}}{\sum_{n=0}^{\infty}(-z)^n\varphi(e_n)}.\end{aligned}$$

Just as in the proof of Theorem 4.8, the denominator of this expression is $e^{-x\mathcal{P}(z)}$. This, along with the definition of ν and φ, shows that the above string of equalities is equal to

$$e^{x\mathcal{P}(z)}\sum_{j=1}^{\infty}y^jx|\mathcal{P}_j|\frac{z^{j-1}}{(j-1)!}=xe^{x\mathcal{P}(z)}\frac{d}{dz}\left(\sum_{j=1}^{\infty}\frac{(yz)^j}{j!}|\mathcal{P}_j|\right)=xe^{x\mathcal{P}(z)}\frac{d}{dz}\left(\mathcal{P}(yz)\right).$$

Integrating the extremities in this string of equalities proves the theorem. □

Theorem 4.9 says, for instance, that

$$\sum_{n=1}\frac{z^n}{n!}\sum_{s\text{ is a set partition}}x^{\text{the number of sets in }s}y^{\text{the size of the set with 1}}=\int xye^{x(e^z-1)+yz}\,dz$$

and

$$\sum_{n=1}^{\infty}\frac{z^n}{n!}\sum_{\sigma\in S_n}y^{\mathrm{one}(\sigma)}q_1^{\mathrm{cyc}_1(\sigma)}q_2^{\mathrm{cyc}_2(\sigma)}\cdots=\int y\left(q_1+q_2(yz)+\cdots\right)e^{q_1z+q_2z^2/2+\cdots}\,dz.$$

If we take $q_1=q_2=\cdots=x$ in this last equation, we find the specialization

$$\sum_{n=1}^{\infty}\frac{z^n}{n!}\sum_{\sigma\in S_n}x^{\mathrm{cyc}(\sigma)}y^{\mathrm{one}(\sigma)}=\int\frac{xy}{(1-z)^x(1-yz)}\,dz.$$

In the proof of Theorem 4.8, we sorted pictures within each brick in increasing order according to the smallest element. This choice of sorting pictures within bricks was arbitrary—any linear order on the pictures can be used to prove the theorem. The choice of sorting by smallest element was made so that the last brick in a fixed point would contain the element 1, enabling us to prove Theorem 4.9 more easily. Different linear orders of pictures combined with the capacity to change the weight on a last brick can refine the exponential formula in different ways, as shown in our next theorem.

Theorem 4.10. *We have*

$$\sum_{n=1}^{\infty} \frac{z^n}{n!} \sum_{s \in \mathcal{S}_n} x^{pic(s)} z^{min(s)} = e^{x\mathcal{P}(z)} \sum_{j=1}^{\infty} \sum_{n=j}^{\infty} y^j \nu(n) \frac{z^n}{n!},$$

where $min(s)$ is the minimum picture size in $s \in \mathcal{S}_n$,

$$\nu(n) = \begin{cases} 0 & \text{if } n < j, \\ \dfrac{x|P_j|}{f(n)} \displaystyle\sum_{m=1}^{n-j} \sum_{\substack{1 \le i_1, \ldots, i_m \le j \\ i_1 + \cdots + i_m = n-j}} \binom{n}{i_1, \ldots, i_m, j} \frac{|\mathcal{P}_{i_1}| \cdots |\mathcal{P}_{i_m}|}{(m+1)!} (-x)^m & \text{if } n \ge j, \end{cases}$$

and $f(n)$ is the function in (4.6).

Proof. Let φ be the homomorphism defined in the proof of Theorem 4.8 and let ν be the function in the statement of the theorem. Applying φ to $n!p_{\nu,n}$ gives

$$n!\varphi(p_{\nu,n}) = \sum_{\lambda \vdash n} \sum_{\substack{T \in B_{\lambda,(n)} \text{ has} \\ \text{bricks } b_1, \ldots, b_\ell}} \binom{n}{b_1, \ldots, b_\ell} f(b_1) \cdots f(b_\ell) \nu(b_\ell). \tag{4.8}$$

From this we can create combinatorial objects which, except for the last brick, are similar to those found in the proof of Theorem 4.8. However, instead of ordering the pictures in each brick according to minimum integers, sort the pictures in increasing order according to size. Then if pictures within a brick have the same size, sort them in increasing order according to minimum element.

The division by $f(n)$ in the definition of ν erases all pictures in the last brick. Then in the same way that $f(n)$ placed pictures in the other bricks, the remaining portion of ν places an ordered list of pictures in the last brick such that the last brick must have a maximum size picture with exactly j integers. For example, when $j = 4$, one such combinatorial object is shown below:

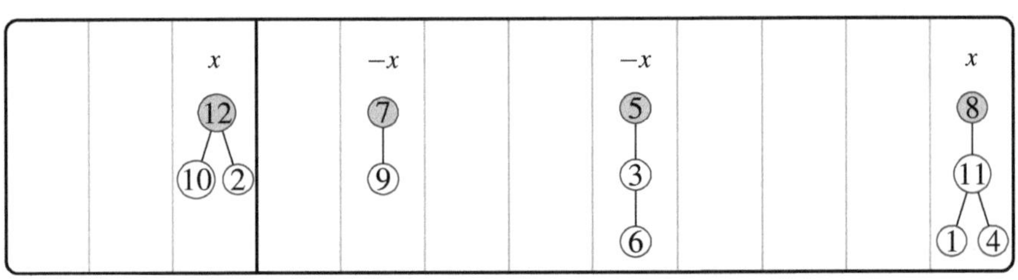

Apply the involution which breaks a brick at the first $-x$ or combines two bricks when the specified order on pictures is preserved. Fixed points must have pictures which weakly decrease according to size and the last brick must contain a picture of size j. This implies that the minimum sized picture in a fixed point must have size j. In other words, fixed points correspond to $\sum x^{\mathrm{pic}(s)}$ where the sum runs over all $s \in \mathcal{S}_n$ which have $\min(s) = j$. The generating function in the statement of the theorem follows from applying φ to 4.1 and then summing over all $j \geq 1$. □

In a similar way as in the proof of this last theorem, the exponential theorem can be refined to keep track of the size of the maximum picture.

We end this section by showing how an unlabeled version of the exponential formula also can be proved using involutions on brick tabloids. Let $\mathcal{P}_n$ to be a set of unlabeled pictures of size n—pictures which are like those described above but with any integer labels erased. Let $\mathcal{U}_n$ be the collection of sets of the form $\{p_1, \ldots, p_k\}$ such that each p_i is an unlabeled picture and the sum of the sizes of the pictures is n. For instance, in the case where $\mathcal{P}_n$ contains unlabeled rooted trees on n nodes, $\mathcal{U}_n$ contains the unlabeled rooted forests on n nodes. Just as in the case of labeled objects, for $u \in \mathcal{U}_n$, $\mathrm{pic}(u)$ denotes the number of pictures used to create u.

Theorem 4.11. *We have* $\sum_{n=0}^{\infty} z^n \sum_{u \in \mathcal{U}_n} x^{pic(u)} = \prod_{i=1}^{\infty} \frac{1}{(1-xz^i)^{|\mathcal{P}_i|}}$.

Proof. Define the homomorphism φ such that

$$\varphi(e_n) = (-1)^n \sum_{\substack{i_1, i_2, \ldots, i_n \geq 0 \\ i_1 + 2i_2 + \cdots + ni_n = n}} \binom{|\mathcal{P}_1|}{i_1} \cdots \binom{|\mathcal{P}_n|}{i_n} (-x)^{i_1 + \cdots + i_n}.$$

Applying φ to h_n gives $\sum_{\lambda \vdash n} (-1)^{n-\ell(\lambda)} |B_{\lambda,(n)}| \varphi(e_{\lambda_1}) \cdots \varphi(e_{\lambda_\ell})$ from which we create combinatorial objects by first selecting $T \in B_{\lambda,(n)}$ for some $\lambda \vdash n$ and then using the -1 sign in this equation and in the definition of φ to assign exactly one -1 sign to each brick in T.

The definition of φ tells us that for each brick of length k we should choose $i_1, \ldots, i_k$ nonnegative integers such that $i_1 + 2i_2 + \cdots + ki_k = k$. Using the binomial coefficients in the definition of φ, select i_j different pictures in $\mathcal{P}_{i_j}$ to be placed in the brick for $j = 1, \ldots, k$. Sort these pictures in increasing order first according to size, and then sort according to some arbitrary linear order given to $\mathcal{P}_n$. Suppose that when this is done, the pictures are $p_{k_1}, \ldots, p_{k_m}$. Place p_{k_j} in cell number $k_1 + \cdots + k_j$ reading left to right. In each one of the cells which now contains a picture, place one factor of $-x$. Since there must be a picture in the last cell in any brick, so must there be a $-x$. Use the factor -1 given to each brick to change this terminal $-x$ to an x.

For example, one possible object created in this way is

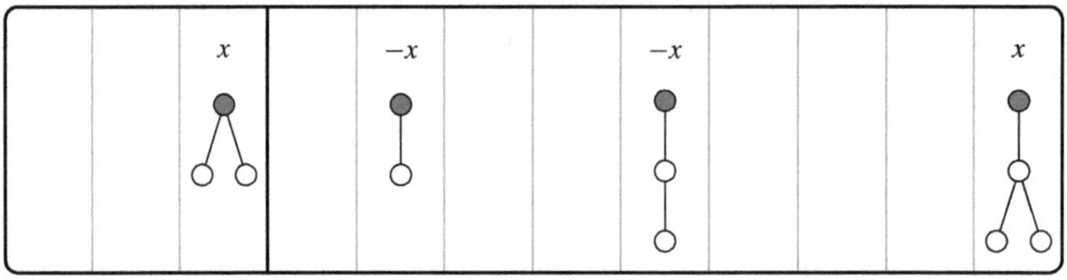

The weighted sum over all such combinatorial objects is $\varphi(h_n)$.

Scan the bricks from left to right looking for a $-x$ or two consecutive bricks which may be combined to preserve the order of the pictures. Break or combine the bricks accordingly, changing the power on x in the process. This involution is sign reversing and weight preserving. The fixed points cannot have any $-x$ labels and hence all bricks must contain one and only one picture. Furthermore, in a fixed point, the pictures when read from left to right must be written in decreasing order.

These fixed points correspond to objects in $\mathcal{U}_n$ with powers of x counting the number of pictures. Applying φ to Theorem 2.5,

$$\sum_{n=0} z^n \sum_{u\in\mathcal{U}_n} x^{\mathrm{pic}(u)} = \left(\sum_{n=0} (-z)^n(-1)^n \sum_{i_1+\cdots+ni_n=n} \binom{|\mathcal{P}_1|}{i_1}\cdots\binom{|\mathcal{P}_n|}{i_n}(-x)^{i_1+\cdots+i_n}\right)^{-1}$$

$$= \left(\prod_{i=1}^{\infty}\sum_{j=0} \binom{|\mathcal{P}_j|}{j}(-x)^j(z^i)^j\right)^{-1},$$

which by the binomial theorem is equal to the desired expression. □

As an example of how Theorem 4.11 can be applied, consider the generating function for the number of partitions of n refined by length. Let $\mathcal{P}_n$ be a set with one picture, the picture consisting of a horizontal strip of n cells. Elements in $\mathcal{U}_n$ therefore correspond to Young diagrams. Theorem 4.11 gives

$$\sum_{n=0} z^n \sum_{\lambda\vdash n} x^{\ell(\lambda)} = \prod_{i=1}^{\infty}\frac{1}{(1-xt^i)},$$

which is a refinement of Theorem 1.8.

4.5 Weighting Multiple Bricks

As we have seen in the previous sections, the ability to weight the last brick in a brick tabloid gives us greater versatility in our method of using ring homomorphisms on Λ_n to find generating functions. In this section we extend this idea further and allow more refined weights of brick tabloids.

Let $v_1,\ldots,v_r$ be functions defined on the set of nonnegative integers. If T is a brick tabloid with bricks of lengths $b_1,\ldots,b_k$ reading left to right, then let

$$w_{v_1,\ldots,v_r}(T) = v_1(b_1)\cdots v_r(b_r)$$

if $k \geq r$ and define $w_{v_1,\ldots,v_r}(T) = 0$ otherwise. Then we can define a new class of symmetric functions by

$$p_{v_1,\ldots,v_r,n} = \sum_{\lambda \vdash n} \sum_{T \in B_{\lambda,(n)}} (-1)^{n-\ell(\lambda)} w_{v_1,\ldots,v_r}(T) e_\lambda.$$

We can first count brick tabloids $T = (b_1,\ldots,b_k)$ with $k = r$ and then count the brick tabloids with $k > r$ by sorting them by the size of the last brick. This gives that $p_{v_1,\ldots,v_r,n}$ is equal to

$$\sum_{\substack{b_1+\cdots+b_r=n \\ b_i \geq 1}} (-1)^{n-r} v_1(b_1)\cdots v_r(b_r) e_{b_1}\cdots e_{b_r} + \sum_{k=1}^{n-r} (-1)^{k-1} e_k p_{v_1,\ldots,v_r,n-k}.$$

From here it follows that

$$\sum_{k=0}^{n-r} (-1)^k e_k p_{v_1,\ldots,v_r,n-k} = \sum_{\substack{b_1+\cdots+b_r=n \\ b_i \geq 1}} (-1)^{n-r} v_1(b_1)\cdots v_r(b_r) e_{b_1}\cdots e_{b_r}$$

which implies that

$$\left(\sum_{n=0}^{\infty} (-1)^k e_k z^k\right)\left(\sum_{n=r}^{\infty} p_{v_1,\ldots,v_r,n} z^n\right) = \prod_{i=1}^{r}\left(\sum_{n=1}^{\infty} (-1)^{n-1} v_i(n) e_n z^n\right).$$

Therefore we have

$$\sum_{n=r}^{\infty} p_{v_1,\ldots,v_r,n} z^n = \frac{\prod_{i=1}^{r}\left(\sum_{n=1}^{\infty} (-1)^{n-1} v_i(n) e_n z^n\right)}{\sum_{n=0}^{\infty} (-1)^n e_n z^n}. \tag{4.9}$$

This equation is analogous to (4.1). This means we can define ring homomorphisms φ on the elementary symmetric functions, apply φ on $p_{v_1,\ldots,v_r,n}$ to get an expansion in terms of weighted brick tabloids, define an involution which leaves interesting fixed points, and then use (4.9) to get a generating function for a permutation statistic. The rest of this section gives an example of this technique, following an approach first shown in [90].

Let $\mathcal{D}_{i+kn+j}^{(i,k,j)}$ denote set of all permutations in S_{i+kn+j} such that the descents in σ can only appear at indices in $\{i+k, i+2k, \ldots, i+nk\}$ where i, j, k, and n are nonnegative integers which satisfy $k \geq 2$, $0 \leq i \leq k-1$, and $0 \leq j \leq k-1$. We have seen special cases of this set of permutations before; the case of $k = 1$ gives our old friend the alternating permutations. For $\sigma \in \mathcal{D}_{i+kn+j}^{(i,k,j)}$, we let $\mathrm{des}_{i,k}(\sigma)$ be the permutation statistic counting the number of integers of the form $i+ks$ such that $0 \le s \le n-1$ and $\sigma_{i+ks} > \sigma_{i+ks+1}$. The next theorem is similar to Theorem 4.6.

Theorem 4.12. *If $k \ge 2$ and $0 < i, j < k$, then*

$$\sum_{n=2}^{\infty} \frac{z^{kn}}{[kn-i-j]_q!} \sum_{\sigma \in \mathcal{D}_{i+kn+j}^{(i,k,j)}} q^{inv(\sigma)} x^{ris_{i,k}(\sigma)}$$
$$= \frac{\left(\sum_{n=1}^{\infty} \frac{(x-1)^k z^{nk}}{[nk-i]_q!}\right)\left(\sum_{n=1}^{\infty} \frac{(x-1)^k z^{nk}}{[nk-j]_q!}\right)}{x-1-\sum_{n=1}^{\infty} \frac{(x-1)^k z^{nk}}{[nk]_q!}}$$
$$-\sum_{n=2}^{\infty} \frac{(n-1)x^{n-2} z^{kn}}{[kn-i-j]_q!} q^{\binom{nk-i-j}{2}} + \sum_{n=2}^{\infty} \frac{x^{(n-1)} z^{kn}}{[kn-i-j]_q!} q^{\binom{nk-i-j}{2}}.$$

Proof. Define a ring homomorphism φ and a function v_i by setting $\varphi(e_n) = 0$ and $v_i(n) = 0$ if n is not a multiple of k and

$$\varphi(e_{kn}) = (-1)^{kn-1} \frac{(x-1)^{k-1}}{[kn]_q!} \qquad \text{and} \qquad v_i(kn) = \frac{[kn]_q!}{[kn-i]_q!}$$

otherwise. Then applying φ to $p_{v_i,v_j,n}$ gives

$$\varphi(p_{v_i,v_j,n}) = \sum_{\lambda \vdash n} \sum_{T \in B_{\lambda,(n)}} (-1)^{n-\ell(\lambda)} w_{v_i,v_j}(T) \varphi(e_\lambda).$$

Since $\varphi(e_\lambda)$ is equal to 0 if any of the parts in λ are not multiples of k, we may assume that all parts in λ and n are multiples of k (and thus $n \ge 2$). Furthermore, if there are fewer than 2 bricks in a brick tabloid T, then the weight $w_{v_i,v_j}(T)$ is defined equal to 0.

In our definition of $p_{v_i,v_j,n}$, the function v_i weights the first brick in a brick tabloid and the function v_j weights the second brick. However, by moving this second brick to the end of the brick tabloid, we can instead choose to apply the function v_i to the first brick and v_j to the last brick in the brick tabloid. Putting this idea together with the same type of logic as found in the proof of Theorem 4.6, we have $[nk-i-j]_q! \varphi(p_{v,v,nk})$ is equal to

$$[nk-i-j]_q! \sum_{\substack{\lambda \vdash nk \\ \ell(\lambda) \ge 2}} \sum_{T \in B_{\lambda,(n)}} (-1)^{nk-\ell(\lambda)} w_{v_1,\ldots,v_r}(T) \varphi(e_\lambda)$$
$$= \sum_{\substack{\lambda \vdash kn \\ \ell(\lambda) \ge 2}} \sum_{\substack{T \in B_{\lambda,(kn)} \text{ has bricks with} \\ \text{lengths } b_1,\ldots,b_\ell \text{ divisible by } k}} \begin{bmatrix} nk-i-j \\ b_1-i, b_2, \ldots, b_{\ell-1}, b_\ell - j \end{bmatrix}_q (x-1)^{\frac{b_1}{k}+\cdots+\frac{b_\ell}{k}-\ell}$$

This last sum tells us to create combinatorial objects by selecting a brick tabloid in $T \in B_{\lambda,(kn)}$ such that there are at least two bricks and each brick length is a multiple of k. In every nonterminal kth cell in a brick, place a choice of x or -1 in a brick, using the power of $(x-1)$. With the q-multinomial coefficient, place a sequence of integers in each brick such that

1. the first i cells of the first brick are left blank,
2. the last j cells of the last brick are left blank,
3. all other cells except the first i and last j contain integers such that each brick contains a decreasing sequence of integers and every integer from $1,\ldots,$ $nk-i-j$ is used exactly once in T, and
4. a power of q in each cell records the number of smaller integers in T appearing to the right.

For example, one such combinatorial object when $k=3$, $i=2$, and $j=1$ is

		x			1			1			-1			1
		q^{11}	q^7	q^6	q^5	q^6	q^4	q^1	q^4	q^3	q^2	q^1	q^0	
		12	8	7	6	10	5	2	11	9	4	3	1	

We next apply a slight variation on our usual brick breaking and combining involution. We want to apply the involution as usual, but run into a problem when T contains exactly two bricks, there are no -1s anywhere in T, and there is a decrease in the integer labeling between the two bricks. In this situation we would want to combine these two bricks but cannot do so since each combinatorial object must have at least two bricks by construction. Define such objects to be fixed points and then proceed otherwise with the usual involution.

Fixed points now come in two varieties. First, there are fixed points which contain no decreases between bricks and there is a power of x for each descent of the form $\sigma_{i+ks} > \sigma_{i+ks+1}$. These are the fixed points we want because these fixed points correspond to $\sum_{\sigma \in \mathcal{D}_{i+kn+j}^{(i,k,j)}} x^{\mathrm{des}_{i,k}(\sigma)} q^{\mathrm{inv}(\sigma)}$.

Second, there are fixed points which contain exactly two bricks which contain a decrease between them. The underlying permutation σ in such a fixed point must be the permutation $\sigma = (kn-j-i)\cdots 2\,1$, that is, all integers must appear in decreasing order. One example when $n=5$, $k=3$, $i=2$, and $j=1$ is below:

		x			x			1			x			1
		q^{11}	q^{10}	q^9	q^8	q^7	q^6	q^5	q^4	q^3	q^2	q^1	q^0	
		12	11	10	9	8	7	6	5	4	3	2	1	

The weight on this type of a fixed point is $x^{n-2}q^{\binom{nk-i-j}{2}}$. There are $n-1$ such fixed points because the two bricks can be partitioned in $n-1$ ways.

Counting the fixed points, we have shown

$$[nk-i-j]_q!\varphi(p_{v_i,v_j,n}) = (n-1)x^{n-2}q^{\binom{nk-i-j}{2}} + \sum_{\sigma\in\mathfrak{D}_{i+kn+j}^{(i,k,j)}} x^{\mathrm{des}_{i,k}(\sigma)}q^{\mathrm{inv}(\sigma)}.$$

Applying φ to both sides of (4.9) with $r=2$ gives

$$\sum_{n=2}^{\infty}\frac{z^{kn}}{[kn-i-j]_q!}\left((n-1)x^{n-2}q^{\binom{nk-i-j}{2}} - x^{n-1}q^{\binom{nk-i-j}{2}} + \sum_{\sigma\in\mathfrak{D}_{i+kn+j}^{(i,k,j)}} x^{\mathrm{des}_{i,k}(\sigma)}q^{\mathrm{inv}(\sigma)}\right)$$
$$= \frac{\left(\sum_{n=1}^{\infty}(-1)^{n-1}v_i(n)\varphi(e_n)z^n\right)\left(\sum_{n=1}^{\infty}(-1)^{n-1}v_j(n)\varphi(e_n)z^n\right)}{1+\sum_{n=1}^{\infty}(-1)^n\varphi(e_n)z^n}$$
$$= \frac{\left(\sum_{n=1}^{\infty}(x-1)^{k-1}z^{nk}/[nk-i]_q!\right)\left(\sum_{n=1}^{\infty}(x-1)^{k-1}z^{nk}/[nk-j]_q!\right)}{1-\sum_{n=1}^{\infty}(x-1)^{k-1}z^{nk}/[nk]_q!},$$

which in turn can be rearranged to be the statement in the theorem. □

Exercises

4.1. Let $\mathrm{fi}(\sigma)$ be the length of the final increasing sequence in a permutation σ. Define a homomorphism φ by $\varphi(e_n) = (-1)^{n-1}(1-x)^{n-1}x/n!$ and a function v by

$$v(n) = \begin{cases} 0 & \text{if } n<j, \\ \dfrac{1}{(1-x)^{n-1}x} & \text{if } n=j, \\ \dfrac{(1-x)^{n-j-1}(-x)}{(1-x)^{n-1}x} & \text{if } n>j, \end{cases}$$

where j is a positive integer. Use φ and v to find a generating function for

$$\sum_{n=0}^{\infty}\frac{z^n}{n!}\sum_{\sigma\in S_n} x^{\mathrm{des}(\sigma)}y^{\mathrm{fi}(\sigma)}.$$

4.2. Let $a_{n,k}$ be the number of weak alternating words in $\{0,\ldots,k-1\}_n^*$, that is, $a_{n,k}$ is the number of words $w_1,\ldots,w_n$ which have $w_i < w_{i+1}$ if and only if i is even. Find a generating function for $\sum_{n=1}^{\infty} a_{n,k}z^n$.

4.3. The Fibonacci sequence is defined by $F_1 = F_2 = 1$ and $F_n = F_{n-1} + F_{n-2}$ for $n \geq 3$. Using the combinatorial interpretation for recurrences given in section 4.3, prove the identities $F_1 + \cdots + F_{n-1} = F_{n+1} - F_2$ and $3F_n = F_{n+2} + F_{n-2}$.

4.4. A path in a labeled graph is a finite sequence of distinct vertices such that consecutive vertices are connected by an edge. A labeled tree on n vertices is a labeled graph with n vertices such that there is a unique path connecting every pair of vertices. This exercise proves there are n^{n-2} labeled trees with n vertices, a result known as Cayley's formula.

Let A_n be the set of objects created by selecting a labeled tree on n vertices, shading one vertex gray, and then labeling the $n-1$ edges with $1,\ldots,n-1$ in some way. For example, one element in A_6 is

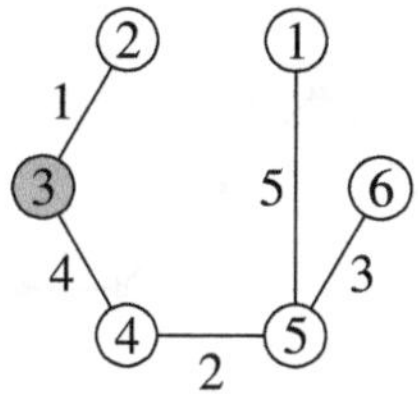

By double counting the elements in A_n, prove that there are n^{n-2} labeled trees on n vertices. Deduce the unusual identity

$$\sum_{n=0}^{\infty}(n+1)^{(n-1)}\frac{z^n}{n!} = exp\left(\sum_{n=1}^{\infty}n^{(n-1)}\frac{z^n}{n!}\right)$$

by considering rooted labeled trees, i.e., labeled trees with one distinguished vertex.

4.5. Let $\mathcal{L}_n(z)$ be the set of lists of the form $(p_1,\ldots,p_k)$ such that each p_i is a picture in $\mathcal{P}_i$, the sizes of $p_1,\ldots,p_k$ sum to n, and the integer labels in the pictures have been replaced so that the total list contains the integers $1,\ldots,n$. This is a similar situation as the exponential formula except that now we are considering ordered lists, not unordered sets.

Define a ring homomorphism φ in order to prove that

$$\sum_{n=0}^{\infty}\frac{z^n}{n!}\sum_{L\in\mathcal{L}_n}x^{\text{pic}(L)} = \frac{1}{1-x\mathcal{P}(z)},$$

where $\mathcal{P}(z) = \sum_{n=1}^{\infty}|\mathcal{P}_n|z^n/n!$.

If z_0 is the unique solution to $\mathcal{P}(z) = 1$ with smallest magnitude and if $\mathcal{P}'(z_0) \neq 0$, find an approximation for $|\mathcal{L}_n|$. As special cases, approximate the number of ordered disjoint cycles with union n and the number of ordered set partitions of n.

4.6. Find a generating function for the number of permutations in S_{2n} with only even sized cycles. Exactly how many such permutations are there?

4.7. Find a generating function for the number of permutations in S_n with cycles which must be an odd length. When taking n is even, this should be the same generating function as found in Exercise 4.6, implying that the number of permutations in S_{2n} with only even cycles must equal the number of permutations in S_{2n} with only odd cycles. Prove this identity bijectively.

4.8. A Motzkin path of length n is a path in the plane starts at $(0,0)$, ends at $(n,0)$, uses steps of the form $(1,1)$, $(1,-1)$, or $(1,0)$, and never travels below (but may touch) the x-axis. A labeled Motzkin path is a Motzkin path where

1. each $(1,1)$ step is not labeled with an integer,
2. each $(1,-1)$ step going from height $y=k$ to height $y=k-1$ is labeled with an integer in $\{1,\ldots,k\}$, and
3. each $(1,0)$ step at height $y=k$ is labeled with an integer in $\{1,\ldots,k+1\}$.

For example, one labeled Motzkin path is

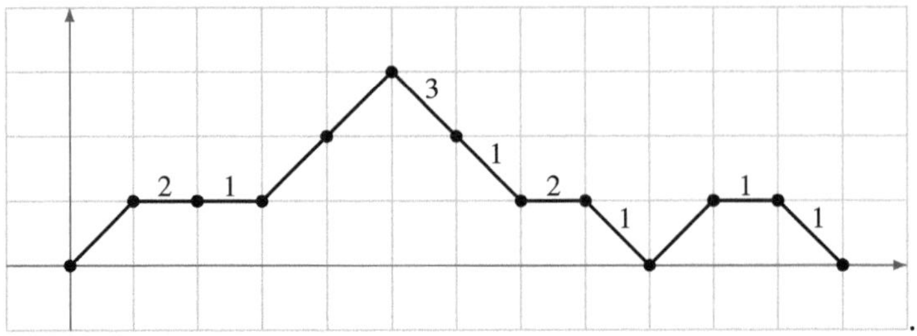

Use a bijection to show that the number of labeled Motzkin paths of length n is equal to the number of set partitions of n.

4.9. Use Exercise 4.8 and the approach in Exercise 3.10 to show that

$$\sum_{n=0}^{\infty} |\{\text{the set partitions of } n\}| z^n = \cfrac{1}{1-z-\cfrac{z^2}{1-2z-\cfrac{2z^2}{1-3z-\cfrac{3z^2}{1-4z-\cdots}}}}.$$

Solutions

4.1 Applying our choices for φ and ν give us

$$n!\varphi(p_{\nu,n}) = \sum_{\lambda\vdash n} \binom{n}{\lambda} w_\nu(B_{\lambda,(n)})(1-x)^{\lambda_1-1}x(1-x)^{\lambda_2-1}x\cdots.$$

From this sum we select a brick tabloid $T \in B_{\lambda,(n)}$ for some $\lambda \vdash n$. Use the $\binom{n}{\lambda}$ term to select a permutation in which to fill the cells of T such that each brick contains an *increasing* sequence. With the powers of x, place an x in the terminal cell of each brick and a choice of either 1 or $-x$ in every other cell.

The function ν tells us that the length of the last brick must be at least j. If the final brick has length j, erase all of the x or 1 weights in the last brick. If the final brick has length greater than j, erase all of the weights in the last brick and replace them $n-j-1$ choices of $-x$ or 1 and then a $-x$.

One example of such a combinatorial object when $j = 6$ is below:

$-x$	x	x	1	$-x$	$-x$						
6	7	11	1	2	3	4	5	8	9	10	12

Modify our usual brick breaking and combining involution by scanning from left to right looking for either the first $-x$ or two consecutive bricks with an increase between them. Break or combine the bricks accordingly, reversing the sign on the resulting middle x or terminal $-x$. The fixed points under this involution tell us that $n!\varphi(p_{\nu,n}) = \sum x^{\mathrm{des}(\sigma)}$ where the sum runs over $\sigma \in S_n$ with $\mathrm{fi}(\sigma) = j$.

A generating function follows from summing over all positive j and then applying φ to equation 4.1. Using similar steps as in the proof of Theorem 4.3,

$$\sum_{n=0}^{\infty} \frac{z^n}{n!} \sum_{\sigma \in S_n} x^{\mathrm{des}(\sigma)} y^{\mathrm{fi}(\sigma)} = 1 + \sum_{j=1}^{\infty} y^j \frac{\sum_{n=1}^{\infty} (-1)^{n-1} \nu(n) \varphi(e_n) z^n}{\sum_{n=0} (-1)^n \varphi(e_n) z^n},$$

which, by using the definitions of φ and ν together with routine simplification steps, can be shown to equal

$$\left(\frac{x-1}{x-e^{(x-1)z}}\right)\left(\frac{x-(1-y)e^{(x-(1-y))z}}{x-(1-y)}\right).$$

4.2 Combining the ideas in Exercise 3.13 with that in Theorem 4.6, we let

$$\varphi(e_n) = (-1)^{n-1}\binom{n+k-1}{k-1}\begin{cases} (-1)^{n/2-1} & \text{if } n \text{ is even,} \\ 0 & \text{if } n \text{ is odd.} \end{cases}$$

Let m be either 0 or 1 and define $\nu(n) = \binom{n-m+k-1}{k-1}/\binom{n+k-1}{k-1}$. Then applying φ to $p_{\nu,2n-m}$ gives

$$\sum_{\lambda \vdash 2n} \sum_{\substack{T \in B_{\lambda,(2n)} \text{ has bricks with} \\ \text{even lengths } b_1,\ldots,b_\ell}} \binom{b_1+k-1}{k-1} \cdots \binom{b_\ell - m + k - 1}{k-1} (-1)^{\frac{b_1}{2}+\cdots+\frac{b_\ell}{2}-\ell}.$$

The binomial coefficients allow us to fill each brick with a weakly decreasing sequence and so from this sum we create combinatorial objects which look like the following (when $k = 4$, $n = 6$, and $m = 1$):

	-1		-1		1		1		1		1
3	2	2	2	1	0	2	1	1	1	1	

Breaking or combining bricks at the first -1 or the first weak decrease between bricks leaves fixed points corresponding to weak alternating words. Applying φ to equation 4.1 and adding together the cases $m = 0$ and $m = 1$, we find that

$$\sum_{n=1}^{\infty} a_{n,k}z^n = \frac{1}{1-\sum_{n=1}^{\infty}(-1)^{n-1}\binom{2n+k-1}{k-1}z^{2n}} + \frac{\sum_{n=1}^{\infty}(-1)^{n-1}\binom{2n-1+k-1}{k-1}z^{2n-1}}{1-\sum_{n=1}^{\infty}(-1)^{n-1}\binom{2n+k-1}{k-1}z^{2n}}.$$

It can be shown that this expression simplifies to

$$(1+z^2)^{k/2}\sec(k\arctan z)+\tan(k\arctan z),$$

which nicely parallels the result for alternating permutations in Theorem 3.5.

4.3 Section 4.3 tells us that F_n is the number of brick tabloids T such that

1. each brick is of length 1 or 2 and has a weight of 1,
2. T either ends with a brick of length 2 or a brick of length 2 followed by a brick of length 1.

For example, $F_5 = 5$ because there are 5 such tabloids when $n = 5$:

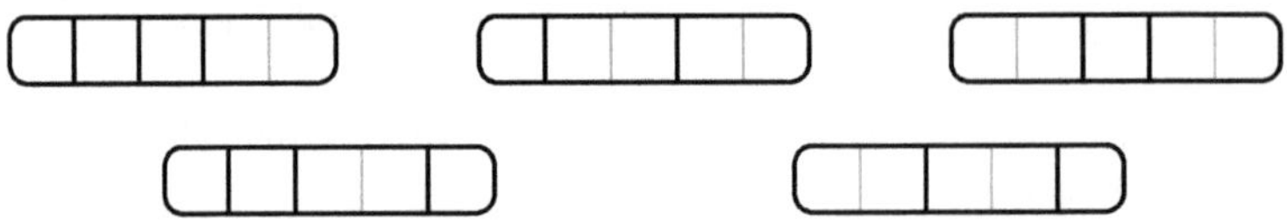

Let $\mathfrak{F}_n$ be the collection of the brick tabloids with these properties so that $|\mathfrak{F}_n| = F_n$.

To show $F_1 + \cdots + F_{n-1} = F_{n+1} - F_2$, take $T \in \mathfrak{F}_k$ for some $1 \le k \le n-1$ and create an element in $\mathfrak{F}_n$ by prepending a sequence of bricks with lengths $1,1,\ldots,1,2$ to T. This operation is reversible and creates every element in $\mathfrak{F}_{n+1}$ except for the one brick tabloid which only contains bricks of lengths $1,1,\ldots,1,2$.

To show $3F_n = F_{n+2} + F_{n-2}$, take $T \in \mathfrak{F}_n$. If T begins with a brick of length 1, replace this brick in T three ways: with a brick of length 1 followed by a brick of length 2, with a brick of length 2 then 1, and with three bricks of length 1. If T begins with a brick of length 2, then change T in three ways: by removing this brick of length 2, by prepending T with a brick of length 2, and by prepending two bricks of length 1. In pictures,

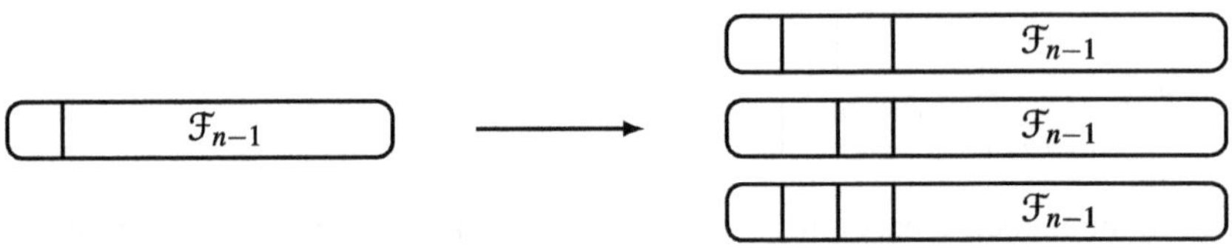

and

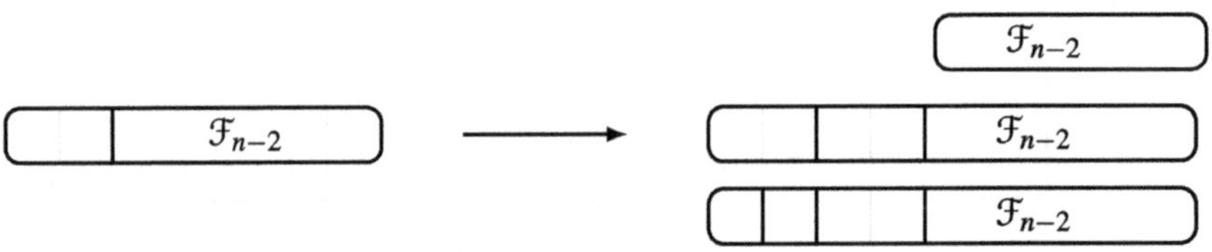

These identities also hold for any initial conditions F_1, F_2 since we have left the final segment of the brick tabloid unchanged.

4.4 Let T_n be the number of labeled trees on n vertices. There are T_n ways to select a labeled tree, n ways to select a vertex to shade gray, and $(n-1)!$ ways to label the $n-1$ edges in T. Therefore there are $T_n n(n-1)!$ elements in A_n.

For a second way to count the number of elements in A, consider constructing an element in A by following this algorithm:

1. Begin with a graph with n labeled gray vertices and no edges.
2. Select any vertex u.
3. Select a gray vertex v such that there is no path from u to v (which, initially, can be any vertex other than u).
4. Connect u and v with an edge and change v from a gray vertex to a white vertex. If this is the ith edge added to the graph, label the edge with i.
5. If there are two or more gray vertices in the graph, go back to step 2.

As an example of this process, we display the sequence of graphs created by algorithm in order to create the element in A_6 displayed in the statement of this exercise:

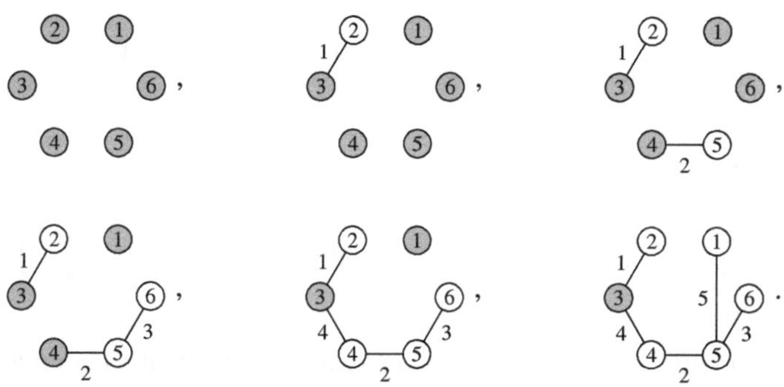

There are always n choices for the vertex u in step 2 of this algorithm and there are $n-i$ choices for vertex v in iteration i. Therefore the number of elements in A_n is $n(n-1)n(n-2)\cdots n(1) = n^{n-1}(n-1)!$. Since this must also equal $T_n n(n-1)!$, we find $T_n = n^{n-2}$, as desired.

There are n^{n-1} rooted labeled trees. There are $(n+1)^{n-1}$ forests of rooted labeled trees since, by connecting the vertex with label "$n+1$" to each root, these forests are in one-to-one correspondence with labeled trees on $n+1$ vertices. The identity follows from an application of the exponential formula and taking $x=1$.

4.5 If $\varphi(e_n) = (-1)^{n-1}x|\mathcal{P}_n|/n!$, then $n!\varphi(h_n) = \sum_{\lambda\vdash n}\binom{n}{\lambda}|B_{\lambda,(n)}|x^{\ell(\lambda)}|\mathcal{P}_{\lambda_1}||\mathcal{P}_{\lambda_2}|\cdots$, from which (in the case of taking pictures to be rooted labeled trees) we create combinatorial objects which look like this:

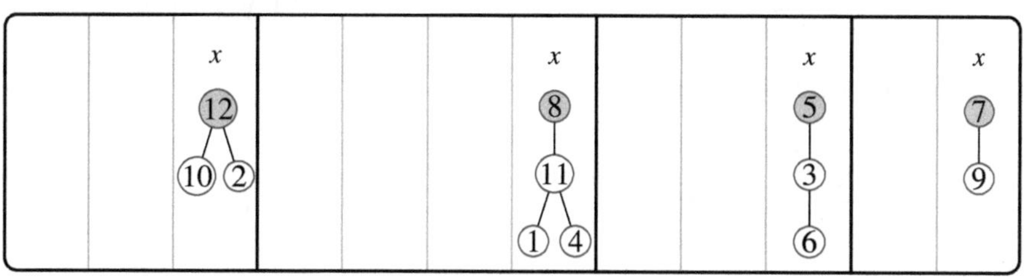

These objects immediately correspond with $\sum_{L\in\mathcal{L}_n} x^{\text{pic}(s)}$. Applying $\varphi(n)$ to Theorem 2.5 gives $\sum_{n=0}^{\infty}\frac{z^n}{n!}\sum_{L\in\mathcal{L}_n} x^{\text{pic}(s)} = 1/(1-x\mathcal{P}(z))$.

Taking $x=1$, we find a singularity at $|z_0|$. Since $\lim_{z\to z_0}\frac{z-z_0}{1-\mathcal{P}(z)} = -\frac{1}{\mathcal{P}'(z_0)}$,

$$\left|\frac{|\mathcal{L}_n|}{n!} - \frac{1}{\mathcal{P}'(z_0)|z_0|^{n+1}}\right| < \left(\frac{1}{|z_1|}+\varepsilon\right)^n$$

for all $\varepsilon > 0$ and large enough n where z_1 denotes the solution to $\mathcal{P}(z)=1$ with second smallest magnitude. This means $|\mathcal{L}_n|$ is approximately $n!/(\mathcal{P}'(z_0)|z_0|^{n+1})$.

In the special case of ordered cycles, $\mathcal{P}(z) = \sum_{n=1}^{\infty} z^n/n = -\ln(1-z)$. Here we find $z_0 = 1-1/e$ and $\mathcal{P}'(z_0) = e$, so the approximate number of ordered cycles of size n is $n!e^n/(e-1)^{n+1}$. In the special case of ordered set partitions, $\mathcal{P}(z) = \sum_{n=1}^{\infty} z^n/n! = e^z - 1$. Here we find $z_0 = \ln 2$ and $\mathcal{P}'(z_0) = 2$, so the approximate number of ordered set partitions of size n is $n!/(2(\ln 2)^{n+1})$.

4.6 Let $\mathcal{P}_n$ be empty if n is odd and the set of n cycles if n is even. Then

$$\mathcal{P}(z) = \sum_{n=1}^{\infty}\frac{z^{2n}}{(2n)!} = \frac{\ln\left(\frac{1}{1-z}\right)+\ln\left(\frac{1}{1+z}\right)}{2} = \ln\left(\frac{1}{\sqrt{1-z^2}}\right).$$

Using the exponential formula,

$$\sum_{n=0}^{\infty}\frac{z^n}{n!}\sum_{\substack{\sigma\in S_n \text{ has only}\\ \text{even sized cycles}}} x^{\text{cyc}(\sigma)} = e^{x\ln\left(\frac{1}{\sqrt{1-z^2}}\right)} = (1-z^2)^{-x/2} = \sum_{k=0}^{\infty}\binom{-x/2}{k}(-1)^k z^{2k},$$

where the last equality comes from an application of Newton's binomial theorem. Taking $x=1$ and extracting the coefficient of $z^{2n}/(2n!)$ from this sum gives an exact formula for the number of permutations in S_{2n} with only even sized cycles:

$$\begin{aligned}(-1)^n\binom{-1/2}{n}(2n)! &= (-1)^n\frac{(-1/2)(-1/2-1)\cdots(-1/2-n+1)}{n!}(2n)!\\ &= \frac{1}{2}\cdot\frac{3}{2}\cdots\frac{2n-1}{2}\cdot(1\cdot 3\cdots(2n-1))\,2^n\\ &= 1^2\cdot 3^2\cdot 5^2\cdots(2n-1)^2.\end{aligned}$$

4.7 Let $\mathcal{P}_n$ be empty if n is even and the set of n cycles if n is odd. Then

$$\mathcal{P}(z) = \sum_{n=1}^{\infty}\frac{z^{2n-1}}{(2n-1)!} = \frac{\ln\left(\frac{1}{1-z}\right)-\ln\left(\frac{1}{1+z}\right)}{2} = \ln\left(\sqrt{\frac{1+z}{1-z}}\right).$$

Using the exponential formula,

$$\sum_{n=0}^{\infty}\frac{z^n}{n!}\sum_{\substack{\sigma\in S_n \text{ has only}\\ \text{odd sized cycles}}} x^{\text{cyc}(\sigma)} = e^{x\ln\left(\sqrt{\frac{1+z}{1-z}}\right)} = \left(\frac{1+z}{1-z}\right)^{x/2}.$$

Taking $x = 1$, this last function can be written as $(1-z^2)^{-1/2} + z(1-z^2)^{-1/2}$. The even terms in this series are given by the function $(1-z^2)^{-1/2}$, which is the same generating function found in Exercise 4.6.

To show that the number of permutations in S_{2n} with only even cycles must equal the number of permutations in S_{2n} with only odd cycles bijectively, begin with a permutation $\sigma \in S_{2n}$ with only odd sized cycles.

Suppose that the cycles of σ are $c_1, \ldots, c_{2k}$ where the cycles are written in decreasing order according to maximum element and this maximum element is found at the end of the cycle. Change each pair of cycles c_{2i-1} and c_{2i} by removing the first element from c_{2i} and make it as the first element of c_{2i-1}. For example,

$$(5\ 7\ 12)\ (11)\ (9\ 6\ 10)\ (1\ 4\ 2\ 3\ 8)$$

would be changed to

$$(11\,5\ 7\ 12)\ (1\ 9\ 6\ 10)\ (4\ 2\ 3\ 8).$$

This process is reversible: working with pairs of cycles from left to right, remove the first element in each cycle and place it into the cycle immediately to the right. However, if doing so creates a permutation without the maximum element at the end of each cycle, instead make the removed integer its own cycle of length 1. This process is therefore a bijection, as desired.

4.8 In order to turn a labeled Motzkin path M into a set partition, begin by coloring each step in M blue.

Suppose the most left blue $(1,-1)$ step in M is in position j and has label ℓ. Locate the ℓth blue $(1,1)$ step in M, say this occurs at position i. The set

$$s = \{i, j\} \cup \{i < m < j : \text{the } m^{th} \text{ step in } M \text{ is a blue } (1,0) \text{ step with label } \ell\}$$

is one set in the set partition created from M. Recolor each step in M with a position in s black. Iterate this process until all $(1,-1)$ and $(1,1)$ steps are black.

After this iteration, some blue $(1,0)$ may remain. For each blue $(1,0)$ step in position j, place the set $\{j\}$ into the set partition. This completes our description of how to turn a labeled Motzkin path M into a set partition.

For example, the Motzkin path displayed in the statement of this exercise corresponds to the set partition $\{\{5,6\},\{1,3,7\},\{4,9\},\{10,11,12\},\{2\},\{8\}\}$.

This method of turning a Motzkin path M into a set partition S is a bijection since the inverse function can be described. Indeed, each set of size 1 in S gives the position of a $(1,0)$ step in M with maximum possible label. For $s \in S$ with size at least 2, the minimum number m in s gives the position of a $(1,1)$ step in M, the maximum number n in s gives the position of a $(1,-1)$ step in M, and all other numbers in s correspond to $(1,0)$ steps in M. This gives us the Motzkin path without the labels. The labels on the $(1,-1)$ step and the $(1,0)$ steps coming from s can be found from the number of $(1,1)$ steps between positions m and n in M.

4.9 Let $a_{n,k}$ be the number of labeled Motzkin paths which start at $(0,k)$ and end at (n,k). Let $A_k(z) = \sum_{n=0}^{\infty} a_{n,k} z^n$.

Suppose that the first time after $(0,k-1)$ that a labeled Motzkin path counted by $a_{i,k-1}$ returns to the line $y = k-1$ is at $(i,k-1)$. The path might begin with a step of the form $(1,0)$. In this case there are $ka_{n-1,k-1}$ ways to complete the path since there are k ways to label the $(1,0)$ step and $a_{n-1,k-1}$ ways to draw a labeled Motzkin path from $(1,k-1)$ to $(n,k-1)$.

Otherwise, if $i \geq 2$, the underlying Motzkin path must look like this:

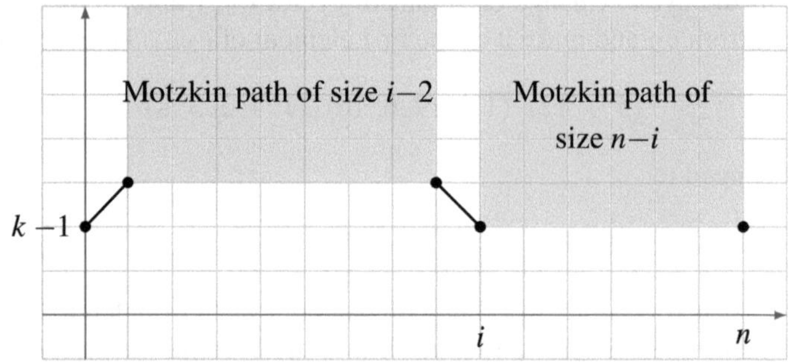

Since there are k ways to label the $(1,-1)$ step ending at $x = i$, the number of labeled Motzkin paths if $i \geq 2$ is $ka_{i-2,k}a_{n-i,k-1}$.

Summing over all possible i gives $a_{n,k-1} = ka_{n-1,k-1} + k\sum_{i=2}^{n} a_{i-2,k}a_{n-i,k-1}$ for $n \geq 2$. Therefore

$$\begin{aligned} A_{k-1}(z) - 1 - kz &= \sum_{n=2}^{\infty} a_{n,k-1} z^n \\ &= \sum_{n=2}^{\infty} a_{n-1,k-1} z^n + \sum_{n=2}^{\infty} \sum_{i=2}^{n} a_{i-2,k} a_{n-i,k-1} z^n \\ &= kz(A_{k-1}(z) - 1) + kz^2 A_k(z) A_{k-1}(z). \end{aligned}$$

Solving for $A_{k-1}(z)$ gives $A_{k-1}(z) = 1/(1 - kz - kz^2 A_k(z))$.

Exercise 4.8 says $A_0(z) = \sum_{n=0}^{\infty} |\{\text{the set partitions of } n\}| z^n$ and so the continued fraction follows from repeatedly applying $A_{k-1}(z) = 1/(1 - kz - kz^2 A_k(z))$ starting with $k = 1$.

Notes

The first use of the symmetric functions $p_{v,n}$ in enumeration is found in [76, 87]. The first use of the symmetric functions $p_{v_1,v_2,n}$ in enumeration is found in [90].

The generating function for the number of valleys in a permutation was first proved by Roger Entringer by solving differential equations [38]. Leonard Carlitz published a few papers containing the result [17, 20, 21] and Ira Gessel showed how this generating function fit into his framework [51]. In these publications the connection to the set of permutations $\sigma \in S_n$ without 2-descents and $\sigma_{n-1} < \sigma_n$ was not noted.

Carl Gustav Jacobi indicated a formal version of the exponential formula and special cases of the exponential formula were given for permutations by Jacques Touchard in 1939 and graphs by Robert Riddell and George Uhlenbeck in 1953 [62, 101, 111]. The full generality we give in Theorem 4.8 was first published in the early 1970s in papers by Edward Bender with Jay Goldman, Peter Doubilet with Gian-Carlo Rota and Richard Stanley, and Dominique Foata with Marcel-Paul Schützenberger [11, 30, 45]. Since that time there have been a number of extensions of the theory [10, 13, 49, 60, 68].

The proof of Cayley's formula given in Exercise 4.4 is due to Jim Pitman [1].

Chapter 5
Counting with RSK

The RSK algorithm is a bijection from the set of matrices with nonnegative integer entries to pairs of the form (P, Q) where P and Q are column strict tableaux of the same shape. The simple algorithm has amazing relationships to both symmetric functions and enumeration—no book on these topics is complete without it.

5.1 Row Insertion

Let T be a column strict tableau and j an integer. We define the row insertion of j into T, denoted $T \leftarrow j$, to be the tableau found by following these rules:

I0. If T is the empty tableau, then $T \leftarrow j$ is the column strict tableaux with 1 cell which contains the integer j.

If T is not empty, assume T is of the form shown here:

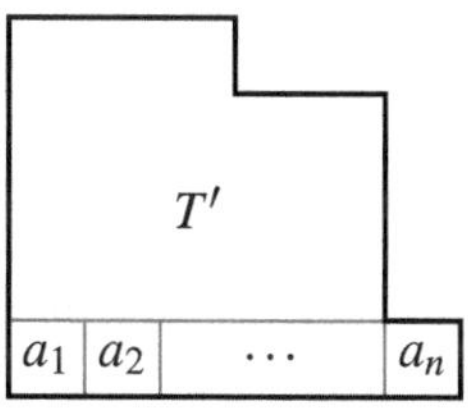

That is, the first row of T is $a_1 \leq \cdots \leq a_n$ and T' is the column strict tableau found by removing the first row of T.

I1. If $a_n \leq j$, then $T \leftarrow j$ results from T by adding a cell containing j to the end of the bottom row of T.

I2. If $j < a_n$, then let a_k be the leftmost entry in bottom row of T that is larger than j. Replace a_k with j and insert a_k into T'. In this case we say that j bumps a_k.

A. Mendes, J. Remmel, *Counting with Symmetric Functions*, Developments in Mathematics 43, DOI 10.1007/978-3-319-23618-6_5

For example, to row insert a 2 into the column strict tableau

4			
2	3	5	
1	1	3	5

,

we replace the 3 in the bottom row with a 2, the 5 in the second row with a 3, and then place a new cell with a 5 into the third row. Graphically, this row insertion looks like

4				
2	3	5		
1	1	3	5	$\leftarrow 2$

,

4				
2	3	5	$\leftarrow 3$	
1	1	2	5	

,

4	$\leftarrow 5$		
2	3	3	
1	1	2	5

,

4	5		
2	3	3	
1	1	2	5

.

Shading the path of the replaced cells in the above row insertion process gives

4	5		
2	3	3	
1	1	2	5

.

We will call this path the bumping path when i is inserted into T. As we will show in Theorem 5.1, such a path must start in the bottom row and, since T is a column strict tableau, move up and weakly to the left. Furthermore, the integers in this path are strictly increasing when read from bottom to top.

Theorem 5.1. *If T is a column strict tableau and j is a positive integer, then $T \leftarrow j$ is a column strict tableau and the bumping path for the insertion of j in T moves weakly to the left as one proceeds from bottom to top. Moreover, if $sh(T)$ is shape of T and $sh(T \leftarrow j)$ is the shape of $T \leftarrow j$, then $sh(T) \subseteq sh(T \leftarrow j)$.*

Proof. We proceed by induction on $|T|$, with the theorem holding true if applying rule I0 or I1.

Thus assume that rule I2 was applied and suppose j bumps a_k in the bottom row of T. The first row of $T \leftarrow j$ is therefore $a_1 \cdots a_{k-1} j a_{k+1} \cdots a_n$. By our choice of a_k, we must have $a_{k-1} \leq j < a_k \leq a_{k+1} \leq \cdots \leq a_n$, meaning that the first row of $T \leftarrow j$ is weakly increasing.

By induction, $T' \leftarrow a_k$ is a column strict tableaux. Thus to show that $T \leftarrow j$ is a column strict tableau, we need to only show that $T \leftarrow j$ is strictly increasing in columns in the first two rows. Suppose that $b_1 \leq \cdots \leq b_s$ is the first row of T'. There are two cases.

Case 1. Suppose $s \geq k$. In this case we know $b_k > a_k$, so $T' \leftarrow a_k$ either bumps b_k or it must bump some b_s with $s \leq k$. If b_k is bumped, then the first two cells of the kth column contains j in the first row and a_k in the second row. Since $j < a_k$,

$T \leftarrow j$ satisfies the column strict condition in kth column as the elements in the first two rows of the remaining columns are the same as in T. Thus $T \leftarrow j$ will be column strict.

If a_k bumps b_s for $s < k$, then we know that in column k, the first two elements are j and b_k, but $j < a_k < b_k$ so that we satisfy the column strict condition in column k. In column s, the first two elements are a_s and a_k, but we know that $a_s \leq a_{k-1} \leq j < a_k$ so that we satisfy the column strict condition in column s. In the remaining columns, the first two elements are the same as in T. Thus $T \leftarrow j$ will be column strict.

Case 2. Suppose $s < k$. In this case, we know that either a_k sits at the end of the first row in T' or it must bump some b_s with $s \leq k$. If a_k sits at the end of the first row in T', then either a_k sits on top of j if $s = k-1$ (in which case $j < a_k$) or a_k sits on top of some a_{s+1} where $s+1 \leq k-1$, in which case $a_{s+1} \leq a_{k-1} \leq j < a_k$. In either case, $T \leftarrow j$ is column strict in the first two rows of the column that contains a_k. The first two elements in the remaining columns are the same as in T and so $T \leftarrow j$ will be column strict.

If a_k bumps b_s for $s < k$, then we know that in column k, there is only one element. In column s, the first two elements are a_s and a_k, but we know that $a_s \leq a_{k-1} \leq j < a_k$ so that we satisfy the column strict condition in column s. In the remaining columns, the first two elements are the same as in T. Thus $T \leftarrow j$ will be column strict.

Lastly, the bumping path moves weakly to the left in the first two rows. By induction, the bumping moves weakly to the left in $T' \leftarrow a_k$. Thus the entire bumping path moves weakly to the left. □

If we are given $T \leftarrow j$ and the location of the cell c_1 in $sh(T \leftarrow j)$ but not $sh(T)$, then we can find both T and j. To reconstruct T and find j, follow these steps to reverse the row insertion process:

1. If the final cell inserted into T is on the bottom row, then this cell contains j and the removal of this cell leaves T.
2. If k is the integer in the final cell inserted into T, then find the rightmost entry in the row below k that is smaller than k, say ℓ. Remove the cell with k, replace ℓ with k, and repeat the process, moving down one row with each step.

Given a word $w_1 w_2 \cdots w_n$, let $T \leftarrow w_1 \cdots w_n = (\cdots((T \leftarrow w_1) \leftarrow w_2)\cdots) \leftarrow w_n$ so that $T \leftarrow w_1 \cdots w_n$ is the result of successively inserting $w_1, w_2, \ldots, w_n$ into T. In such a situation, we denote $T_0 = T$ and $T_i = T \leftarrow w_1 \ldots w_i$ for $i = 1, \cdots, n$. It follows from Theorem 5.1 that

$$sh(T_0) \subset sh(T_1) \subset sh(T_2) \subset \cdots \subset sh(T_n)$$

where the notation $\mu \subseteq \lambda$ means that the Young diagram for μ fits inside that of λ. We also let c_i to be the cell in $sh(T_i)$ but not $sh(T_{i-1})$ for $i = 1, \ldots, n$.

Theorem 5.2. *If $w = w_1 \cdots w_n$ is a word of length $n \geq 2$ with letters in $\{1, 2, \ldots\}$, then the following two facts are true:*

1. If $w_1 \le \cdots \le w_n$, then c_{i+1} is strictly to the right and weakly below c_i for all i.
2. If $w_1 > \cdots > w_n$, then c_{i+1} is strictly above and weakly to the left of c_i for all i.

These conditions are pictured below:

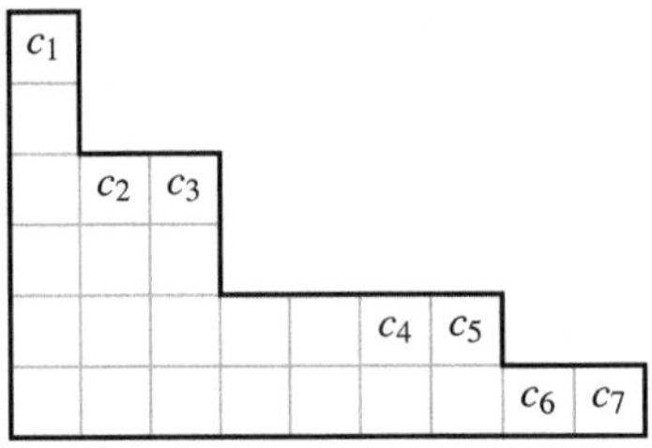

The case of $w_1 \le w_2 \le \cdots \le w_7$,

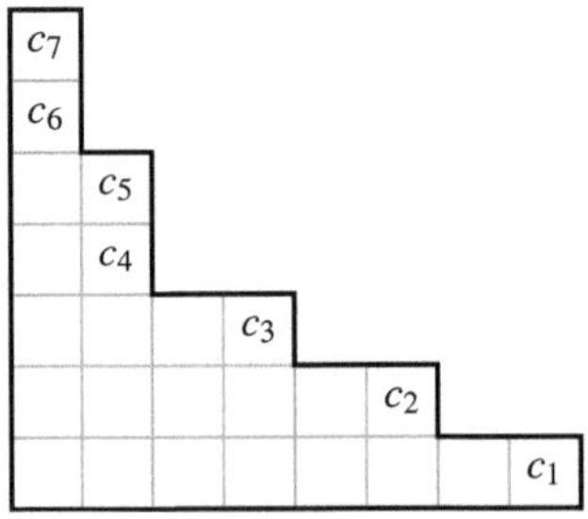

The case of $w_1 > w_2 > \cdots > w_7$.

Proof. To prove either statement 1 or 2, it is enough to consider only the case of $n = 2$. If this is done, the results will follow by induction.

We prove the first statement by induction on the size of T. If c_1 is row 1 (which would be the case only if insertion rules I0 and I1 were applied to $T \leftarrow w_1$), then w_1 is at the end of the row in T_1. Since $w_1 \le w_2$, the integer w_2 is placed at the end of the first row of T_1 in the insertion $T_1 \leftarrow w_2$. In this case, c_2 is clearly strictly to the right and weakly below c_1.

Suppose the bottom row of T contains $a_1 \le \cdots \le a_j$ and w_1 bumps a_k in $T \leftarrow w_1$. By our choice of a_k, we have $a_{k-1} \le w_1 < a_k$ and w_1 is in the kth cell of row 1 in T_1. This means that in the insertion $T_1 \leftarrow w_2$, either (i) w_2 is placed in the end of row 1 or (ii) w_2 must bump a_s where $s > k$. In case (i), our result follows since the cell c_1 is the new cell created by $T' \leftarrow a_k$ and we know that bumping paths move weakly to the left by Theorem 5.2. In case (ii), the result follows by induction since c_1 and c_2 are the cells created by the insertion $T' \leftarrow a_k a_s$ and $a_k \le a_s$. This completes the proof of the first statement.

The proof of the second statement is very similar, also following by induction on the size of T.

If $T = \emptyset$, then T_1 is the tableau with one cell containing w_1. As $w_1 > w_2$, this means that w_2 will bump w_1 in the insertion $T_1 \leftarrow w_2$ so c_2 will sit directly on top of c_1 in this case.

Suppose the bottom row of T contains $a_1 \le \cdots \le a_j$. If c_1 is row 1 and we applied rule I1 to find $T \leftarrow w_1$, then w_1 is at the end of the bottom row in T_1. As $w_1 > w_2$, then either w_2 bumps w_1 or it bumps some a_k. In either case, our result immediately follows from the fact that bumping paths move weakly to the left as we go up.

Finally, suppose that w_1 bumps a_k in $T \leftarrow w_1$. Hence $a_{k-1} \le w_1 < a_k$. Then w_1 is in the kth cell of row 1 in T_1. This means that either (i) w_2 bumps w_1 in the insertion $T_1 \leftarrow w_2$ or (ii) w_2 must bump a_i where $i < k$. In case (i), our result follows because cell c_1 is the new cell created by $T' \leftarrow a_k w_1$ and we know that $a_k > w_1$. In case (ii), the result follows by induction since c_1 and c_2 are the cells created by the insertion

$T' \leftarrow a_k a_s$ and $a_k > a_{k-1} \geq a_s$. This proves the second statement and completes the proof of the theorem. □

The Pieri rules, found in Theorem 5.3, give a nice description of how to expand the products $h_n s_\mu$ and $e_n s_\mu$ into a sum of Schur symmetric functions. One nice consequence of Theorems 5.1 and 5.2 is that they allow us to give a combinatorial proof of the Pieri rules.

For integer partitions μ, λ, we let λ/μ the cells in the Young diagram for λ but not those in μ. This type of object is called a skew shape. We say that skew shape λ/μ is a skew row if λ/μ has no two cells in the same column and λ/μ is a skew column if no two cells of λ/μ lie in the same row.

For an alternative proof of the Pieri rule giving the expansion of $e_n s_\mu$ using labeled abaci, see Exercises 2.10, 2.11, and 2.12 in Chapter 2.

Theorem 5.3 (The Pieri rules). *For any partition μ and for all $n \geq 1$,*

$$h_n s_\mu = \sum_{\substack{\lambda/\mu \text{ is a skew row} \\ \text{with } n \text{ cells}}} s_\lambda \qquad \text{and} \qquad e_n s_\mu = \sum_{\substack{\lambda/\mu \text{ is a skew column} \\ \text{with } n \text{ cells}}} s_\lambda.$$

Proof. We begin with the expansion of $h_n s_\mu$. Let $\mathcal{H}_{n,\mu}$ denote set of all pairs (S,T) where S is a column strict tableau of shape (n) and T is a column strict tableau of shape μ. Let $SR_{n,\mu}$ denote the set of all column strict tableaux P such that $sh(P)/\mu$ is a skew row of size n. The expansion of $h_n s_\mu$ in the statement of the theorem is equivalent to

$$\sum_{(S,T)\in\mathcal{H}_{n,\mu}} w(S)w(T) = \sum_{P\in SR_{n,\mu}} w(P),$$

where $w(P)$ is the usual weight for column strict tableaux as defined in Chapter 1.

We claim that row insertion algorithm allows us to give a weight preserving bijection $\theta : \mathcal{H}_{n,\mu} \to SR_{n,\mu}$. That is, given a pair (S,T), let $a_1 \leq \cdots \leq a_n$ be the elements of S, reading from left to right. We define $\theta(S,T) = P = T \leftarrow a_1 \ldots a_n$. By the first statement in Theorem 5.2, we know that if $\lambda = sh(P)$, then λ/μ is skew row. Moreover, we know that in the insertion of $a_1 \ldots a_n$ into T, the new cells where created from left to right. This allows us to reverse our steps, showing that θ is one-to-one.

To show that θ is bijection, it remains to be shown that θ is surjective, that is, we must show that if P is any column strict tableau such that $sh(P)/\mu$ is skew row of size n, then there exists a sequence $a_1 \leq \cdots \leq a_n$ and a column strict tableau T of shape μ such that $P = T \leftarrow a_1 \ldots a_n$.

The idea is that given P, we can look at the cells in $sh(P)/\mu$. If P did come from an insertion of the form $T \leftarrow a_1 \cdots a_n$ where $a_1 > \cdots > a_n$, then we know that new cells were created from left to right by Theorem 5.2. Thus we just try to reverse the process but first doing the reverse row insertion on the rightmost cell of $sh(P)/\mu$. Thus $P = P_1 \leftarrow a_n$ for some column strict tableau P_1 such that $sh(P_1)/\mu$ is skew row. Next we reverse the row insertion for P_1 using the rightmost cell of $sh(P_1)/\mu$. Thus $P_1 = P_2 \leftarrow a_{n-1}$ so that $P = P_2 \leftarrow a_{n-1}a_n$. Continuing on in this way, we will obtain a sequence $a_1 \cdots a_n$ and a column strict tableau of shape μ such that $P = T \leftarrow a_1 \cdots a_n$.

For example, suppose $\mu = (4,2,1)$, $\lambda = (5,4,2,1)$, and P is the column strict tableau shown below

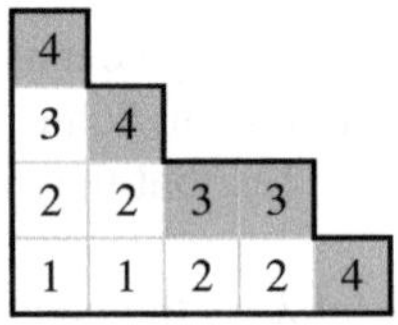

where we have shaded the cells in λ/μ. We illustrate the idea of reversing the row insertion process by first undoing the row insertion of the bottom 4, then undoing the row insertion of the right most 3, and so on.

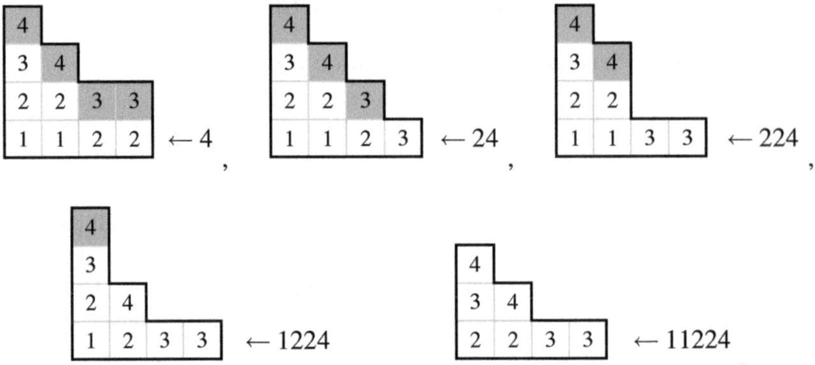

The question of why $a_1, \cdots, a_n$ is weakly increasing remains. Appealing to the second statement in Theorem 5.2, if $a_j > a_{j+1}$ for some j, then the new cell c_{j+1} created by the insertion of a_{j+1} into $T \leftarrow a_1 \cdots a_j$ is strictly above and weakly to left of the new cell c_j created by the insertion of a_j into $T \leftarrow a_1 \cdots a_{j-1}$. But this is not what happens in our process—in our reverse process, c_{i+1} is strictly to the right and weakly below c_i. Hence there can be no such j and so $a_1, \cdots, a_n$ is weakly increasing.

This now proves that θ is a bijection, proving the Pieri rule involving the homogeneous symmetric functions.

The proof of the Pieri rule involving the elementary symmetric functions is similar. Let $\mathcal{E}_{n,\mu}$ be the set of all pairs (S,T) where S is a column strict tableau of shape (1^n) and T is a column strict tableau of shape μ. Let $SC_{n,\mu}$ denote the set of all column strict tableaux P such that $sh(P)/\mu$ is a skew column of size n. Then we wish to show

$$\sum_{(S,T)\in\mathcal{E}_{n,\mu}} w(S)w(T) = \sum_{P\in SC_{n,\mu}} w(P).$$

Row insertion again allows us to give a weight preserving bijection $\Gamma : \mathcal{E}_{n,\mu} \to SC_{n,\mu}$. Given a pair (S,T), let $a_1 > \cdots > a_n$ be the elements of S, reading from top to bottom. Define $\Gamma(S,T) = P = T \leftarrow a_1 \cdots a_n$. By Theorem 5.2, if $\lambda = sh(P)$, then λ/μ is skew column. Moreover, we know that in the insertion of $a_1 \cdots a_n$ into T, the new cells were created from bottom to top. This allows us to reverse our steps, implying that Γ is one-to-one.

We show that Γ is also surjective in a similar manner as when we showed θ is a surjection earlier in the proof. Take and column strict tableau P such that $sh(P)/\mu$ is a skew column with n cells. Look at the cells in $sh(P)/\mu$. If P did come from an insertion of the form $T \leftarrow a_1 \cdots a_n$ where $a_1 > \cdots > a_n$ then we know that new cells were created from bottom to top.

Thus to reverse the process, we first apply reverse row insertion on the topmost cell of $sh(P)/\mu$, moving down the skew column, undoing row inserting at each step. Continuing in this way, we will obtain a sequence $a_1 \cdots a_n$ and a column strict tableau of shape μ such that $P = T \leftarrow a_1 \cdots a_n$.

For example, suppose $\mu = (3,2,1)$, $\lambda = (4,3,2,1,1)$, and P is the column strict tableau shown below:

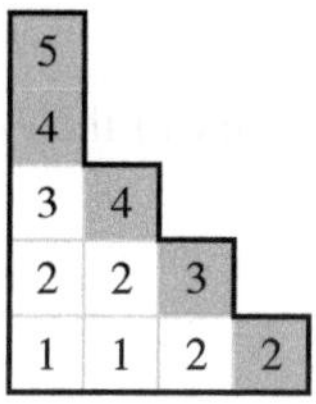

where we have shaded the cells in λ/μ. We illustrate the idea of this reverse row insertion here:

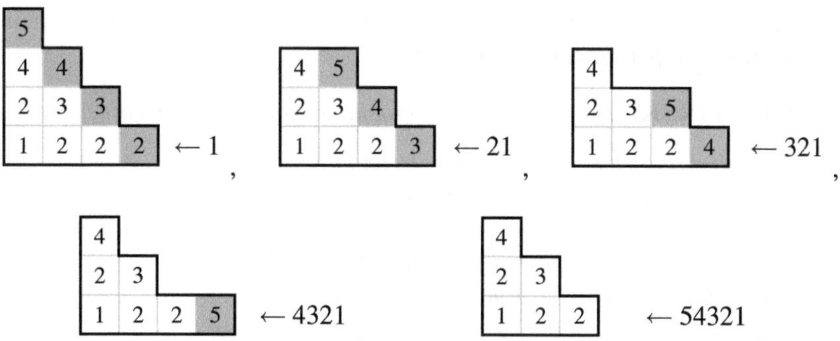

Why is $a_1 \ldots a_n$ strictly decreasing? If $a_j \leq a_{j+1}$ for some j, then the new cell c_{j+1} created by the insertion of a_{j+1} into $T \leftarrow a_1 \cdots a_j$ is strictly to the right and weakly below the new cell c_j created by the insertion of a_j into $T \leftarrow a_1 \cdots a_{j-1}$. But this is not what happens in our process. That is, in our reverse process, c_{i+1} is strictly above and weakly to the left of c_i. Hence there can be no such j and $a_1 \ldots a_n$ is strictly decreasing. □

By iterating the Pieri rule for the homogeneous symmetric functions, it can be seen that if $\mu \vdash n$, then

$$h_\mu = \sum_{\lambda \vdash n} K_{\lambda,\mu} s_\lambda, \tag{5.1}$$

thereby giving the entries of the h-to-s transition matrix.

That is, suppose $\mu = (\mu_1, \ldots, \mu_k)$. Then $h_{\mu_1} = s_{(\mu_1)}$; we can place 1s in the cells of the shape (μ_1). Using the Pieri rule to multiply $h_{\mu_2} s_{(\mu_1)}$, we find all s_λ such that $\lambda/(\mu_1)$ is a skew row. For each such λ, place 2s in the cells of $\lambda/(\mu_1)$. It follows that $h_{\mu_2} s_{(\mu_1)}$ equals the sum over all s_λ such that the shape λ is the shape of column strict tableau T of weight $x_1^{\mu_1} x_2^{\mu_2}$.

For each such λ, we can use the Pieri rule again to find $h_{\mu_3} s_\lambda$. We mark each new cell added to λ with a 3. Then $h_{\mu_3} h_{\mu_2} s_{(\mu_1)}$ equals the sum over all s_δ such that the shape δ is the shape of column strict tableau T of weight $x_1^{\mu_1} x_2^{\mu_2} x_3^{\mu_3}$. Continuing in this way proves (5.1).

Item 9 in the Notes in Chapter 2 provides an alternative route to (5.1).

Applying the ω transformation to both sides of (5.1) gives

$$e_\mu = \sum_{\lambda \vdash n} K_{\lambda,\mu} s_{\lambda'},$$

an identity which says that the λ, μ entry of the e-to-s transition matrix is $K_{\lambda',\mu}$.

5.2 The RSK Algorithm

We now present the RSK algorithm. It is named after Gilbert de Beauregard Robinson, who first described an algorithm equivalent to bumping, Craige Schensted who described the algorithm for permutations, and Donald Knuth who extended the algorithm from permutations to matrices.

Algorithm 5.4 (RSK). *The input is a nonzero, nonnegative integer-valued matrix A.*

1. *Begin with P and Q as empty column strict tableaux.*
2. *Let (i, j) be the topmost and then the leftmost nonzero entry in A.*
3. *Change P to $P \leftarrow j$, thereby adding one cell to P. Add a cell containing i to Q in the same position as the cell that was added to P.*
4. *Change A by subtracting 1 from the (i, j) entry.*
5. *If A is the zero matrix, stop. Otherwise, go back to step 2.*

The output is the pair (P, Q).

For example, consider applying RSK to $A = \begin{bmatrix} 1 & 2 & 0 \\ 3 & 0 & 1 \end{bmatrix}$. Starting at the top left entry of A and moving across each row, the first nonzero entry is at $(1,1)$. After initializing P and Q to be empty, steps 3 and 4 give

$$P = \boxed{1} \qquad Q = \boxed{1} \qquad A = \begin{bmatrix} 0 & 2 & 0 \\ 3 & 0 & 1 \end{bmatrix}.$$

Iterating until A is the zero matrix, we have

$$P = \boxed{1 \; 2} \qquad Q = \boxed{1 \; 1} \qquad A = \begin{bmatrix} 0 & 1 & 0 \\ 3 & 0 & 1 \end{bmatrix},$$

$$P = \boxed{1 \; 2 \; 2} \qquad Q = \boxed{1 \; 1 \; 1} \qquad A = \begin{bmatrix} 0 & 0 & 0 \\ 3 & 0 & 1 \end{bmatrix},$$

$$P = \begin{array}{|l|l|l|} \hline 2 & & \\ \hline 1 & 1 & 2 \\ \hline \end{array} \qquad Q = \begin{array}{|l|l|l|} \hline 2 & & \\ \hline 1 & 1 & 1 \\ \hline \end{array} \qquad A = \begin{bmatrix} 0 & 0 & 0 \\ 2 & 0 & 1 \end{bmatrix},$$

$$P = \begin{array}{|l|l|l|} \hline 2 & 2 & \\ \hline 1 & 1 & 1 \\ \hline \end{array} \qquad Q = \begin{array}{|l|l|l|} \hline 2 & 2 & \\ \hline 1 & 1 & 1 \\ \hline \end{array} \qquad A = \begin{bmatrix} 0 & 0 & 0 \\ 1 & 0 & 1 \end{bmatrix},$$

$$P = \begin{array}{|l|l|l|l|} \hline 2 & 2 & & \\ \hline 1 & 1 & 1 & 1 \\ \hline \end{array} \qquad Q = \begin{array}{|l|l|l|l|} \hline 2 & 2 & & \\ \hline 1 & 1 & 1 & 2 \\ \hline \end{array} \qquad A = \begin{bmatrix} 0 & 0 & 0 \\ 0 & 0 & 1 \end{bmatrix},$$

$$P = \begin{array}{|l|l|l|l|l|} \hline 2 & 2 & & & \\ \hline 1 & 1 & 1 & 1 & 3 \\ \hline \end{array} \qquad Q = \begin{array}{|l|l|l|l|l|} \hline 2 & 2 & & & \\ \hline 1 & 1 & 1 & 2 & 2 \\ \hline \end{array} \qquad A = \begin{bmatrix} 0 & 0 & 0 \\ 0 & 0 & 0 \end{bmatrix}.$$

The output of applying RSK to A is the pair (P, Q).

An alternative way of thinking about RSK is to consider the word of pairs

$$w(A) = \begin{matrix} q_1 & q_2 & \cdots & q_n \\ p_1 & p_2 & \cdots & p_n \end{matrix}$$

obtained by reading the rows of the matrix A from left to right starting at the top row and ending at the bottom row where for each $a_{i,j} > 0$, we write down $a_{i,j}$ pairs of the form $\begin{matrix} i \\ j \end{matrix}$. We call $w(A)$ the bi-word of A. For example, if A is the matrix of our example, then

$$w(A) = \begin{matrix} 1 & 1 & 1 & 2 & 2 & 2 & 2 \\ 1 & 2 & 2 & 1 & 1 & 1 & 3. \end{matrix}$$

Then P is found from $\emptyset \leftarrow p_1 \cdots p_n$. The matrix P is sometimes called the insertion tableau. We then use the elements of $q_1 \ldots q_n$ to record the growth of P. That is, if c_i is the new cell created by the insertion $(\emptyset \leftarrow p_1 \cdots p_{i-1}) \leftarrow p_i$, then we place q_i in cell c_i of Q. The matrix Q is sometimes called the recording tableau.

Theorem 5.5. *The RSK algorithm is a bijection between nonnegative integer valued matrices A and pairs of the form (P, Q) where P and Q are column strict tableau of the same shape.*

Proof. Suppose that

$$w(A) = \begin{matrix} q_1 & q_2 & \cdots & q_n \\ p_1 & p_2 & \cdots & p_n. \end{matrix}$$

The shapes of P and Q are the same by construction. By Theorem 5.1, P is a column strict tableau. Moreover, our construction ensures Q is weakly increasing in rows and columns because at any stage, the new cell attached to Q contains a letter i which is greater than or equal to all the previous letters added to Q.

Thus to prove Q is column strict, we need to only show that for any letter i, no two is in the Q can lie in the same column. But this is an immediate consequence of the first statement in Theorem 5.2. That is, the elements corresponding to the ith row of A were inserted in weakly increasing order because when we created the word of A, we read the elements from left to right. This insures that if $q_s q_{s+1} \cdots q_t$ is a block of is in $q_1 \ldots q_n$, then $p_s \leq \cdots \leq p_t$. But then by the first statement in Theorem 5.2, new cells which were created by the insertion

$$(\emptyset \leftarrow p_1 \cdots p_{s-1}) \leftarrow p_s \cdots p_t$$

were created from left to right and form a skew row. Thus no two is in Q can be in the same column so that Q is column strict.

It follows that we can read off the order in which the cells were created in P. That is, for any fixed i, we know by the first statement in Theorem 5.2, the cells were created from left to right. Thus the last cell c that was created in P corresponds to the right-most cell which contains the largest element in Q. Because we can reverse the row insertion algorithm starting at cell c, it follows that we can reconstruct A from P and Q.

At this point we know that the correspondence $A \to (P,Q)$ is an injection, so it remains to be shown that it is also a surjection. That is, we must show that if P and Q are column strict tableaux of the same shape, then there is a nonnegative-valued matrix A such that RSK sends A to (P,Q).

The idea is that given (P,Q), we can reverse the bumping process as it came from inserting a bi-word of a matrix A. That is, for each i in Q, we assume that the cells containing i were created from left to right. This allows us to reconstruct a bi-word

$$\begin{matrix} q_1 & q_2 & \cdots & q_n \\ p_1 & p_2 & \cdots & p_n \end{matrix}$$

by successively reversing the bumping process in P using the cell that contains the largest and then right-most element of Q.

For example, suppose we wish to undo the bumping process for the pair (P,Q) shown below:

$$P = \begin{array}{|c|c|cc} \hline 2 & 4 & & \\ \hline 1 & 1 & \multicolumn{1}{|c|}{3} & \multicolumn{1}{c|}{4} \\ \hline \end{array} \qquad Q = \begin{array}{|c|c|cc} \hline 2 & 2 & & \\ \hline 1 & 1 & \multicolumn{1}{|c|}{2} & \multicolumn{1}{c|}{2} \\ \hline \end{array}$$

We begin by locating the largest integer in P. If there are ties, select the rightmost. In this example, the rightmost 4 in P is chosen. Then we undo row insertion in P, remove the cell in the same position as Q, and then record this move in the bi-word. This process looks like this:

$$P = \begin{array}{|c|c|c|} \hline 2 & 4 & \\ \hline 1 & 1 & 3 \\ \hline \end{array} \qquad Q = \begin{array}{|c|c|c|} \hline 2 & 2 & \\ \hline 1 & 1 & 2 \\ \hline \end{array} \qquad w(A) = \begin{matrix} 2 \\ 4 \end{matrix}$$

$$P = \begin{array}{|c|c|} \hline 2 & 4 \\ \hline 1 & 1 \\ \hline \end{array} \qquad Q = \begin{array}{|c|c|} \hline 2 & 2 \\ \hline 1 & 1 \\ \hline \end{array} \qquad w(A) = \begin{matrix} 2\,2 \\ 3\,4 \end{matrix}$$

$$P = \begin{array}{|c|c|} \hline 2 & \\ \hline 1 & 4 \\ \hline \end{array} \qquad Q = \begin{array}{|c|c|} \hline 2 & \\ \hline 1 & 1 \\ \hline \end{array} \qquad w(A) = \begin{matrix} 2\,2\,2 \\ 1\,3\,4 \end{matrix}$$

$$P = \begin{array}{|c|c|} \hline 2 & 4 \\ \hline \end{array} \qquad Q = \begin{array}{|c|c|} \hline 1 & 1 \\ \hline \end{array} \qquad w(A) = \begin{matrix} 2\,2\,2\,2 \\ 1\,1\,3\,4 \end{matrix}$$

$$P = \begin{array}{|c|} \hline 2 \\ \hline \end{array} \qquad Q = \begin{array}{|c|} \hline 1 \\ \hline \end{array} \qquad w(A) = \begin{matrix} 1\,2\,2\,2\,2 \\ 4\,1\,1\,3\,4 \end{matrix}$$

$$P = \varnothing \qquad Q = \varnothing \qquad w(A) = \begin{matrix} 1\,1\,2\,2\,2\,2 \\ 2\,4\,1\,1\,3\,4 \end{matrix}$$

It only remains to be shown that for each i, if $q_s q_{s+1} \cdots q_t$ is a block of is in $q_1 \cdots q_n$, then $p_s \leq \cdots \leq p_t$. Just like in the proof of Theorem 5.3, we can use the second statement in Theorem 5.2. That is, if $p_j > p_{j+1}$ for some $s \leq j < j+1 \leq t$, then by Theorem 5.2, the cell c_{j+1} created by inserting p_{j+1} in $\emptyset \leftarrow p_1 \cdots p_j$ would be strictly above cell c_i created by p_j in $\emptyset \leftarrow p_1 \cdots p_{j-1}$. But by construction, c_{j+1} is strictly to the right and weakly below c_j so that there can be no such j.

This shows that RSK is bijection from the set of all nonnegative integer-valued matrices A onto the collection of pairs of column strict tableaux (P, Q) of the same shape. □

The Schur symmetric functions s_λ were defined in Chapter 2 using column strict tableaux, so it should not be a complete surprise that information about the Schur symmetric functions can be deduced from the RSK algorithm. The product in the next theorem is called the Cauchy kernel.

Theorem 5.6. *We have* $\displaystyle\prod_{i,j\geq 1} \frac{1}{1 - x_i y_j z} = \sum_\lambda s_\lambda(x_1, x_2, \ldots) s_\lambda(y_1, y_2, \ldots) z^{|\lambda|}$.

Proof. Write each $1/(1 - x_i y_j z)$ term in the infinite product as a geometric series and expand. Record the choice of $x_i^m y_j^m z^m$, selected from the geometric series $1/(1 - x_i y_j z)$ when creating a monomial in the expansion, by placing an m in the (i, j) entry of a matrix A. By defining the weight of $A = \|a_{i,j}\|$ to be $\prod_{i,j\geq 1} x_i^{a_{i,j}} y_j^{a_{i,j}}$, the coefficient of z^n in the infinite product is the weighted sum of all nonnegative integer-valued matrices A which have entries that sum to n.

For every (i,j) entry in A, RSK inserts $a_{i,j}$ copies of i into Q and $a_{i,j}$ copies of j into P. By defining the weight of the pair (P,Q) to be

$$\prod_{i,j\geq 1} x_i^{\text{the number of times } i \text{ is in } Q} y_j^{\text{the number of times } j \text{ is in } P},$$

the weighted sum over all nonnegative integer-valued matrices A is equal to the weighed sum over all pairs (P,Q) of column strict tableau of shape λ where λ ranges over all integer partitions of n.

The Schur symmetric function $s_\lambda(x_1,x_2,\ldots)$ is the weighted sum of all column strict tableau Q of shape λ and $s_\lambda(y_1,y_2,\ldots)$ gives P, so the coefficient of z^n on the right-hand side of the equality is also the weighted sum over all pairs (P,Q) of column strict tableaux of shape $\lambda \vdash n$. □

Taking $z=1$ and applying Theorems 2.25 and 2.26, Theorem 5.6 tells us that the basis $\{s_\lambda : \lambda \vdash n\}$ is self-dual in Λ_n.

We will prove a second Cauchy identity, namely

$$\prod_{i,j\geq 1}(1+x_iy_jz) = \sum_{\lambda} s_\lambda(x_1,x_2,\ldots)s_{\lambda'}(y_1,y_2,\ldots)z^{|\lambda|},$$

in Theorem 5.10 by slightly modifying the RSK algorithm to an algorithm known as dual RSK. Dual RSK involves an analogue of row insertion for row strict tableaux.

We say that T is a row strict tableau of shape λ if T is a tableau of shape λ such that integers strictly increase in rows, reading from left to right, and weakly increase in columns, reading from bottom to top.

Let T be a row strict tableau and j an integer. We define the dual row insertion of j into T, denoted $T \overset{d}{\leftarrow} j$, to be the tableau found by the following three rules:

D0. If T is the empty tableau, then $T \overset{d}{\leftarrow} j$ is the column strict tableaux with 1 cell which contains the integer j.

Otherwise we assume that the first row of T contains the integers $a_1 < \cdots < a_n$ and T' is the row strict tableau that is found by removing the first row of T.

D1. If $a_n < j$, then $T \overset{d}{\leftarrow} j$ results from T by adding a cell containing j at the end of the bottom row of T.

D2. If $a_k \geq j$, then let a_k be the leftmost entry in bottom row of T that is greater than or equal to j. Replace a_k with j and then dual row insert a_k into T'.

For example, to dual row insert a 3 into the row strict tableau

$$\begin{array}{|c|c|c|c|} \hline 2 \\ \cline{1-1} 1 & 2 & 4 \\ \cline{1-3} 1 & 2 & 3 & 5 \\ \hline \end{array},$$

Graphically, this dual row insertion looks like

$$\begin{array}{|l|l|l|l|}\hline 2 \\ \cline{1-1} 1 & 2 & 4 \\ \cline{1-3} 1 & 2 & 3 & 5 \\ \hline \end{array} \overset{d}{\leftarrow} 3, \quad \begin{array}{|l|l|l|l|}\hline 2 \\ \cline{1-1} 1 & 2 & 4 & \overset{d}{\leftarrow} 3 \\ \cline{1-3} 1 & 2 & 3 & 5 \\ \hline \end{array}, \quad \begin{array}{|l|l|l|l|}\hline 2 & \overset{d}{\leftarrow} 4 \\ \cline{1-1} 1 & 2 & 3 \\ \cline{1-3} 1 & 2 & 3 & 5 \\ \hline \end{array}, \quad \begin{array}{|l|l|l|l|}\hline 2 & 4 \\ \cline{1-2} 1 & 2 & 3 \\ \cline{1-3} 1 & 2 & 3 & 5 \\ \hline \end{array}.$$

Theorems 5.7 and 5.8 are the dual versions of Theorems 5.1 and 5.2. Their proofs are so similar to the proofs of Theorems 5.1 and 5.2 that they are left to the reader.

Theorem 5.7. *If T is a row strict tableau and j is a positive integer, then $T \overset{d}{\leftarrow} j$ is a row strict tableau and the bumping path for the insertion of j in T moves weakly to the left as one proceeds from bottom to top. Moreover, if $sh(T)$ is shape of T and $sh(T \overset{d}{\leftarrow} j)$ is the shape of $T \overset{d}{\leftarrow} j$, then $sh(T) \subseteq sh(T \overset{d}{\leftarrow} j)$.*

Given a word $w_1 \cdots w_n$, let $T \overset{d}{\leftarrow} w_1 \cdots w_n = (\cdots((T \overset{d}{\leftarrow} w_1) \overset{d}{\leftarrow} w_2) \cdots) \overset{d}{\leftarrow} w_n$. Further, let $T_i = T \overset{d}{\leftarrow} w_1 \cdots w_i$ for all i and let c_i be the single cell in $sh(T_i)/sh(T_{i-1})$.

Theorem 5.8. *If $w = w_1 \cdots w_n$ is a word of length $n \geq 2$ with letters in $\{1,2,\ldots\}$, then these two facts are true:*

1. *If $w_1 < \cdots < w_n$, then c_{i+1} is strictly to the right and weakly below c_i for all i.*
2. *If $w_1 \geq \cdots \geq w_n$, then c_{i+1} is strictly above and weakly to the left of c_i for all i.*

Algorithm 5.9 (Dual RSK). *The input is a matrix A with entries either 0 or 1.*

1. *Begin with P and Q the empty tableaux.*
2. *Let (i,j) be the topmost and then the leftmost nonzero entry in A.*
3. *Change P to $P \overset{d}{\leftarrow} j$, thereby adding one cell to P. Add a cell containing i to Q in the same position as the cell that was added to P.*
4. *Change A by subtracting 1 from the (i,j) entry.*
5. *If A is the zero matrix, stop. Otherwise, go back to step 2.*

The output is the pair (P,Q).

Like RSK, dual RSK can be phrased in terms of the bi-word of A instead of the matrix A itself. In this way, we would apply dual RSK by successively dual row inserting the pairs $\begin{smallmatrix} i \\ j \end{smallmatrix}$ for every i,j pair in the bi-word of A.

From here, Theorems 5.7 and 5.8 can be combined to prove that P is a row strict tableau, Q is a column strict tableau, and we can recover the order in which the cells were created in P from Q. Mimicking the proof of Theorem 5.5, the reader can verify that dual RSK is a bijection between such pairs (P,Q) and $0,1$ valued matrices A.

Theorem 5.10. $\prod_{i,j\geq 1}(1+x_iy_jz) = \sum_{\lambda} s_{\lambda}(x_1,x_2,\dots)s_{\lambda'}(y_1,y_2,\dots)z^{|\lambda|}.$

Proof. Expand each term $(1+x_iy_jz)$ in the infinite product and record the choice of x_iy_jz by placing a 1 in the (i,j) entry of a matrix A and the choice of 1 by placing a 0 in the (i,j) entry of a matrix A.

By defining the weight of $A = \|a_{i,j}\|$ to be $\prod_{i,j\geq 1} x_i^{a_{i,j}} y_j^{a_{i,j}}$, the coefficient of z^n in the infinite product is the weighted sum of all $\{0,1\}$-valued matrices A which have entries that sum to n. The dual RSK row insertion algorithm associates A with a pair (P,Q) of the same shape where P is a row strict tableau and Q is a column strict tableau. Define the weight of (P,Q) to be

$$\prod_{i,j\geq 1} x_i^{\text{the number of times } i \text{ is in } Q} y_j^{\text{the number of times } j \text{ is in } P},$$

the weighted sum over all nonnegative $\{0,1\}$-valued matrices A is equal to the weighed sum over all pairs (P,Q) of tableau of shape λ where λ ranges over all partitions of n, P is a row strict tableau, and Q is a column strict tableau.

The Schur symmetric function $s_{\lambda}(x_1,x_2,\dots)$ is the weighted sum of all column strict tableau Q of shape λ and $s_{\lambda'}(y_1,y_2,\dots)$ is the weighted sum of all row strict tableau P so the coefficient of z^n on the right-hand side of the equality is also the weighted sum over all pairs (P,Q) of shape $\lambda \vdash n$ where P is row strict tableau and Q is a column strict tableau. □

Dual row insertion can be rephrased in terms of column insertion. Given a column strict tableau P, we define a column insertion by first transposing P, doing dual row insertion, and then transposing again. Then instead of repeating row insertion to define the RSK algorithm, we can repeat column insertion to define the column RSK algorithm.

As a graphical example of column insertion, we have

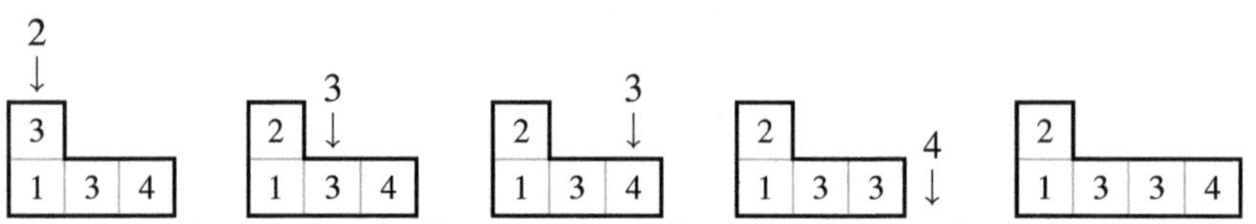

Algorithm 5.11 (Column RSK). *The input is a matrix A with entries either* 0 *or* 1. *Apply RSK to A except replace each instance of row insertion with an instance of column insertion. The output is the pair* (P,Q).

For example, column RSK sends $A = \begin{bmatrix} 1 & 0 & 1 & 1 \\ 0 & 1 & 1 & 0 \end{bmatrix}$ to the pair (P,Q) where

$$P = \begin{array}{|c|c|} \hline 3 \\ \cline{1-1} 2 & 4 \\ \hline 1 & 3 \\ \hline \end{array}, \qquad Q = \begin{array}{|c|c|} \hline 1 \\ \cline{1-1} 1 & 2 \\ \hline 1 & 2 \\ \hline \end{array}.$$

In a similar way that row insertion maintains row strictness, the operation of column insertion maintains column strictness. Therefore the output P from RSK column insertion algorithm is a column strict tableau and one can use Theorems 5.7 and 5.8 to prove that Q is a row strict tableau.

The inverse column RSK algorithm can be described: starting with the pair (P,Q) of tableau of the same shape such that P and Q' are column strict, the inverse column RSK algorithm is the same as the inverse RSK algorithm (found in the proof of Theorem 5.5) with the exception that the topmost i is selected if there is more than one largest integer i in Q. Therefore column RSK is a bijection between $0,1$ matrices A and pairs (P,Q) of tableau of the same shape such that P and Q' are column strict.

5.3 Weakly Increasing Subsequences in Words

Matrices can be used to represent words and permutations. The matrix representing the word $w = w_1 \cdots w_n \in \{1,\ldots,k\}_n^*$ is the $n \times k$ matrix that has a 1 in entry (i,w_i) for all i and 0s elsewhere. If the word happens to be a permutation, this is a permutation matrix. For example, the permutation matrix for 1 4 2 5 3 is

$$\begin{bmatrix} 1\,0\,0\,0\,0 \\ 0\,0\,0\,1\,0 \\ 0\,1\,0\,0\,0 \\ 0\,0\,0\,0\,1 \\ 0\,0\,1\,0\,0 \end{bmatrix}.$$

Define the RSK algorithm (or the dual RSK algorithm) on words $w = w_1 \cdots w_n$ to be the result when the matrix representing w is inputted into the RSK algorithm. This amounts to row inserting $w_1,\ldots,w_n$ into P while recording the placement of new cells with $1,\ldots,n$ in Q. For example, RSK sends 1 4 2 5 3 to (P,Q) where

$$P = \begin{array}{|c|c|c}\hline 4 & 5 & \\ \hline 1 & 2 & \multicolumn{1}{c|}{3} \\ \hline \end{array}\,, \qquad Q = \begin{array}{|c|c|c}\hline 3 & 5 & \\ \hline 1 & 2 & \multicolumn{1}{c|}{4} \\ \hline \end{array}\,.$$

If w is a word or a permutation, we let $P(w)$ and $Q(w)$ denote the output tableaux P and Q that come from applying RSK to w. We define the words w and v to be P-equivalent if $P(w) = P(v)$. For example, the words 1 2 1 2 3 3 2 and 2 3 1 1 2 2 3 are P-equivalent because

$$P(1\,2\,1\,2\,3\,3\,2) = P(2\,3\,1\,1\,2\,2\,3) = \begin{array}{|c|c|ccc}\hline 2 & 3 & & & \\ \hline 1 & 1 & \multicolumn{1}{c|}{2} & \multicolumn{1}{c|}{2} & \multicolumn{1}{c|}{3} \\ \hline \end{array}\,.$$

Our next goal is to characterize P-equivalent words because doing so will allow us to provide an amazing relationship between the RSK algorithm and increasing subsequences in words.

An elementary Knuth operation on a word w is one of these two operations:

1. If $y\,z\,x$ appears consecutively in w and $x < y \leq z$, then change the order of these three letters to $y\,x\,z$ and leave the rest of w unchanged.
2. If $x\,z\,y$ appears consecutively in w and $x \leq y < z$, then change the order of these three letters to $z\,x\,y$ and leave the rest of w unchanged.

Both of these two actions interchange consecutive integers on one side of the middle-valued integer y and both of these actions are invertible. The reason for this definition is because these operations (and their inverses) preserve the row insertion process. That is, the first and second elementary Knuth operations give

$$P(y\,z\,x) = P(y\,x\,z) = \begin{array}{|c|c}\hline y & \\ \hline x & \multicolumn{1}{c|}{z} \\ \hline \end{array} \qquad \text{and} \qquad P(x\,z\,y) = P(z\,x\,y) = \begin{array}{|c|c}\hline z & \\ \hline x & \multicolumn{1}{c|}{y} \\ \hline \end{array},$$

respectively.

We define two words w and v to be Knuth equivalent if w can be transformed into v by a series of elementary Knuth operations or inverse elementary Knuth operations. For example, the words 1 2 1 2 3 3 2 and 2 3 1 1 2 2 3 are Knuth equivalent; we display the sequence of elementary Knuth operations below with a boldfaced x and z to be interchanged:

1 2 1 2 3 3 2	(operation 2)
2 1 1 2 3 **3 2**	(operation 1)
2 1 1 **2 3** 2 3	(operation 2)
2 1 **1 3** 2 2 3	(operation 2)
2 **1 3** 1 2 2 3	(inverse operation 1)
2 3 1 1 2 2 3	

The row word of a tableau T is a canonical example of a word w with $P(w) = T$. It is created by listing the elements in each row of T, starting with the top row and moving down. For example, the row word of

$$\begin{array}{|c|c|ccc}\hline 2 & 3 & & & \\ \hline 1 & 1 & \multicolumn{1}{c|}{2} & \multicolumn{1}{c|}{2} & \multicolumn{1}{c|}{3} \\ \hline \end{array}$$

is 2 3 1 1 2 2 3.

Theorem 5.12. *Two words are P-equivalent if and only if they are Knuth equivalent.*

Proof. Assume that $w = w_1 \cdots w_n$ and v are P-equivalent. We will show that word w is Knuth equivalent to the row word r for $P(w)$. This will imply that v is also Knuth equivalent to r, from which this direction of the if and only if statement will follow from the transitivity of Knuth equivalence.

We proceed by induction on the length of the word w. Let r' be the row word for $P(w_1 \cdots w_{n-1})$ so that, by induction, r' and $w_1 \cdots w_{n-1}$ are Knuth equivalent. We need to show that $r' w_n$ and r are Knuth equivalent.

How are r and r' related? Suppose the bottom row of $P(w_1 \cdots w_{n-1})$ contains the weakly increasing sequence $r_1, \ldots, r_\ell$. This is the sequence at the tail end of r'. Let i be the integer that satisfies

$$r_1 \le r_2 \le \cdots \le r_{i-1} < w_n \le r_i \le \cdots \le r_\ell.$$

Then the bottom row of $P(w_1 \cdots w_{n-1} w_n)$ contains the weakly increasing sequence $r_1 \cdots r_{i-1} w_n r_{i+1} \cdots r_\ell$. This is the sequence at the tail end of r.

Using $\sim$ to denote Knuth equivalence, we have

$$\begin{aligned}
r' w_n &= r''\ r_1\ \cdots\ r_{i-1}\ r_i\ \cdots\ r_{\ell-1}\ \mathbf{r_\ell}\ \mathbf{w_n} && \text{(operation 1)}\\
&\sim r''\ r_1\ \cdots\ r_{i-1}\ r_i\ \cdots\ \mathbf{r_{\ell-1}}\ \mathbf{w_n}\ r_\ell && \text{(operation 1)}\\
&\ \vdots \\
&\sim r''\ r_1\ \cdots\ r_{i-1}\ \mathbf{r_i}\ \mathbf{w_n}\ \cdots\ r_{\ell-1}\ r_\ell && \text{(operation 1)}\\
&\sim r''\ r_1\ \cdots\ \mathbf{r_{i-1}}\ \mathbf{r_i}\ w_n\ r_{i+1}\ \cdots\ r_{\ell-1}\ r_\ell && \text{(operation 2)}\\
&\ \vdots \\
&\sim r''\ \mathbf{r_1}\ \mathbf{r_i}\ \cdots\ r_{i-1}\ w_n\ r_{i+1}\ \cdots\ r_{\ell-1}\ r_\ell && \text{(operation 2)}\\
&\sim r''\ r_i\ r_1\ \cdots\ r_{i-1}\ w_n\ r_{i+1}\ \cdots\ r_{\ell-1}\ r_\ell
\end{aligned}$$

This shows that the tail end of $r' w_n$ and r are the same. The resulting words found by removing these tail ends from $r' w_n$ and r are P-equivalent since they both are the row words for $P(w_1 \cdots w_n)$ with the bottom row removed. By induction we have shown that $r' w_n$ and r are Knuth equivalent, as needed.

For the reverse implication, assume that words w and v are Knuth equivalent. It is enough to assume that w and v differ by a single elementary Knuth operation. We will assume that v can be found from applying the first elementary Knuth operation to w; the case where they differ by the second elementary Knuth operation is similar and left to the reader.

It is enough to show that

$$((T \leftarrow y) \leftarrow z) \leftarrow x = ((T \leftarrow y) \leftarrow x) \leftarrow z \tag{5.2}$$

whenever $x < y \le z$ and T is an arbitrary column strict tableau T.

By Theorem 5.2, the path of replaced cells in $T \leftarrow y$ appears strictly to the left of the path of placed cells in $(T \leftarrow y) \leftarrow z$. Similarly, the path of replaced cells in $T \leftarrow y$ appears strictly to the right of the path of placed cells in $(T \leftarrow y) \leftarrow x$.

Since row inserting z occurs on the other side of the path of replaced cells from y, the path of replaced cells when row inserting x into $T \leftarrow y$ is the same as the path of replaced cells when row inserting x into $(T \leftarrow y) \leftarrow z$. Similarly, since row inserting z does not cross the barrier created by the path of replaced cells from y, row inserting z commutes with row inserting x. This shows 5.2, as needed. $\square$

A weakly increasing subsequence in a word $w = w_1 \cdots w_n \in \{1,\ldots,k\}_n^*$ is a sequence $w_{i_1} \le w_{i_2} \le \cdots \le w_{i_k}$ for some $1 \le i_1 < \cdots < i_k \le n$. Let $I_k(w)$ be the maximum number of integers in a union of k disjoint weakly increasing sequences in w.

For example, if $w = 1\ 2\ 3\ 2\ 3\ 3\ 2\ 2\ 3\ 1\ 1$, then $I_1(w) = 6$; a longest weakly increasing subsequence in w is boldfaced:

1 2 3 **2** 3 3 **2 2 3** 1 1.

There are other weakly increasing subsequences of length 6. To show that $I_2(w) = 9$, a second weakly increasing subsequence in w is italicized:

1 2 *3* **2** *3 3* **2 2 3** 1 1.

Although it worked this time it may not always be the case that $I_2(w)$ can be found by examining the integers not used in the subsequence found for $I_1(w)$. From here we find $I_k(w) = 11$ for all $k \ge 3$ since it takes three disjoint weakly increasing subsequences to use every integer in w.

Continuing this example, we have

$$P(1\ 2\ 3\ 2\ 3\ 3\ 2\ 2\ 3\ 1\ 1) = \begin{array}{|c|c|c|c|c|c|}\cline{1-2} 3 & 3 \\ \cline{1-3} 2 & 2 & 3 \\ \hline 1 & 1 & 1 & 2 & 2 & 3 \\ \hline \end{array}\,.$$

The shape of this tableau is the integer partition $(6,3,2)$. As our next theorem shows, it is not a coincidence that $I_1(w) = 6$, $I_2(w) = 6+3$, and $I_3(w) = 6+3+2$.

Theorem 5.13. *Let w be a word and let $\lambda = (\lambda_1, \lambda_2, \ldots)$ be the shape of $P(w)$ where λ_ℓ is the last nonzero entry in λ. Then $I_k(w) = \lambda_1 + \cdots + \lambda_k$ for all $k \ge 1$.*

Proof. Let r be the row word for $P(w)$. Integers in a weakly increasing subsequence in r must come from different columns in $P(w)$. The longest weakly increasing subsequence in r therefore uses every column. This means that $I_1(r) = \lambda_1$ and that the longest weakly increasing subsequence in r consists of the λ_1 integers at the tail end of r. Removing the bottom row of $P(w)$ and continuing inductively, we see that $I_k(r) = \lambda_1 + \cdots + \lambda_k$ for the row word r.

It is enough to show that $I_k(w) = I_k(v)$ whenever w and v are Knuth equivalent—after all, we have just proved the theorem true for row words and, by the proof of Theorem 5.12, every word w is Knuth equivalent to the row word for $P(w)$. Further, we only need to show that $I_k(w) = I_k(v)$ when w and v differ by an elementary Knuth operation. We will assume that w and v differ by the first elementary Knuth operation and leave the similar case of the second elementary Knuth operation to the reader.

Suppose w and v are the same except for three consecutive letters $x < y \le z$ which appear as $y\ z\ x$ in w but appear as $y\ x\ z$ in v. The inequality $I_k(w) \le I_k(v)$ follows immediately because both x and z can be a part of a weakly increasing subsequence in v but not in w.

Suppose the $v_1,\ldots,v_k$ are disjoint weakly increasing subsequences in v which contain a total of $I_k(v)$ integers. To show $I_k(w) \geq I_k(v)$, we will construct weakly increasing subsequences in w with the same lengths as $v_1,\ldots,v_k$. If x and z do not both appear in the same subsequence v_i for some i, then $v_1,\ldots,v_k$ are the desired disjoint weakly increasing subsequences in w. For the remainder of this proof we now assume that x and z both appear in v_i for some i.

If y is not in $v_1,\ldots,v_k$, then replace x in v_i with y to find the desired weakly increasing subsequences in w.

If y appears in v_j for $i \neq j$, then let v_i' be the initial portion of v_i that appears up to and including the x concatenated with the tail end of v_j that appears after the y. Similarly, let v_j' be the initial portion v_i that appears up to and including the y concatenated with the tail end of v_i that appears after the x. Then $v_1,\ldots,v_k$ with v_i and v_j changed to v_i' and v_j' are the desired disjoint weakly increasing subsequences.

We have now shown $I_k(w) \leq I_k(v)$ and $I_k(w) \geq I_k(v)$, so these two integers are equal, as needed. □

5.4 Paths in Permutation Matrices

Drawing certain paths connecting the 1s in a permutation matrix can give a visual representation of the RSK algorithm. This new understanding will allow us to prove that if RSK sends the matrix A to (P,Q), then RSK sends A^{T} to (Q,P).

Let M be a permutation matrix and let (i,j) be an entry containing a 1 in M. We define an upright path segment to be a path created by starting at an entry in column j below the 1, moving up column j until reaching the 1, and then moving right to the first column which has a 1 above row i (or the last column of M if no such 1 exists). An upright path in M is created by chaining together upright path segments, staring at the bottom entry of the first column in M.

For example, below is an upright path in the permutation matrix for 6 2 5 7 1 4 3:

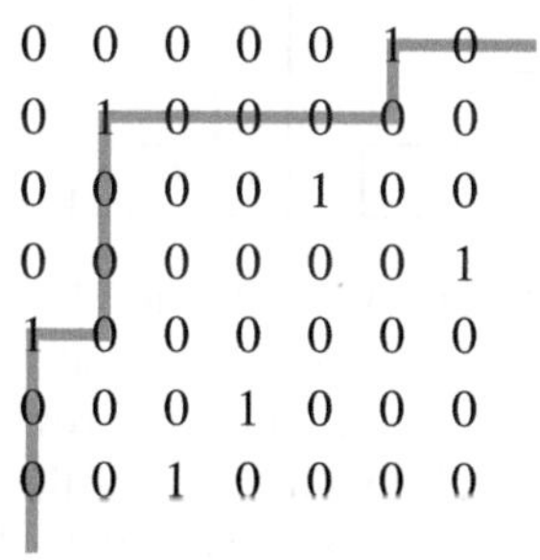

A path diagram for the permutation matrix M is the matrix M together with nested upright paths. Create a path diagram by drawing an upright path staring in the bottom left entry of M. Ignore any 1s appearing in this path and draw another upright path, this time starting at the bottom entry of the leftmost column which contains a 1. Iterate, continuing to draw upright paths until every 1 is in a path.

For example, here is the path diagram for the permutation matrix for 6 2 5 7 1 4 3:

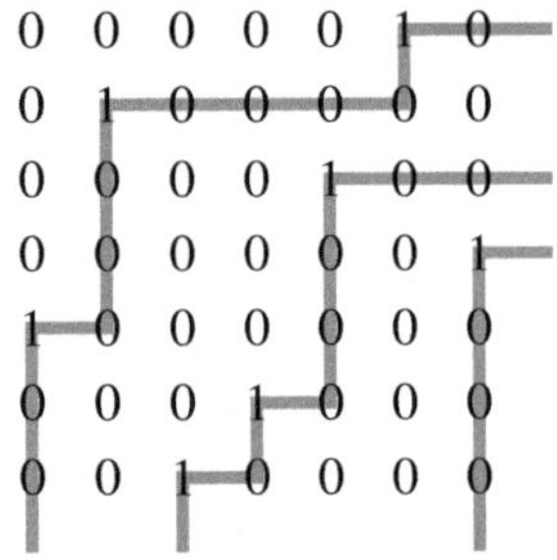

The path diagram for the permutation matrix representing σ is related to the column strict tableau $P(\sigma)$ and $Q(\sigma)$. We will soon show that an upright path begins in column j exactly when there is a j in the bottom row of $P(\sigma)$ and an upright path ends in row i exactly when there is an i in the bottom row of $Q(\sigma)$.

To verify this with our above example, the upright paths begin in columns 1, 3, and 7 and end in rows 1, 3, and 4. These are indeed the bottom rows of

$$P(6\,2\,5\,7\,1\,4\,3) = \begin{array}{|c|c|c|}\hline 6 \\ \cline{1-1} 5 \\ \cline{1-2} 2 & 4 \\ \hline 1 & 3 & 7 \\ \hline \end{array} \quad \text{and} \quad Q(6\,2\,5\,7\,1\,4\,3) = \begin{array}{|c|c|c|}\hline 7 \\ \cline{1-1} 5 \\ \cline{1-2} 2 & 6 \\ \hline 1 & 3 & 4 \\ \hline \end{array}\,.$$

Going one step further, the sequence of column strict tableaux found when building $P(\sigma)$ with the RSK algorithm is

$$\begin{array}{|c|}\hline 6 \\ \hline \end{array}\,,\ \begin{array}{|c|}\hline 6 \\ \hline 2 \\ \hline \end{array}\,,\ \begin{array}{|c|c|}\hline 6 \\ \cline{1-1} 2 & 5 \\ \hline \end{array}\,,\ \begin{array}{|c|c|c|}\hline 6 \\ \cline{1-1} 2 & 5 & 7 \\ \hline \end{array}\,,\ \begin{array}{|c|c|c|}\hline 6 \\ \cline{1-1} 2 \\ \cline{1-1} 1 & 5 & 7 \\ \hline \end{array}\,,\ \begin{array}{|c|c|c|}\hline 6 \\ \cline{1-1} 2 & 5 \\ \cline{1-2} 1 & 4 & 7 \\ \hline \end{array}\,,\ \begin{array}{|c|c|c|}\hline 6 \\ \cline{1-1} 5 \\ \cline{1-1} 2 & 4 \\ \cline{1-2} 1 & 3 & 7 \\ \hline \end{array}\,.$$

The bottom rows are 6, 2, 25, 257, 157, 147, and 137. Look at the top 1 row of M only: the upright path begins in column 6. Look at the top 2 rows of M only: the upright path begins in column 2. Look at the top 3 rows of M only: the upright paths begin in columns 2, 5. In general, looking only at the top k rows of M, an upright path begins in column j exactly when there is a j in the bottom row of the kth column strict tableau found when building $P(\sigma)$. This is the content of our next theorem.

Theorem 5.14. *Let M be the permutation matrix for the permutation $\sigma = \sigma_1 \cdots \sigma_n$. Then the bottom row of $\varnothing \leftarrow \sigma_1 \leftarrow \cdots \leftarrow \sigma_k$ contains a j exactly when the top k rows of the path diagram for M contains an upright path beginning in column j.*

Proof. The case $k = 0$ is vacuously true. We proceed by induction.

Suppose σ_{k+1} is larger than each entry in the bottom row of $\varnothing \leftarrow \sigma_1 \leftarrow \cdots \leftarrow \sigma_k$. Row inserting σ_{k+1} simply appends σ_{k+1} to the bottom row. In this case the 1 in row $k+1$ in the permutation matrix M must appear farther to the right than all above 1s. This means that a new upright path in the first k rows of M begins in the σ_{k+1} column, as needed.

Suppose σ_{k+1} is not larger than each entry in the bottom row of $\varnothing \leftarrow \sigma_1 \leftarrow \cdots \leftarrow \sigma_k$. Let x be the smallest entry in the bottom row which is larger than σ_{k+1}. Then row inserting σ_{k+1} into $\varnothing \leftarrow \sigma_1 \leftarrow \cdots \leftarrow \sigma_k$ replaces the x in the first row with σ_{k+1}.

By our choice of x, the 1 in column x in M is the nearest 1 northeast of the 1 in column σ_{k+1}. This means that the 1 in column x and the 1 in column σ_{k+1} are in the same upright path. Since the 1 in column σ_{k+1} appears to the left of the x, this 1 now marks the start of this upright path, as needed. □

Theorem 5.15. *Let σ be a permutation with permutation matrix M. An upright path in the path diagram for M begins in column j and ends in row i if and only if j is in the bottom row of $P(\sigma)$ and i is in the bottom row of $Q(\sigma)$.*

Proof. Taking $k = n$ in Theorem 5.14 says that $P(\sigma)$ contains j whenever an upright path begins in column j.

An element $k+1$ appears in the first row of Q whenever σ_{k+1} is greater than each entry in the bottom row of $\varnothing \leftarrow \sigma_1 \leftarrow \cdots \leftarrow \sigma_k$. As seen in the proof of Theorem 5.14, there is an up-down path with a final 1 in row $k+1$ each time this happens. Therefore each time there is a $k+1$ in the first row of Q, there is an up-down path which ends in row $k+1$. □

The proof of Theorem 5.14 says that an integer x is placed into the second row of $P(\sigma)$ each time an up-down path has a new up-down path segment. In other words, a 0 appearing in a corner of an up-down path appears in column j exactly when the second row of $P(\sigma)$ contains a j. Therefore to find the second row of $P(\sigma)$ and $Q(\sigma)$, we can create a second path diagram by connecting the corner 0s from the first path diagram with upright path segments. For example, below we show this second path diagram in our running example:

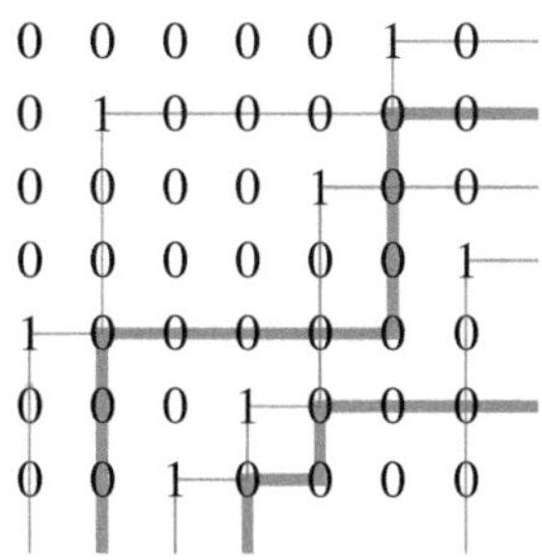

This second path diagram has upright paths which begin in columns 2 and 4 and end in rows 2 and 6, so the second rows of $P(\sigma)$ and $Q(\sigma)$ are $2,4$ and $2,6$.

To find the third row of $P(\sigma)$ and $Q(\sigma)$ and beyond, iterate this procedure. Create a path diagram using the corner 0s from the previous iteration of the path diagram to form new upright paths. The columns and rows which begin these paths give the desired row in $P(\sigma)$ and $Q(\sigma)$.

Theorem 5.16. *We have* $P(\sigma^{-1}) = Q(\sigma)$ *and* $Q(\sigma^{-1}) = P(\sigma)$ *for all* $\sigma \in S_n$.

Proof. The columns in the path diagram of the permutation matrix M tell us $P(\sigma)$ and the rows tell us $Q(\sigma)$. Therefore the rows in the path diagram of M^{T} tell us $P(\sigma)$ and the columns tell us $Q(\sigma)$. Since the permutation matrix for σ^{-1} is M^{T}, we have the desired result. □

Another way to phrase Theorem 5.16 is to say that RSK sends M to (P,Q) if and only if RSK sends M^{T} to (Q,P) for all permutation matrices M. We can extend this result to any nonnegative integer-valued matrix A by carefully changing A into a permutation matrix.

Let $A = \|a_{i,j}\|$ be an $m \times n$ nonnegative integer-valued matrix. Let r_i be the sum of row i and let c_j be the sum of column j in A. The permutation matrix $st(A)$, called the standardization of A, is the block matrix defined in the following way. Change the i,j entry in A to the $r_i \times c_j$ block matrix with j,i block equal to the $a_{i,j} \times a_{i,j}$ identity matrix and all other blocks equal to 0. For example,

$$st\left(\begin{bmatrix} 3 & 1 & 2 \\ 1 & 2 & 1 \end{bmatrix}\right) = \left[\begin{array}{cc|cc|cc} I_3 & 0 & 0 & 0 & 0 & 0 \\ 0 & 0 & I_1 & 0 & 0 & 0 \\ 0 & 0 & 0 & 0 & I_2 & 0 \\ \hline 0 & I_1 & 0 & 0 & 0 & 0 \\ 0 & 0 & 0 & I_2 & 0 & 0 \\ 0 & 0 & 0 & 0 & 0 & I_1 \end{array}\right]. \tag{5.3}$$

This definition of standardization gives that transposition and standardization commute. In symbols, $st(A^{\mathrm{T}}) = st(A)^{\mathrm{T}}$.

The reason for this definition of $st(A)$ comes from the order in which entries in A are used in RSK. For example, if A is the matrix shown in (5.3), then we would use the (i,j) pairs in the order

$$(1,1),(1,1),(1,1),(1,2),(1,3),(1,3),(2,1),(2,2),(2,2),(2,3)$$

when creating P and Q. The effect of standardization is that the 1s in the first coordinate of this list of ordered pairs are relabeled from left to right with $1,2,\ldots,\ell_1$. Then the 2s are relabeled from left to right with $\ell_1+1,\ldots,\ell_1+\ell_2$, and so on. Applying this relabeling procedure twice, once for each coordinate, we find

$$(1,1),(2,2),(3,3),(4,5),(5,8),(6,9),(7,4),(8,6),(9,7),(10,10).$$

Standardization makes the integers in both the first and second coordinates distinct. The matrix $st(A)$ is the permutation matrix for the permutation created by reading the second coordinates off of this list of ordered pairs.

If we keep track of $r_1r_2\cdots$ and $c_1c_2\cdots$, then standardization is reversible. These row and column sums tell us the size of each block in the block matrix $st(A)$ from which we can reconstruct A. In other words, the function which sends A to $(st(A), r_1r_2\cdots, c_1c_2\cdots)$ is a bijection.

We define a standard tableau to be a column strict tableau T of size n such that each of the integers $1,\ldots,n$ appear exactly once in T. If w is a word, the tableau $Q(w)$ will be standard. Similarly, both $P(\sigma)$ and $Q(\sigma)$ are standard tableau if and only if σ is a permutation.

Just as the standardization of a matrix A makes the integers inserted in RSK distinct, we can define the standardization of a column strict tableau T in order to make the integers in T distinct. Suppose T is a column strict tableau with ℓ_1 appearances of 1, ℓ_2 appearances of 2, and so on. We can take T and create a standard tableau $st(T)$ by relabeling the 1s in T from left to right with $1,\ldots,\ell_1$, then relabeling the 2s in T from left to right with $\ell_1+1,\ldots,\ell_1+\ell_2$, and so on. Below we display an example of T and $st(T)$:

$T =$

3	3				
2	2	3			
1	1	1	2	2	3

$st(T) =$

8	9				
4	5	10			
1	2	3	6	7	11

If we let t_i the number of is in T, then we can reconstruct T from the pair $(st(T), t)$. In other words, the function which sends T to the pair $(st(T), t_1t_2\cdots)$ is a bijection.

Theorem 5.17. *Let A be a nonnegative integer-valued matrix. If RSK sends A to (P,Q), then RSK sends A^{T} to (Q,P).*

Proof. Suppose that applying RSK to A involves the sequence of row insertions $\varnothing \leftarrow j_1 \leftarrow \cdots \leftarrow j_n$. When changing to the standardized matrix $st(A)$, the RSK algorithm will instead row insert $\varnothing \leftarrow j'_1 \leftarrow \cdots \leftarrow j'_n$ where $j'_1,\ldots,j'_n$ is a sequence such that $j'_i < j'_k$ whenever $j_i \leq j_k$. In other words, the relative magnitude of any two terms in the sequences $j_1,\ldots,j_n$ and $j'_1,\ldots,j'_n$ is the same.

This means that the row insertions when RSK is applied to A match exactly the row insertions when RSK is applied to $st(A)$, with the exception that the integers $j'_1,\ldots,j'_n$ are used instead of $j_1,\ldots,j_n$. RSK inserts equal integers from left to right, which matches our method of standardizing a column strict tableau, and so we can conclude that RSK sends $st(A)$ to $(st(P), st(Q))$.

We can now use Theorem 5.16 on the permutation matrix $st(A)$ to find that RSK sends $st(A)^{\mathrm{T}} = st(A^{\mathrm{T}})$ to $(st(Q), st(P))$.

Let $r_1r_2\cdots$ and $c_1c_2\cdots$ be the row and column sums of A. We can undo the standardization process on $st(A^{\mathrm{T}})$, using $c_1c_2\cdots$ as row sums and $r_1r_2\cdots$ as column sums, to find A^{T}. Similarly, we can undo the standardization of $st(Q)$ using $r_1r_2\cdots$ to find Q and undo the standardization of $st(P)$ using $c_1c_2\cdots$ to find P. We have now found that RSK sends A^{T} to (Q,P), as desired. □

5.5 Permutation Statistics from the Cauchy Kernel

Let $\sigma = \sigma_1 \cdots \sigma_n$ be a permutation. If we let $\mathrm{Des}(\sigma)$ the set of indices i for which $\sigma_i > \sigma_{i+1}$, then the descent and major index statistics can be written as $\mathrm{des}(\sigma) = |\mathrm{Des}(\sigma)|$ and $\mathrm{maj}(\sigma) = \sum_{i\in\mathrm{Des}(\sigma)} i$.

We can adapt these definitions for standard tableaux. If T is a standard tableau, we define $\mathrm{Des}(T)$ to be the set of indices i such that $i+1$ appears in a row above i. The descent and major index statistic for standard tableau are defined by $\mathrm{des}(T) = |\mathrm{Des}(T)|$ and $\mathrm{maj}(T) = \sum_{i\in\mathrm{Des}(T)} i$. The relationship between the set Des for permutations and the set Des for the output of RSK is in our next theorem.

Theorem 5.18. *For all $\sigma \in S_n$, $Des(\sigma) = Des(Q(\sigma))$ and $Des(\sigma^{-1}) = Des(P(\sigma))$.*

Proof. Let T be a column strict tableau. If $\sigma_i < \sigma_{i+1}$, Theorem 5.2 says that the last cell inserted in $T \leftarrow \sigma_i$ appears strictly to the left of the last cell inserted into $(T \leftarrow \sigma_i) \leftarrow \sigma_{i+1}$. This implies that $i+1$ does not appear above the row containing i in $Q(\sigma)$.

If $\sigma_i > \sigma_{i+1}$, then a similar argument as found in the proof of Theorem 5.2 shows that the path of replaced cells in $T \leftarrow \sigma_i$ appears strictly to the right of the path of replaced cells in $(T \leftarrow \sigma_i) \leftarrow \sigma_{i+1}$. Therefore the last cell inserted into $T \leftarrow \sigma_i$ appears strictly to the right of the last cell inserted into $(T \leftarrow \sigma_i) \leftarrow \sigma_{i+1}$. This implies that $i+1$ appears above the row containing i in $Q(\sigma)$.

We have now shown that $\mathrm{Des}(\sigma) = \mathrm{Des}(Q(\sigma))$. Using Theorem 5.16, we also have $\mathrm{Des}(\sigma^{-1}) = \mathrm{Des}(Q(\sigma^{-1})) = \mathrm{Des}(P(\sigma))$. □

Theorem 5.19. *For all $\lambda \vdash n$,*

$$s_\lambda(1,q,q^2,\ldots) = \frac{1}{(1-q)(1-q^2)\cdots(1-q^n)} \sum_{T\in ST_\lambda} q^{maj(T)}$$

where ST_λ is the set of standard tableaux of shape λ.

Proof. Let $\mathrm{sum}(T)$ denote the sum of the integers in a tableau T. In Exercise 2.9 it is seen that $s_\lambda = \sum_{R\in RCS_\lambda} w(R)$ where RCS_λ is the set of reverse column strict tableaux. Therefore $s_\lambda(1,q^1,q^2,\ldots)$ is equal to $\sum q^{\mathrm{sum}(R)-n}$ where the sum runs over all $R \in RCS_\lambda$.

Each $R \in RCS_\lambda$ can be turned into a $T \in CS_\lambda$ by a reverse standardization procedure: working from the largest integer to the smallest in R and moving from left to right for repeated integers, replace the integers in R with the integers $1,2,\ldots,n$. An example of an $R \in RCS_\lambda$ and the tableau $T \in ST_\lambda$ found by the reverse standardization procedure is below:

$$R = \begin{array}{|c|c|c|c|c|c|} \hline 1 & 1 \\ \cline{1-2} 2 & 2 & 1 & 1 \\ \cline{1-4} 4 & 3 & 3 & 2 & 2 & 1 \\ \hline \end{array} \qquad T = \begin{array}{|c|c|c|c|c|c|} \hline 8 & 9 \\ \cline{1-2} 4 & 5 & 10 & 11 \\ \cline{1-4} 1 & 2 & 3 & 6 & 7 & 12 \\ \hline \end{array}.$$

Let $r = r_1 \cdots r_n$ be the word created by listing the integers in R in weakly decreasing order. Define a word $a = a_1 \cdots a_n$ by $a_i = r_i - r_{i+1}$ for $i = 1, \ldots, n-1$ and $a_n = r_n$. For example, the table below shows r and a for the example R and T shown earlier in this proof:

	1	2	3	4	5	6	7	8	9	10	11	12
$r =$	4	3	3	2	2	2	2	1	1	1	1	1
$a =$	1	0	1	0	0	0	1	0	0	0	0	1

Then we have that $r_i = a_i + a_{i+1} + \cdots + a_n$, and so $\mathrm{sum}(R) = r_1 + \cdots + r_n = 1a_1 + 2a_2 + \cdots + na_n$. Additionally, we have $a_n \geq 1$ and $a_i \geq \chi(i \in \mathrm{Des}(st(T)))$ for $i = 1, \ldots, n-1$ where $\chi(A)$ is 1 if the statement A is true and 0 if A is false. This fact comes from the observation that an element in $\mathrm{Des}(\sigma)$ can only exist when the reverse standardization process changes from relabeling the ks in R to relabeling the $(k+1)$s in R for some k.

Therefore $s(1, q, q^2, \ldots) = \sum_{R \in RCS_\lambda} q^{\mathrm{sum}(R)-n}$ is equal to

$$q^{-n} \sum_{T \in ST_\lambda} \sum_{a_1 \geq \chi(1 \in \mathrm{Des}(T))} \cdots \sum_{a_{n-1} \geq \chi(n-1 \in \mathrm{Des}(T))} \sum_{a_n \geq 1} q^{1a_1 + 2a_2 + \cdots + na_n}$$

$$= q^{-n} \sum_{T \in ST_\lambda} \sum_{a_1 \geq \chi(1 \in \mathrm{Des}(T))} (q^1)^{a_1} \cdots \sum_{a_{n-1} \geq \chi(n-1 \in \mathrm{Des}(T))} (q^{n-1})^{a_{n-1}} \sum_{a_n \geq 1} (q^n)^{a_n}$$

$$= q^{-n} \sum_{T \in ST_\lambda} \frac{(q^1)^{\chi(1 \in \mathrm{Des}(T))} \cdots (q^{n-1})^{\chi(n-1 \in \mathrm{Des}(T))} q^n}{(1-q^1) \cdots (1-q^{n-1})(1-q^n)}$$

where the last line came from summing the multiple geometric series. This last equation can be seen to equal $\sum_{T \in ST_\lambda} q^{\mathrm{maj}(T)} / ((1-q) \cdots (1-q^n))$, as needed. □

By making small modifications to the above proof, we can refine Theorem 5.19 by descents. This refinement will be used to find generating functions for permutation statistics, as we will see in Theorem 5.21.

Theorem 5.20. *Let $(x;q)_{n+1}$ denote the product $(1 - xq^0)(1 - xq^1) \cdots (1 - xq^n)$. Then for all $\lambda \vdash n$ we have*

$$\sum_{k=0}^{\infty} x^k s_\lambda(1, q^1, \ldots, q^k) = \frac{1}{(x;q)_{n+1}} \sum_{T \in ST_\lambda} x^{des(T)} q^{maj(T)},$$

where ST_λ is the set of standard tableaux of shape λ.

Proof. Let $\max(T)$ denote the maximum integer in the tableau T. For reasons given in the proof of Theorem 5.19, $s_\lambda(1, q^1, \ldots, q^k)$ is equal to $\sum q^{\mathrm{sum}(R)-n}$ where the sum runs over all $R \in RCS_\lambda$ with $\max(R) \leq k+1$.

Each $R \in RCS_\lambda$ with $\max(R) \leq k+1$ can be turned into a column strict tableau $C \in CS_\lambda$ with $\max(C) \leq k+1$ by replacing each integer i in R with $k+2-i$. For instance, if $k = 3$, the R on the left turns into the C on the right in the diagram below:

$$R=\begin{array}{|c|c|c|c|c|c|}\hline 1&1\\\cline{1-2} 2&2&1&1\\\cline{1-4} 4&3&3&2&2&1\\\hline\end{array}\qquad C=\begin{array}{|c|c|c|c|c|c|}\hline 4&4\\\cline{1-2} 3&3&4&4\\\cline{1-4} 1&2&2&3&3&4\\\hline\end{array}\,.$$

Changing R to C in this way changes the sum of the tableau. Indeed, $\mathrm{sum}(R) = (k+2)n - \mathrm{sum}(C)$.

At this point we have

$$\begin{aligned}\sum_{k=0}^{\infty} x^k s_\lambda(1,q^1,\ldots,q^k) &= \sum_{k=0}^{\infty} x^k \sum_{R\in RCS_\lambda \text{ has } \max(R)\le k+1} q^{\mathrm{sum}(R)-n}\\ &= \sum_{k=0}^{\infty} x^k \sum_{C\in CS_\lambda \text{ has } \max(C)\le k+1} q^{(k+2)n-\mathrm{sum}(C)-n}\\ &= \sum_{k=0}^{\infty} (xq^n)^k \sum_{C\in CS_\lambda \text{ has } \max(C)\le k+1} q^{n-\mathrm{sum}(C)}\\ &= \frac{1}{1-xq^n}\sum_{C\in CS_\lambda} (xq^n)^{\max(C)-1} q^{n-\mathrm{sum}(C)}, \end{aligned} \tag{5.4}$$

where the last equality can be seen by expanding $1/(1-xq^n)$ as a geometric series and then examining the coefficient of x^k.

The usual standardization procedure associates each $C \in CS_\lambda$ with a pair of the form $(st(C), c)$ where $st(C)$ is an element in ST_λ and $c = c_1c_2\cdots$ is a word such that c_i is the number of appearances of i in C. Let $w = w_1\cdots w_n$ be the word created from c by replacing c_i with c_i copies of i for all i. Define a word $a = a_1\cdots a_n$ by $a_i = w_{i+1} - w_i$ for $i = 1,\ldots,n-1$ and $a_n = w_1$. The word a has the following properties:

1. We have $\max(C) = a_1 + \cdots + a_n$.
2. We have $\mathrm{sum}(C) = n\cdot\max(C) - (1a_1 + 2a_2 + \cdots + (n-1)a_{n-1})$.
3. We have $a_n \ge 1$ and $a_i \ge \chi(i \in \mathrm{Des}(st(T)))$ for $i = 1,\ldots,n-1$.

Therefore (5.4) is equal to

$$\begin{aligned}&\frac{x^{-1}}{1-xq^n}\sum_{T\in ST_\lambda}\sum_{a_1\ge\chi(1\in\mathrm{Des}(T))}\cdots\sum_{a_{n-1}\ge\chi(n-1\in\mathrm{Des}(T))}\sum_{a_n\ge 1} x^{a_1+\cdots+a_n} q^{1a_1+\cdots+(n-1)a_{n-1}}\\ &= \frac{x^{-1}}{1-xq^n}\sum_{T\in ST_\lambda}\sum_{a_1\ge\chi(1\in\mathrm{Des}(T))}(xq^1)^{a_1}\cdots\sum_{a_{n-1}\ge\chi(1\in\mathrm{Des}(T))}(xq^{n-1})^{a_{n-1}}\sum_{a_n\ge 1} x^{a_n}\\ &= \frac{x^{-1}}{1-xq^n}\sum_{T\in ST_\lambda}\frac{(xq^1)^{\chi(1\in\mathrm{Des}(T))}\cdots(xq^{n-1})^{\chi(n-1\in\mathrm{Des}(T))}x^1}{(1-xq^1)\cdots(1-xq^{n-1})(1-x)},\end{aligned}$$

where the last line came from summing the multiple geometric series. This last equation can be seen to equal $\sum_{T\in ST_\lambda} x^{\mathrm{des}(T)}q^{\mathrm{maj}(T)}/(x;q)_{n+1}$, as needed. □

The Cauchy kernel can be used to extract information about permutation statistics, as we see in our following theorem and in Exercises 5.3 and 5.6.

Theorem 5.21. *Let* $(z;p,q)_{k+1,\ell+1}$ *denote* $\prod_{i=0}^{k}\prod_{j=0}^{\ell}(1-zp^iq^j)$. *Then we have*

$$\sum_{n=0}^{\infty}\frac{z^n}{(x;p)_{n+1}(y;q)_{n+1}}\sum_{\sigma\in S_n}x^{des(\sigma)}y^{des(\sigma^{-1})}p^{maj(\sigma)}q^{maj(\sigma^{-1})}=\sum_{k,\ell\geq 0}\frac{x^ky^\ell}{(z;p,q)_{k+1,\ell+1}}.$$

Proof. Taking $x_i=p^{i-1}$ for $i=1,\ldots,k+1$ and $y_j=q^{j-1}$ for $j=1,\ldots,\ell+1$ and all other variables $x_i,y_j=0$ in Theorem 5.6, we find

$$\begin{aligned}\frac{1}{(z;p,q)_{k+1,\ell+1}}&=\prod_{i=0}^{k}\prod_{j=0}^{\ell}\frac{1}{1-p^iq^jz}\\&=\sum_{n=0}^{\infty}\sum_{\lambda\vdash n}s_\lambda(1,p,\ldots,p^k)s_\lambda(1,q,\ldots,q^\ell)z^n\\&=\sum_{n=0}^{\infty}z^n\sum_{\lambda\vdash n}\sum_{Q\in ST_\lambda}\frac{x^{\mathrm{des}(Q)}p^{\mathrm{maj}(Q)}}{(x;p)_{n+1}}\sum_{P\in ST_\lambda}\frac{y^{\mathrm{des}(P)}q^{\mathrm{maj}(P)}}{(y;q)_{n+1}}\Bigg|_{x^ky^\ell},\end{aligned}$$

where the last equality comes from taking the coefficient of x^k in Theorem 5.20. When restricted to permutations, the RSK algorithm is a bijection between permutations $\sigma\in S_n$ and pairs $(P(\sigma),Q(\sigma))$. Using this fact together with Theorem 5.18, we have

$$\frac{1}{(z;p,q)_{k+1,\ell+1}}=\sum_{n=0}^{\infty}\frac{z^n}{(x;p)_{n+1}(y;q)_{n+1}}\sum_{\sigma\in S_n}x^{\mathrm{des}(\sigma)}y^{\mathrm{des}(\sigma^{-1})}p^{\mathrm{maj}(\sigma)}q^{\mathrm{maj}(\sigma^{-1})}\Bigg|_{x^ky^\ell}.$$

The result follows from multiplying by x^ky^ℓ and summing over all $k,\ell\geq 0$. □

5.6 Hooks

Let λ be an integer partition of n. The hook of a cell c is the "L"-shaped subset of cells in the Young diagram of λ consisting of the cell c, all cells to the right of c and in the same row, and all cells above c and in the same column. For example, the hook of a cell c is shaded in the Young diagram below:

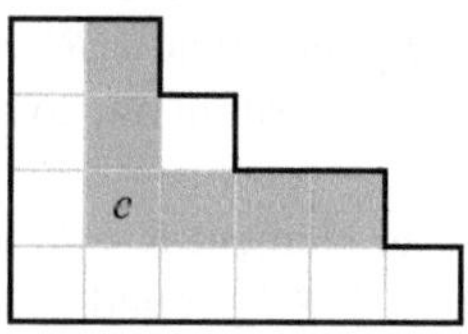

We define the hook length of the cell c, denoted $h(c)$, to be the number of cells in the hook of c. Below we have filled each cell c in the Young diagram for the integer partition $(6,5,3,2)$ with its hook length $h(c)$:

2	1				
4	3	1			
7	6	4	2	1	
9	8	6	4	3	1

Theorem 5.22. *For any* $\lambda = (\lambda_1, \lambda_2, \ldots) \vdash n$,

$$q^{-0\cdot\lambda_1-1\cdot\lambda_2-\cdots} s_\lambda(1,q,q^2,\ldots) = \prod_{c\in\lambda} \frac{1}{1-q^{h(c)}},$$

where the notation $c \in \lambda$ *means that* c *is a cell in the Young diagram for* λ.

Proof. As seen in the proof of Theorem 5.19, $s_\lambda(1,q,q^2,\ldots) = \sum_{T\in CS_\lambda} q^{\mathrm{sum}(T)-n}$ where $\mathrm{sum}(T)$ denotes the sum of the elements in T. Use the factor of q^{-n} in this sum to subtract 1 from each integer in $T \in CS_\lambda$ and use the $q^{-0\cdot\lambda_1-1\cdot\lambda_2-\cdots}$ term to subtract $(i-1)$ from each integer in row i of T. This process changes each $T \in CS_\lambda$ into a tableau filled with nonnegative integers such that

1. the integers weakly increase when reading bottom to top within columns and
2. the integers weakly increase when reading left to right within rows.

A tableau which satisfies the above conditions is called a reverse plane partition.

Let RPP_λ be the set of reverse plane partitions of shape λ. Given any $T \in RPP_\lambda$, let T_c be the nonnegative integer found in cell c. By defining the weight of $T \in RPP_\lambda$ to be $w(T) = \prod_{\text{cells } c \text{ in } T} q^{T_c}$, we now have

$$q^{-0\cdot\lambda_1-1\cdot\lambda_2-\cdots} s_\lambda(1,q,q^2,\ldots) = \sum_{R\in RPP_\lambda} w(R). \tag{5.5}$$

Let T_λ denote the set of tableaux of shape λ, filled freely with nonnegative integers. Looking at the right side of the equality in the statement of the theorem, write each term in the product $\prod_{c\in\lambda} 1/(1-q^{h(c)})$ as a geometric series and expand, selecting a term of the form $(q^{h(c)})^i$ for each $c \in \lambda$. Record the choices of i made for each cell by placing an i into cell c in a tableau $T \in T_\lambda$. By defining the hook weight of $T \in T_\lambda$ to be $hw(T) = \prod_{\text{cells } c \text{ in } T} (q^{h(c)})^{T_c}$, we now have

$$\prod_{c\in\lambda} \frac{1}{(1-q^{h(c)})} = \sum_{T\in T_\lambda} hw(T). \tag{5.6}$$

Comparing equations (5.5) and (5.6), we see that the theorem can be proved by defining a bijection $\varphi : RPP_\lambda \to T_\lambda$ such that $w(R) = hw(\varphi(R))$ for all $R \in RPP_\lambda$. The bijection we will describe is an algorithm due to Abraham Hillman and Richard Grassl [61].

The input to the algorithm is an element $R \in RPP_\lambda$ with not every entry equal to 0. The image of R is the element $T \in T_\lambda$ found by the following these steps:

1. Begin with T the tableaux of shape λ with all entries 0.
2. Locate the most north west nonzero element in R, say it lies in cell c.
3. Create a path P which moves down and to the right in R by starting at cell c. The next step in P will move down if the cell below c contains the same integer as c and the next step in P will move to the right otherwise. Continue creating P by moving down and to the right in this manner until no more moves are possible.

4. Subtract 1 from each cell in R that lies on the path P.
5. If P begins in column j and ends in row i, then add 1 to row i, column j entry of T.
6. If R contains a nonzero entry, go back to step 2. If R contains only 0s, then the output of the algorithm is T.

We give an example of this process below, where the appropriate path P is shaded in each step:

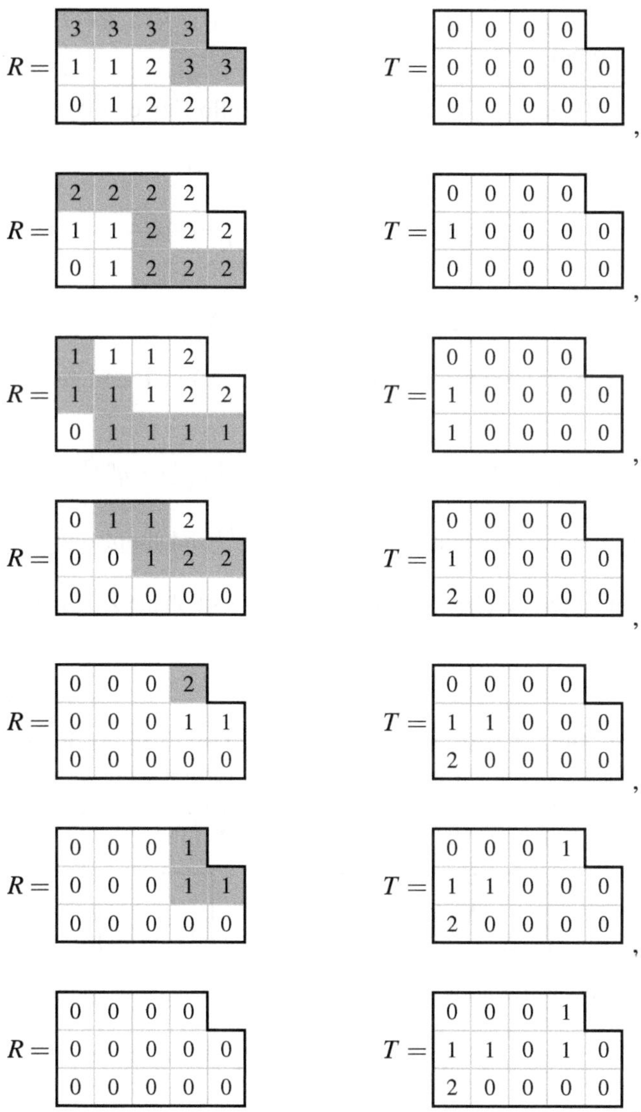

This algorithm moves from left to right down the columns of R, so the last entry added to T lies in the right-most nonzero column of T. Furthermore, the weakly increasing rows and columns in R force the last entry added to T to appear in the bottom entry of this column. Therefore since the last increased cell in T can be identified, this algorithm is reversible: given a pair R,T, recreate the path P by traveling up and to the left, moving up whenever neighboring entries in different rows are equal and moving left otherwise.

The algorithm is a bijection since it is reversible. The algorithm is also weight preserving since at any step in the algorithm we have $w(R) = hw(T)$ by design. This completes the proof. □

Theorems 5.19 and 5.22 combine to give the following corollary.

Corollary 5.23. *For any $\lambda = (\lambda_1, \lambda_2, \ldots) \vdash n$, we have*

$$\sum_{T \in ST_\lambda} q^{maj(T)} = \frac{[n]_q!}{\prod_{c \in \lambda} [h(c)]_q} q^{0\lambda_1 + 1\lambda_2 + \cdots},$$

where ST_λ is the set of standard tableaux of shape λ.

Proof. Comparing the expressions for $s_\lambda(1, q, q^2, \ldots)$ in Theorems 5.19 and 5.22,

$$\frac{1}{(1-q)(1-q^2)\cdots(1-q^n)} \sum_{T \in ST_\lambda} q^{\mathrm{maj}(T)} = \frac{1}{\prod_{c \in \lambda}(1-q^{h(c)})} q^{0\lambda_1 + 1\lambda_2 + \cdots}.$$

The corollary follows by solving for $\sum_{T \in ST_\lambda} q^{\mathrm{maj}(T)}$ and simplifying. □

Let f^λ denote the number of standard tableaux of shape $\lambda \vdash n$. Taking $q = 1$ in Corollary 5.23 gives that

$$f^\lambda = \frac{n!}{\prod_{c \in \lambda} h(c)} \tag{5.7}$$

for any $\lambda \vdash n$. This identity, which gives a wonderful way to find the number of standard tableaux, is known as the hook length formula. For example, instead of calculating $f^{(3,2)}$ by listing the standard tableaux below, we can instead use the hook length formula to see that $f^{(3,2)} = 5!/(4 \cdot 3 \cdot 2 \cdot 1 \cdot 1) = 5$.

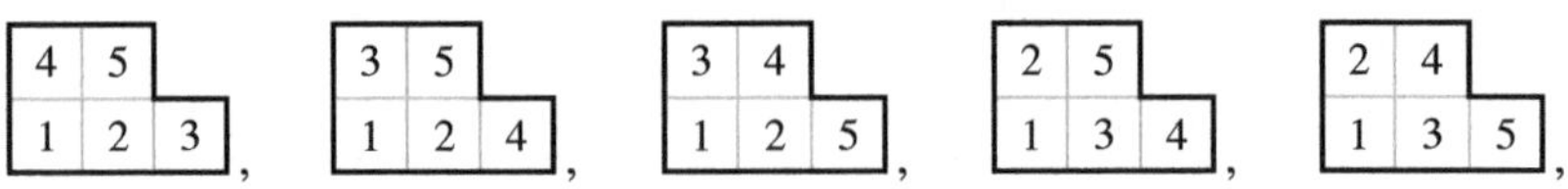

Since we proved the hook length formula in a somewhat roundabout way, the intuition giving why the formula is correct may not have been fully formed. To gain insight, consider the following incorrect, but enlightening "proof" of (5.7):

Proof (incorrect). Randomly place the integers $1,\ldots,n$ into the Young diagram of shape $\lambda \vdash n$ to create a tableau. Such a naïve filling will be a standard tableau exactly when each cell c is the smallest of the integers in its hook. The probability that this happens for any given $c \in \lambda$ is $1/h(c)$. Therefore the probability of creating a standard tableau is $1/\prod_{c\in\lambda} h(c)$.

There are $n!$ random placements of the integers $1,\ldots,n$ into the Young diagram of shape λ, so the number of standard tableaux f^λ is equal to $n!/\prod_{c\in\lambda} h(c)$. □

The error in the proof is that the probability of one cell $c_1 \in \lambda$ being the smallest integer in c_1s hook is not independent from the probability of a second cell $c_2 \in \lambda$ being the smallest in c_2s hook. Because the probabilities are not independent, we cannot simply multiply them together in order to find the probability of creating a standard tableau. Strangely, however, since the hook length formula is indeed true, the probability of creating a standard tableau by randomly placing the integers $1,\ldots,n$ into a Young diagram is $1/\prod_{c\in\lambda} h(c)$ nevertheless!

Exercises

5.1. Let T be a column strict tableau with distinct entries and let $j \neq k$ be integers not found in T. Show that $(T \leftarrow j) \downarrow k = (T \downarrow k) \leftarrow j$, that is, show that row insertion and column insertion commute when all integers involved are distinct.

5.2. For $\sigma = \sigma_1 \cdots \sigma_n \in S_n$, define σ^r to be the reverse permutation $\sigma_n \cdots \sigma_1$. Use exercise 5.1 to show that $P(\sigma^r) = P(\sigma)'$.

5.3. Show that the generating function

$$\sum_{n=0}^{\infty} \frac{z^n}{(x;p)_{n+1}(y;q)_{n+1}} \sum_{\sigma\in S_n} x^{(n-1)-\mathrm{des}(\sigma)} y^{\mathrm{des}(\sigma^{-1})} p^{\binom{n}{2}-\mathrm{maj}(\sigma)} q^{\mathrm{maj}(\sigma^{-1})}$$

is equal to $\sum_{k,\ell\geq 0} x^k y^\ell (-z;p,q)_{k+1,\ell+1}$.

5.4. Let A be a symmetric matrix. By Theorem 5.17, RSK sends A to (P,P) where P is a column strict tableau. Show that the trace of A is equal to the number of columns of an odd length in P.

5.5. Use exercise 5.4 to show that

$$\prod_{i\geq 1} \frac{1}{1-x_i yz} \prod_{i<j} \frac{1}{1-x_i x_j z^2} = \sum_{\lambda} s_\lambda(x_1,x_2,\ldots) y^{\mathrm{odd}(\lambda')} z^{|\lambda|},$$

where $\mathrm{odd}(\lambda')$ is the number of odd parts in the conjugate partition λ'.

5.6. Let I_n be the set of involutions in S_n, that is, the set of permutations in S_n with $\sigma = \sigma^{-1}$. Use exercise 5.5 to find a generating function for

$$\sum_{n=0}^{\infty} \frac{z^n}{(x;p)_{n+1}} \sum_{\sigma \in I_n} x^{\mathrm{des}(\sigma)} y^{\mathrm{fix}(\sigma)} p^{\mathrm{maj}(\sigma)}$$

where $\mathrm{fix}(\sigma)$ is the number of fixed points in the permutation σ.

5.7. Simplify these sums:

$$\sum_{\lambda \vdash n} f^{\lambda}, \qquad \sum_{\lambda \vdash n} (f^{\lambda})^2, \qquad \sum_{\lambda \vdash n} f^{2\lambda}.$$

As usual, f^{λ} denotes the number of standard tableaux of shape λ. The notation 2λ means every part in λ is multiplied by 2. Exercise 5.4 may help with the third sum.

Solutions

5.1 We proceed by induction on the maximum element m found in $(T \leftarrow j) \downarrow k$.

Let $j = j_1, j_2, \ldots, j_\ell$ be the sequence of replaced cells in $T \leftarrow j$. This increasing sequence moves up and weakly to the right in T. Similarly, let $k = k_1, k_2, \ldots, k_n$ be the sequence of replaced cells in $T \downarrow k$. This increasing sequence moves to the right and weakly down in T. The diagram below simultaneously displays how these two sequences must appear in $T \leftarrow j$ and $T \downarrow k$:

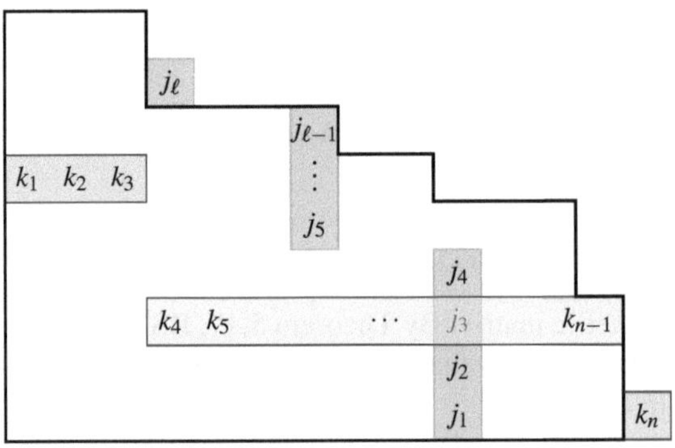

Case 1: m is not in either of the sequences $j_1, \ldots, j_\ell$ or $k_1, \ldots, k_n$. Then m can be removed from T to form T' without influencing either sequence—the operation of adding or removing m is independent of row inserting j or column inserting k. Remove m, invoke induction to find that $(T' \leftarrow j) \downarrow k = (T' \downarrow k) \leftarrow j$, and then reinsert m to see that row and column insertion commute.

Case 2: m is found in the sequence $j_1,\dots,j_\ell$ (this means $j_\ell = m$) but not in the sequence $k_1,\dots,k_n$. Since m is the largest integer involved, row insertion terminated one step after m was bumped from T. In other words, either $j = m$ or the position of m in T is at the cell labeled $j_{\ell-1}$ in $T \leftarrow j$.

If the cell containing m in T is removed to form T', then the sequence of replaced cells in the row insertion $T' \leftarrow j$ is the same sequence $j_1,\dots,j_{\ell-1}$ but with the last integer j_ℓ removed. This means $T' \leftarrow j$ is equal to $T \leftarrow j$ with m removed, showing that removing or replacing m is independent of row inserting j.

Since m is not in $k_1,\dots,k_n$, this sequence cannot involve the cell containing m in T. Therefore removing or replacing m is independent of column inserting k. Now we can remove m to form T', invoke induction $(T' \leftarrow j) \downarrow k = (T' \downarrow k) \leftarrow j$, and then reinsert m to see that row and column insertions commute.

Case 3: m is not found in the sequence $j_1,\dots,j_\ell$ but is in the sequence $k_1,\dots,k_n$. The result follows from an argument very similar to that found in Case 2.

Case 4: m is in both sequences $j_1,\dots,j_\ell$ and $k_1,\dots,k_n$ and $j_{\ell-1} > k_{n-1}$. A diagram simultaneously illustrating the sequences $j_1,\dots,j_\ell$ and $k_1,\dots,k_n$ in $T \leftarrow j$ and $T \downarrow k$ looks like this:

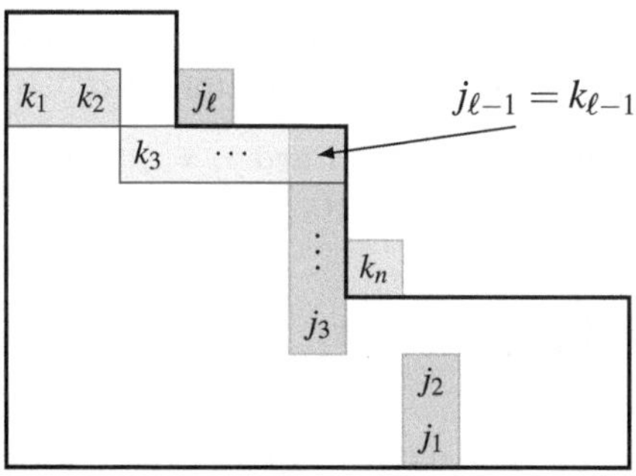

If more than one integer in the sequence $j_1,\dots,j_{\ell-1}$ was found in the column of $T \leftarrow j$ containing $j_{\ell-1}$, then since k_{n-1} appears above $j_{\ell-1}$ in $T \downarrow k$, it would follow that $k_{\ell-1} > j_{\ell-1}$. This means T actually looks something like this:

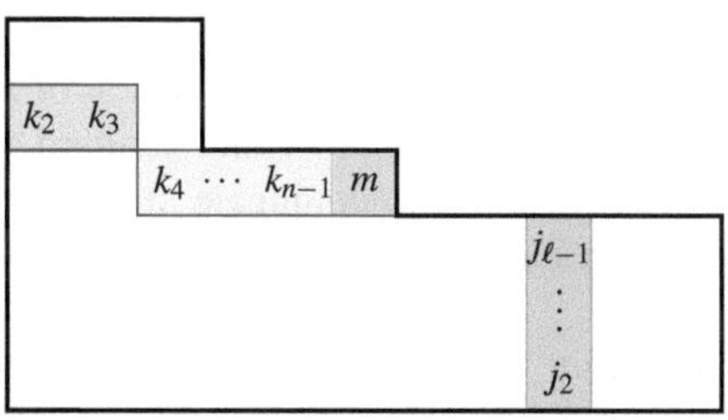

From here it can be seen that both $(T \leftarrow j) \downarrow k$ and $(T \downarrow k) \leftarrow j$ are the same; in particular, illustrating what happens with the above diagram, they are both equal to

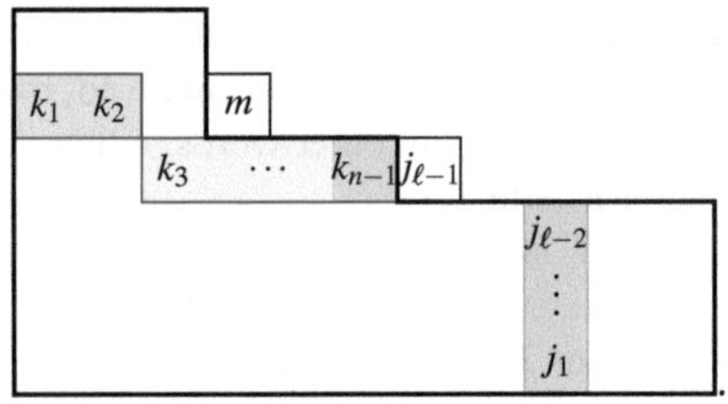

Case 5: m is in both sequences $j_1,\dots,j_\ell$ and $k_1,\dots,k_n$ and $j_{\ell-1} < k_{n-1}$. The result follows from an argument very similar to that found in Case 4.

5.2 We have

$$P(\sigma^r) = \varnothing \leftarrow \sigma_n \leftarrow \sigma_{n-1} \leftarrow \cdots \leftarrow \sigma_1 = \varnothing \downarrow \sigma_n \leftarrow \sigma_{n-1} \leftarrow \cdots \leftarrow \sigma_1. \qquad (5.8)$$

Exercise 5.1 says that (5.8) is equal to $\varnothing \leftarrow \sigma_{n-1} \leftarrow \cdots \leftarrow \sigma_1 \downarrow \sigma_n$, so, by induction, we know that

$$P(\sigma^r) = \varnothing \downarrow \sigma_1 \downarrow \sigma_2 \downarrow \cdots \downarrow \sigma_n.$$

Repeated column insertion of a list of distinct integers gives the conjugate standard Young tableau as repeated row insertion. Therefore $P(\sigma^r) = P(\sigma)'$.

5.3 Taking $x_i = p^{i-1}$ for $i = 1,\dots,k+1$ and $y_j = q^{j-1}$ for $j = 1,\dots,\ell+1$ and all other variables $x_i, y_j = 0$ in Theorem 5.10, we find

$$\begin{aligned}(-z;p,q)_{k+1,\ell+1} &= \prod_{i=0}^{k}\prod_{j=0}^{\ell}(1+p^iq^jz)\\ &= \sum_{n=0}^{\infty}\sum_{\lambda\vdash n} s_{\lambda'}(1,p,\dots,p^k)s_\lambda(1,q,\dots,q^\ell)z^n\\ &= \sum_{n=0}^{\infty} z^n \sum_{\lambda\vdash n}\sum_{Q\in ST_\lambda}\frac{x^{\operatorname{des}(Q')}p^{\operatorname{maj}(Q')}}{(x;p)_{n+1}}\sum_{P\in ST_\lambda}\frac{y^{\operatorname{des}(P)}q^{\operatorname{maj}(P)}}{(y;q)_{n+1}}\Bigg|_{x^ky^\ell},\end{aligned}$$

where the last equality comes from taking the coefficient of x^k in Theorem 5.20.

The integer $i+1$ is in a row above i in the standard tableau T if and only if $i+1$ is not in a row above i in the conjugate standard tableau T'. Therefore $\operatorname{Des}(T) = \{1,\dots,n-1\} \setminus \operatorname{Des}(T')$. By theorem 5.18, $\operatorname{des}(T') = (n-1) - \operatorname{des}(\sigma)$ and $\operatorname{maj}(T') = \binom{n}{2} - \operatorname{maj}(\sigma)$.

Using Theorem 5.18 again for the P tableau and using the observation that the dual RSK algorithm is a bijection between permutations $\sigma \in S_n$ and pairs $(P(\sigma), Q(\sigma)')$, we have that $(-z;p,q)_{k+1,\ell+1}$ is equal to

$$\sum_{n=0}^{\infty}\frac{z^n}{(x;p)_{n+1}(y;q)_{n+1}}\sum_{\sigma\in S_n} x^{(n-1)-\operatorname{des}(\sigma)}y^{\operatorname{des}(\sigma^{-1})}p^{\binom{n}{2}-\operatorname{maj}(\sigma)}q^{\operatorname{maj}(\sigma^{-1})}\Bigg|_{x^ky^\ell}.$$

The result follows from multiplying by x^ky^ℓ and summing over all $k,\ell \geq 0$.

5.4 We proceed by induction on the number of rows in P, with the case of zero rows being vacuously true.

The standardized matrix $st(A)$ is symmetric and the trace of A and $st(A)$ are equal, allowing us to reduce to the case where A is a symmetric permutation matrix.

The path diagram for A is symmetric, so each upright path in the path diagram must either have a 1 on the diagonal or a corner 0 on the diagonal. Let $\tilde{A}$ be the matrix A with any 1s found in the path diagram removed and any corner 0s in the path diagram turned into 1s. Let $\tilde{P}$ be P with the bottom row removed.

Since the second path diagram determines the shape of $\tilde{P}$, we know by induction that $tr(\tilde{A})$ is equal to the number of columns of an odd length in $\tilde{P}$. Therefore $tr(\tilde{A})$ is the number of columns of an even length in P—equivalently, each 0 on the diagonal of the path diagram accounts for a columns of an even length. Since the total number of up-down paths in the path diagram is the total number of columns, it follows that the number of 1s on the diagonal of A is the number of columns of an odd length.

5.5 We use ideas similar to those found in the proof of Theorem 5.6.

The coefficient of z^n on the left-hand side is the weighted sum over all symmetric matrices A with nonnegative integer entries that sum to n. This can be seen by expanding each term in the products as geometric series. The terms in the first product dictate the diagonal entries of A and the terms in the second product dictate both the (i,j) and (j,i) entries in A. Since each choice of diagonal entry comes along with a power of y, the exponent on y is the trace of A.

RSK is a weight preserving bijection which changes the symmetric matrix A into a single-column strict tableau P such that P has size n and, by exercise 5.4, P has $tr(A)$ columns of an odd length. Summing over all possible shapes λ of P and noticing that $\mathrm{odd}(\lambda) = tr(A)$, we can see this is exactly what is counted by the coefficient of z^n on the right side of the equality.

5.6 Take $x_i = p^{i-1}$ for $i = 1, \ldots, k+1$ and all other $x_i = 0$ in exercise 5.5 to find

$$\frac{1}{(yz;p)_{k+1}} \prod_{0\le i<j\le k} \frac{1}{(1-p^{i+j}z^2)} = \sum_{n=0}^{\infty} \sum_{\lambda\vdash n} s_\lambda(1,p,\ldots,p^k) y^{\mathrm{odd}(\lambda')} z^n$$

$$= \sum_{n=0}^{\infty} z^n \sum_{\lambda\vdash n} y^{\mathrm{odd}(\lambda')} \sum_{P\in ST_\lambda} \frac{x^{\mathrm{des}(P)} p^{\mathrm{maj}(P)}}{(x;p)_{n+1}} \Bigg|_{x^k},$$

where the last equality comes from taking the coefficient of x^k in Theorem 5.20.

Each standard tableau P corresponds to an involution $\sigma \in I_n$ by applying the inverse to the RSK algorithm to the pair (P,P). Furthermore, by exercise 5.4, the trace of the permutation matrix representing σ is equal to $\mathrm{odd}(\lambda')$ where λ is the shape of P. Since the trace of the permutation matrix gives the number of fixed points in σ, the above string of equalities is equal to

$$\sum_{n=0}^{\infty} \frac{z^n}{(x;p)_{n+1}} \sum_{\sigma\in I_n} x^{\mathrm{des}(\sigma)} y^{\mathrm{fix}(\sigma)} p^{\mathrm{maj}(\sigma)} \Bigg|_{x^k}$$

by Theorem 5.18. Multiply by x^k and sum over all $k \geq 0$ to find that the desired generating function is

$$\sum_{k=0}^{\infty} \frac{x^k}{(yz;p)_{k+1}} \prod_{0 \leq i < j \leq k} \frac{1}{(1-p^{i+j}z^2)}.$$

5.7 Let I_n be the set of involutions in S_n. RSK is a bijection from I_n to pairs of the form (P,P) where P is a standard tableaux of size n. Therefore

$$\sum_{\lambda \vdash n} f^{\lambda} = (\text{the number of standard tableau of size } n) = |I_n|.$$

The generating function for the number of involutions of n was given in section 4.4.

RSK is a bijection from S_n to pairs of the form (P,Q) where P and Q are both standard tableaux of size n. Therefore

$$\sum_{\lambda \vdash n} (f^{\lambda})^2 = (\text{the number of standard tableau of size } n)^2 = |S_n| = n!.$$

As for the third sum, notice that

$$\sum_{\lambda \vdash n} f^{2\lambda} = \sum_{\substack{\lambda \vdash 2n \\ \text{odd}(\lambda)=0}} f^{\lambda} = \sum_{\substack{\lambda \vdash 2n \\ \text{odd}(\lambda')=0}} f^{\lambda},$$

where odd(λ) denotes the number of odd parts in $\lambda \vdash n$.

The number of fixed points in an involution $\sigma \in I_n$ is the trace of the permutation matrix representing σ. Exercise 5.4 tells us that RSK is a bijection between involutions σ without a fixed point and pairs of the form (P,P) where P is a standard tableau of a shape λ with odd$(\lambda') = 0$. Therefore

$$\begin{aligned}
\sum_{\lambda \vdash n} f^{2\lambda} &= \sum_{\substack{\lambda \vdash 2n \\ \text{odd}(\lambda)=0}} f^{\lambda} \\
&= \sum_{\substack{\lambda \vdash 2n \\ \text{odd}(\lambda')=0}} f^{\lambda} \\
&= (\text{the number of standard tableau of shape } \lambda \vdash 2n \text{ with odd}(\lambda') = 0) \\
&= (\text{the number of } \sigma \in I_{2n} \text{ without a fixed point}).
\end{aligned}$$

The number of $\sigma \in I_{2n}$ without a fixed point is equal to $(2n-1)(2n-3)\cdots 3 \cdot 1$; this can be seen by writing σ in cyclic notation, selecting an integer from $\{2,\ldots,2n\}$ to place in a cycle of length 2 with 1, and then proceeding by induction.

Notes

A straightening algorithm equivalent to the RSK was first published by Robinson [103]. Later, Schensted created the row insertion algorithm [105], which Knuth extended to a map between matrices and pairs of column strict tableaux [72]. Our proof the Cauchy identities come from the ideas in [72]. In this same paper, Knuth defined the notion of Knuth equivalence and proved Theorem 5.12. Theorem 5.13 is due to Green [54].

Craige Schensted changed his name to Ea, after the Sumerian god Enki, in 1995. At the end of 1999, anticipating millennium computer glitches, Ea added a second name, becoming Ea Ea.

The hook length formula is due to Frame, Robinson, and Thrall [46]. In [104], an interesting anecdote about the origins of the hook length formula are given; apparently this identity was independently and simultaneously discovered by mathematicians visiting Michigan State University in 1953.

Section 5.4 is based on of a geometric version of the RSK algorithm due to Viennot [112].

The generating function found in Theorem 5.21 is due to Gessel [50] and first appeared in [48]. This first proof did not use the Cauchy kernel; the connection to the Cauchy kernel is due to Désarménien and Foata [26, 27]; these papers also included Exercise 5.7.

A (k,ℓ) hook Schur function of shape λ is defined by

$$HS_\lambda(x_1,\ldots,x_k;y_1,\ldots,y_\ell) = \sum_{\mu\subseteq\lambda} s_\mu(x_1,\ldots,x_k)s_{\lambda'/\mu'}(y_1,\ldots,y_\ell).$$

The ideas of Désarménien and Foata were extended from identities involving Schur functions to the realm of (k,ℓ) Schur functions in [98]. This work was further extended by Desésarménien and Foata in [28].

The method of proving the hook length formula using symmetric functions is due to Stanley [106].

Chapter 6
Counting Problems That Involve Symmetry

In this chapter we introduce Pólya's enumeration theorem. The theory is designed to solve counting problems which involve symmetry, like these:

1. How many ways are there to create a necklace with n black beads and n red beads? Two necklaces are considered the same if the first necklace can be rotated and/or flipped over to match the second necklace.
2. How many ways are there to color the faces of a cube such that 2 faces are red, 3 faces are black, and one face is fluorescent beige? Two colorings are the same if one cube can be rotated to get the second.

6.1 Pólya's Enumeration Theorem

This chapter uses some beginning concepts in group theory. For our purposes, a finite group G is a nonempty subset of S_n such that $\sigma\tau$ and σ^{-1} are both in G for all $\sigma, \tau \in G$. Let G be a group and let $\lambda(g)$ be the cycle type of $g \in G$. Define the cycle index polynomial for G to be the symmetric function

$$Z_G = \frac{1}{|G|} \sum_{g \in G} p_{\lambda(g)},$$

where, as usual, p_k is the power symmetric function.

For example, the dihedral group D_4, generated by the rotation $(1\ 2\ 3\ 4)$ and the reflection $(1\ 2)(3\ 4)$, is a subgroup of S_4. The group D_4 acts on the set of vertices of the square

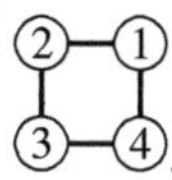

.

A. Mendes, J. Remmel, *Counting with Symmetric Functions*, Developments in Mathematics 43, DOI 10.1007/978-3-319-23618-6_6

The elements in D_4 are

$$\{(1),(1\,2\,3\,4),(1\,3)(2\,4),(1\,4\,3\,2),(1\,2)(3\,4),(1\,4)(2\,3),(1\,3),(2\,4)\},$$

and so the cycle index polynomial is

$$Z_{D_4} = \frac{1}{8}\left(2p_4 + 3p_2^2 + 2p_1^2 p_2 + p_1^4\right).$$

Let $\mathbf{N}$ denote the set of all functions $f : \{1,\dots,n\} \to \{1,2,\dots\}$. This function f is called a coloring because each element $\{1,\dots,n\}$ is assigned a "color" (a positive integer). The weight of such a function is

$$w(f) = \prod_{i=1}^{\infty} x_i^{(\text{the number of } j \text{ with } f(j)=i)}.$$

Given $g \in G$ and $f \in \mathbf{N}$, let gf denote the function defined by $gf(i) = f(gi)$ for all i. In this way, an equivalence relation on $\mathbf{N}$ can be defined such that $f \sim f'$ if and only if there is a $g \in G$ such that $f = gf'$. The symmetric function F_G is defined to be

$$F_G = \sum_{[f] \in \mathbf{N}/\sim} w(f),$$

where $[f]$ is the equivalence class containing $f \in \mathbf{N}$.

Using our running example of the vertices of the square, two colors are equivalent if the first can be rotated and or reflected to find the second. For instance,

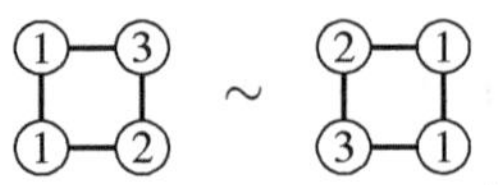

All possible inequivalent colorings which have a weight $x_1^{\lambda_1}x_2^{\lambda_2}x_3^{\lambda_3}x_4^{\lambda_4}$ for some integer partition $\lambda = (\lambda_1,\dots,\lambda_4) \vdash 4$ are shown below:

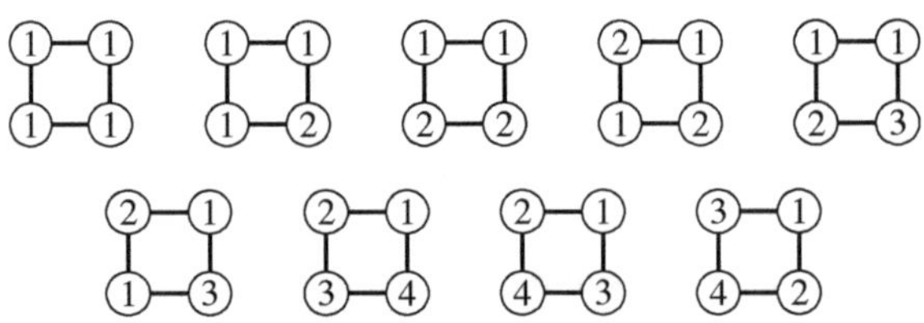

Each one of these colorings represents an equivalence class in $\mathbf{N}/\sim$. There are many more equivalence classes not listed here; the colors in each one of the above equivalence classes can be changed to find more inequivalent colorings. For instance, here are more inequivalent colorings with weight $x_i^2 x_j^2$ for $i \neq j$:

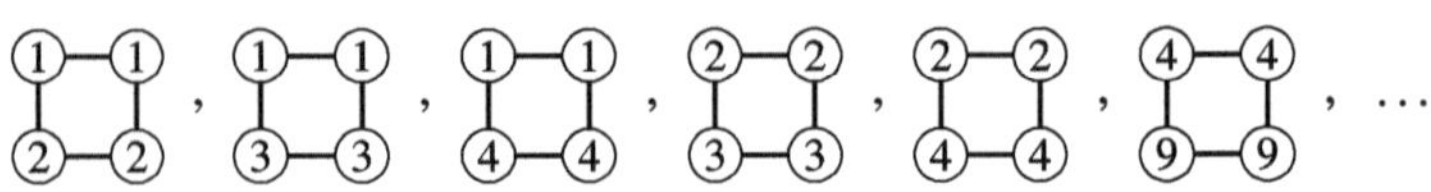

For this example, the symmetric function F_{D_4} is therefore

$$F_{D_4} = m_{(4)} + m_{(3,1)} + 2m_{(2,2)} + 2m_{(2,1,1)} + 3m_{(1,1,1,1)}.$$

Using the m-to-p transition matrix, found by inverting the p-to-m transition matrix given in section 2.3, we see that

$$\begin{bmatrix} 1&1&1&1&1 \\ 0&1&0&2&4 \\ 0&0&2&2&6 \\ 0&0&0&2&12 \\ 0&0&0&0&24 \end{bmatrix}^{-1} \begin{bmatrix} 1\\1\\2\\2\\3 \end{bmatrix} = \frac{1}{8} \begin{bmatrix} 2\\0\\3\\2\\1 \end{bmatrix}$$

and so $F_{D_4} = \frac{1}{8}\left(2p_{(4)} + 3p_{(2,2)} + 2p_{(2,1,1)} + p_{(1,1,1,1)}\right) = Z_{D_4}$. This is no accident, as we see in Theorem 6.1.

Theorem 6.1 (Pólya). *The symmetric polynomials Z_G and F_G are equal.*

Proof. The statement that $Z_G = F_G$ is equivalent to

$$\sum_{g \in G} p_{\lambda(g)} = |G| \sum_{[f] \in \mathbf{N}/\sim} w(f).$$

We will prove this identity bijectively.

Consider the set of objects formed by drawing a strip of k cells all filled with the same positive integer underneath each cycle of length k in an element $g \in G$. For instance, one possible object is

(1 5 3)	(2)	(4 7)	(6 10 9 8 11 12)
2 2 2	3	2 2	1 1 1 1 1 1

.

By defining the weight of such an object to be the product of the weights of the underlying column strict tableau, the weighted sum over all possible objects is equal to $\sum_{g \in G} p_{\lambda(g)}$.

Given such an object, let f be the coloring such that $f(i) = j$ if i appears in a cycle above cells filled with j. Our objects are therefore ordered pairs (g, f) where $g \in G$, $f \in \mathbf{N}$, and—since f is constant on the cycles of g—it must be the case that $gf = f$. Let φ be the function which sends (g, f) to $(g'g, f')$ where f' is the lexicographically least element in $[f]$ and g' is the lexicographically least element in G for which $g'f = f'$.

The function φ is weight preserving. It is also a bijection because we can describe its inverse. Let f' be the lexicographically least element in $[f']$. Given $h \in G$, let g be the lexicographically least element in G for which $gh^{-1}f' = f'$. Then the inverse image of (h, f') is the pair $(g^{-1}h, h^{-1}f')$ because

$$\varphi\left((g^{-1}h, h^{-1}f')\right) = (gg^{-1}h, f') = (h, f').$$

The function φ is the desired weight preserving bijection; it sends ordered pairs (g,f) with $gf = f$ to ordered pairs in $G \times \mathbf{N}/\sim$. The weighed sum over all possible elements in $G \times \mathbf{N}/\sim$ is equal to the right-hand side of the identity, as required. □

To illustrate the utility of Theorem 6.1 we continue our running example with the square. How many ways are there to color the vertices of the square if two colorings are considered the same if one coloring can be rotated and or reflected to find the other? The answer is $F_{D_4}(1,\ldots,1,0,\ldots)$, the number found by taking $x_1 = \cdots = x_k = 1$ and all other variables x_i equal to 0. Since $p_k(1,\ldots,1,0,\ldots) = 1^k + \cdots + 1^k = k$, we have that the number of possible colorings is equal to

$$F_{D_4}(1,\ldots,1,0,\ldots) = Z_{D_4}(1,\ldots,1,0,\ldots) = \frac{1}{8}\left(2k + 3k^2 + 2k^3 + k^4\right).$$

How many ways are there to color the vertices of square if we must color two vertices black and two vertices white? The answer is the coefficient of $x_1^2x_2^2$ in F_{D_4}. We have

$$\begin{aligned}
F_{D_4} &= Z_{D_4} \\
&= \frac{1}{8}\left(2p_{(4)} + 3p_{(2,2)} + 2p_{(2,1,1)} + p_{(1,1,1,1)}\right) \\
&= \frac{1}{8}\left(2\left(x_1^4 + \cdots\right) + 3(x_1^2 + \cdots)^2 + 2(x_1^2 + \cdots)(x_1 + \cdots)^2 + (x_1 + \cdots)^4\right).
\end{aligned}$$

A term of the form $x_1^2x_2^2$ can come from the $p_{(2,2)}$, $p_{(2,1,1)}$, and $p_{(1,1,1,1)}$ terms. Therefore the coefficient is $(3 \cdot 2 + 2 \cdot 2 + 6)/8 = 2$.

Theorem 6.2 (Fermat's little theorem). *If q is a prime number and a a positive integer, then $a^q - a$ is divisible by q.*

Proof. When q is prime, the group $\mathbf{Z}_q$, which is generated by the cycle $(1\ 2\ \cdots\ q)$, has $q-1$ elements of cycle type (q) and one element of cycle type (1^q). Therefore

$$F_{\mathbf{Z}_q} = Z_{\mathbf{Z}_q} = \frac{1}{q}\left(p_1^q + (q-1)p_q\right).$$

It is clear from the definition that F_G must be a polynomial with positive integer coefficients for any group G. In particular, specializing by taking $x_1 = \cdots = x_a = 1$ and all other terms equal to 0, $(a^q + (q-1)a)/q$ must be an integer. Fermat's little theorem follows. □

6.2 The Cycle Index Polynomial and Schur Functions

This section is devoted to understanding the cycle index polynomial when expanded in terms of the Schur basis. It turns out that the coefficient of the Schur symmetric function s_λ in the cycle index polynomial Z_G is always a nonnegative integer. That is, for any group G,

$$Z_G = \frac{1}{|G|} \sum_{g \in G} p_{\lambda(g)} = \sum_{\lambda \vdash n} c_\lambda s_\lambda \tag{6.1}$$

for nonnegative integer constants c_λ. The proof of this fact is not particularly difficult but is beyond the scope of this text, as it involves the representation theory of the symmetric group. See [104] for more details.

Since Z_G is defined in terms of the power symmetric functions, we can use the p-to-s transition matrix to write Z_G in terms of Schur symmetric functions. Let χ_μ^λ be the λ, μ entry of the p-to-s transition matrix. In Exercises 2.13 and 2.14 in Chapter 2, labeled abaci were used to find a combinatorial description of χ_μ^λ in terms of the so-called rim hook tableaux. We begin this section with an alternative proof of this combinatorial interpretation of χ_μ^λ which is based on the Pieri rules found in Chapter 5. That way this section provides a self-contained description on how to expand Z_G into Schur functions. The proof we include below is based on the work of [79] and has never appeared before in a book.

As first described in Exercise 1.3, a skew shape λ/μ is a rim hook of λ if it contains no 2×2 square and any two consecutive cells are connected by an edge. A skew shape λ/μ is a broken rim hook of λ if it is a union of rim hooks. As described in Exercise 2.13, the sign $sgn(\lambda/\mu)$ of a rim hook λ/μ which spans R rows is $(-1)^{R-1}$.

Theorem 6.3 (The Murnaghan–Nakayama rule). *For all integers r and integer partitions μ,*

$$p_r s_\mu = \sum_{\lambda/\mu \text{ is a rim hook with } r \text{ cells}} sgn(\lambda/\mu) s_\lambda.$$

Proof. Before showing the identity involving $p_r s_\mu$ in the statement of the theorem, we first understand how to expand $s_{(r-k,1^k)} s_\mu$ into a sum of Schur functions.

Using the identity shown in Exercise 2.1, we have

$$s_{(r-k,1^k)} s_\mu = \left(\sum_{i=0}^{k} (-1)^{k-i} e_i h_{r-i} \right) s_\mu = \sum_{i=0}^{k} (-1)^{k-i} e_i h_{r-i} s_\mu; \tag{6.2}$$

therefore to understand $s_{(r-k,1^k)} s_\mu$ we will consider terms of the form $e_i h_{r-i} s_\mu$.

By the Pieri rules, our Theorem 5.3, we have

$$e_i h_{r-i} s_\mu = \sum s_\lambda,$$

where the sum runs over all integer partitions λ which can be found by adding a skew row containing $r-i$ cells to μ and then adding a skew column containing i cells to the result. If we put hs (for horizontal) in the cells in the added skew row and vs (for vertical) in the cells in the skew column, we will find a diagram λ which will look like this:

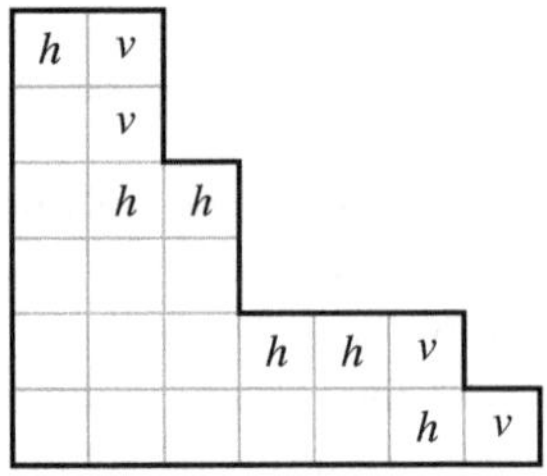

It follows that (6.2) can be interpreted as $\sum \text{sign}(D) s_{sh(D)}$ where D is a diagram formed in the way described above such that D contains at most k cells labeled v and the sign of D is defined to be $(-1)^{k-(\text{the number of } v\text{'s in } D)}$. We now define a sign reversing involution J on these combinatorial objects.

Given any diagram D, find the bottom rightmost cell c in D which can be filled with either an h or a v. If c contains v, then define $J(D)$ to be D with the entry in c changed to h. If c contains h and if D contains fewer than k cells with a v already, then define $J(D)$ to be D with the entry in c changed to h. Otherwise, set $J(D) = D$.

For example, the image of the object displayed earlier in the proof is sent to the object pictured below by J:

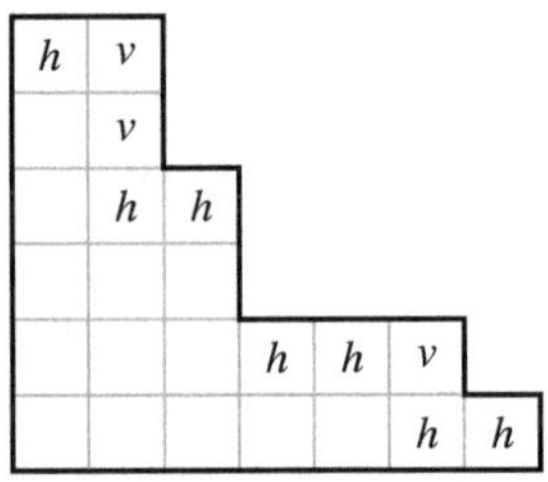

It is not difficult to see that J is an involution and, if D is not a fixed point, J changes the number of vs in D by 1 and therefore changes sign of D. The fixed points D under the involution J are such that D has k cells containing vs (and therefore has sign $+1$) and the bottom rightmost cell which can contain an h or v contains an h. Furthermore, by construction, there are $r-k$ cells filled with hs which form a skew row on the outside of μ and there are k cells filled with vs which form a skew column in λ.

The arrangement of hs and vs in such a fixed point can either be a single rim hook or a broken rim hook (a broken rim hook example is shown in the above diagrams). If the hs and vs form a single rim hook, then the placement of the hs and vs is completely determined since, by construction, two hs cannot appear in the same column and two v's cannot appear in the same row. An example:

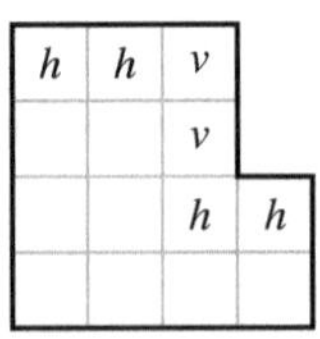

At this point in the proof, we have shown that $s_{(r-k,1^k)}s_\mu = \sum s_{sh(D)}$ where the sum runs over all fixed points D under J. From here, we will prove the Murnaghan–Nakayama rule with the help of the identity found in Exercise 2.2 in Chapter 2. Using this exercise, we have

$$p_r s_\mu = \left(\sum_{k=0}^{r-1} (-1)^k s_{(r-k,1^k)} \right) s_\mu = \sum_{k=0}^{r-1} (-1)^k s_{(r-k,1^k)} s_\mu = \sum \overline{\text{sign}}(D) s_{sh(D)},$$

where the sum runs over all fixed points D under J and where we define $\overline{\text{sign}}(D)$ to be $(-1)^{\text{the number of } v\text{'s in } D}$.

To finish the proof we apply a second sign reversing involution K. If D contains a single rim hook, then define $K(D) = D$. Otherwise, consider the bottom rightmost cell c in the second rim hook reading bottom to top. If c contains an h, then change this h to a v. If c contains a v, then change this v to an h. Thus K is the involution J with two exceptions: we consider the bottommost cell in the *second* bottommost rim hook, and we always change an h to a v with disregard to how many vs are present in D. It can be seen that K is an involution which changes the sign of D unless $K(D) = D$.

Since fixed points D make $sh(D)/\mu$ a single rim hook, and the number of vs in D is equal to (the number of rows spanned by $sh(D)/\mu) - 1$, we have that $\overline{\text{sign}}(D)$ is equal to the sign of the rim hook $sh(D)/\mu$. Therefore by summing the signs over all fixed points D under K, we have

$$p_r s_\mu = \sum_{\lambda/\mu \text{ is a rim hook with } r \text{ cells}} sgn(\lambda/\mu) s_\lambda,$$

as needed. □

The λ, μ entry χ_μ^λ of the p-to-s transition matrix can be found by iterating the Murnaghan–Nakayama rule. That is, if $\mu = (\mu_1, \dots, \mu_\ell)$, we write $p_\mu = p_{\mu_1} \cdots p_{\mu_\ell}$. We expand p_{μ_ℓ} as a sum of Schur functions using the Murnaghan–Nakayama rule to find shapes λ of size μ_ℓ. Using the Murnaghan–Nakayama rule again on each such choice of λ, we find all β such that β/λ is a rim hook in order to find the expansion of $p_{\mu_{\ell-1}} p_{\mu_\ell}$. We continue this process, successively adding rim hooks of lengths given by the parts of μ.

This suggests building a rim hook tableau of shape λ and content $\mu = (\mu_1, \dots, \mu_\ell)$, where a rim hook tableau of shape λ and content μ by filling the cells of the Young diagram of λ with rim hooks of lengths $\mu_1, \dots, \mu_\ell$ labeled with $1, \dots, \ell$ such that the removal of the last i rim hooks leaves the Young diagram of a smaller integer partition (see Exercise 2.14).

We define the sign of a rim hook tableau to be the product of the signs of the rim hooks that it contains and let χ_μ^λ the sum of the signs of all rim hook tableaux of shape λ with content μ. Then we have

$$p_\mu = \sum_{\lambda \vdash n} \chi_\mu^\lambda s_\lambda, \tag{6.3}$$

thereby giving the p-to-s transition matrix. For example, the two rim hook tableaux of shape $(3,2,1)$ with content $(3,2,1)$ are

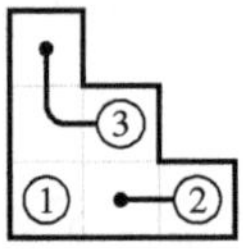

and

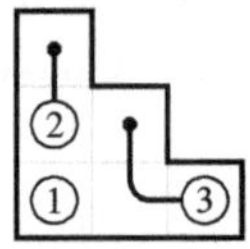

These rim hook tableaux have signs -1 and $+1$, and so $\chi^{(3,2,1)}_{(3,2,1)} = 0$.

The integers χ^{λ}_{μ} appear in other areas of mathematics. Most notably, χ^{λ}_{μ} is the value of the irreducible character corresponding to the integer partition λ on the conjugacy class corresponding to the integer partition μ, see [104]. This means that there are books which contain tables of these values (see [63]) and software packages such as Sage (see `sagemath.org`) and a "SF" Maple package written by John Stembridge that provide the ability to expand power symmetric functions in terms of the Schur basis. This provides a route to finding the constants c_λ in (6.1).

For example, consider the group G of rotations of the faces of a cube. By labeling the faces of the cube in this way

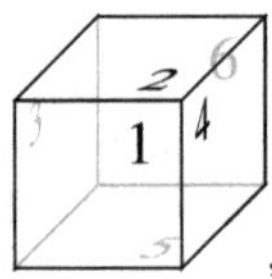

,

we see that G is generated by the quarter turns $(1\ 2\ 6\ 5)(3)(4)$ and $(1\ 4\ 6\ 3)(2)(5)$. From here it can be checked that the cycle index polynomial is

$$Z_G = \frac{1}{24}\left(p_{(1^6)} + 6p_{(4,1^2)} + 3p_{(2^2,1^2)} + 8p_{(3^2)} + 6p_{(2^3)}\right).$$

Either by using software, looking up tables of the values χ^{λ}_{μ}, or computing directly by hand using (6.3), we can expand each power symmetric function above into the Schur basis and then simplify to find

$$Z_G = s_{(6)} + s_{(4,2)} + s_{(3,1^3)} + 2s_{(2^3)}.$$

This last expression can be used to answer counting problems such as this: how many ways are there to color the faces of the cube such that three faces are red, two faces are blue, and one face is neon gray? This question is asking for the coefficient of $x_1^3x_2^2x_1$ in Z_G, and so we consider

$$\begin{aligned} Z_G|_{x_1^3x_2^2x_3} &= \left(s_{(6)} + s_{(4,2)} + s_{(3,1^3)} + 2s_{(2^3)}\right)\Big|_{x_1^3x_2^2x_3} \\ &= K_{(6),(3,2,1)} + K_{(4,2),(3,2,1)} + K_{(3,1^3),(3,2,1)} + 2K_{(2^3),(3,2,1)}, \end{aligned}$$

where $K_{\lambda,\mu}$ is the Kostka number counting the number of column strict tableaux of shape λ and type μ. Thus we have reduced the problem of counting colorings of the cube into the problem of counting column strict tableaux. It is not difficult to find these Kostka numbers by hand; when this is done, we find $Z_G|_{x_1^3x_2^2x_3} = 3$.

This leads to an interesting open problem first mentioned in [24]. For a fixed $\mu = (\mu_1, \ldots, \mu_6) \vdash 6$, is there a natural bijection between column strict tableaux of content μ and inequivalent colorings of the cube such that μ_i faces are color i such that each column strict tableau of shape $(6), (4,2)$, and $(3,1)$ is sent to one coloring and each column strict tableau of shape (2^3) is sent to two colorings?

We end this section with an example of a Schur function expansion of a cycle index polynomial which is more complicated than the cycle index polynomial coming from rotating the faces of the cube. Suppose we want to color the vertices of the complete binary tree of height 2 where the vertices are labeled as below:

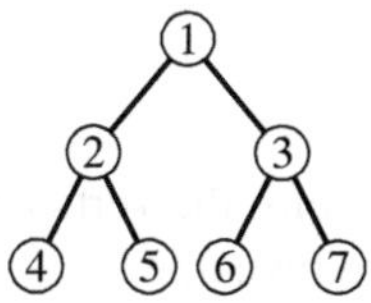

The group of symmetries G is generated by reflections about any one of the internal vertices of the tree. These are listed here:

(1) (4 5) (6 7) (4 5)(6 7)

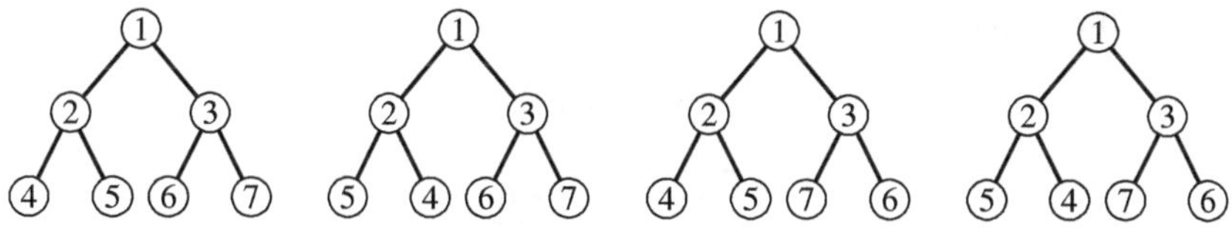

(2 3)(4 6)(5 7) (2 3)(4 7 5 6) (2 3)(4 6 5 7) (2 3)(4 7)(5 6)

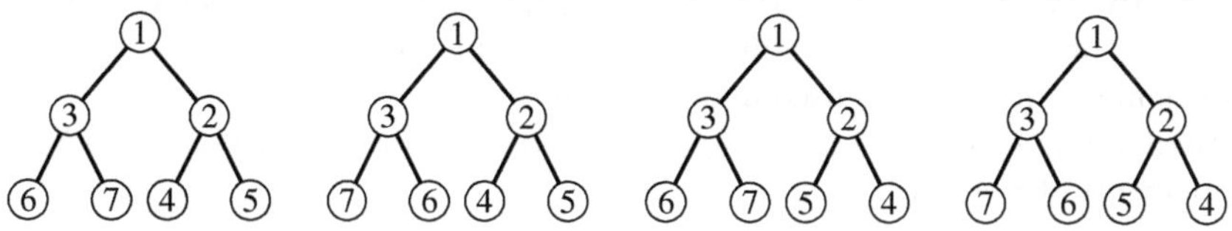

From these figures it can be readily seen that

$$Z_G = \frac{1}{8}\left(p_{(1^7)} + 2p_{(2,1^5)} + p_{(2^2,1^3)} + 2p_{(2^3,1)} + 2p_{(4,2,1)}\right).$$

Doing the calculations involving rim hook tableaux to expand this in terms of the Schur basis, it can be shown that Z_G is equal to

$$s_{(2^3,1)} + s_{(3,1^4)} + 3s_{(3,2,1^2)} + 3s_{(3,2^2)} + 2s_{(3^2,1)} + 2s_{(4,1^3)} \\ + 6s_{(4,2,1)} + 3s_{(4,3)} + 2s_{(5,1^2)} + 4s_{(5,2)} + 2s_{(6,1)} + s_{(7)}.$$

This expansion into the Schur basis is somewhat involved; it is still possible to compute the number of inequivalent colorings by hand. For example, if we want to use color 1 three times, color 2 two times, and color 3 two times, then we calculate

$$\begin{aligned} Z_G\big|_{x_1^3x_2^2x_3^2} &= K_{(2^3,1),(3,2^2)} + K_{(3,1^4),(3,2^2)} + 3K_{(3,2,1^2),(3,2^2)} + 3K_{(3,2^2),(3,2^2)} \\ &\quad + 2K_{(3^2,1),(3,2^2)} + 2K_{(4,1^3),(3,2^2)} + 6K_{(4,2,1),(3,2^2)} + 3K_{(4,3),(3,2^2)} \\ &\quad + 2K_{(5,1^2),(3,2^2)} + 4K_{(5,2),(3,2^2)} + 2K_{(6,1),(3,2^2)} + K_{(7),(3,2^2)} \\ &= 42. \end{aligned}$$

Exercises

6.1. How many ways are there to color the vertices of a cube such that four vertices are red, two are black, and two are invisible?

6.2. Let E be the set of two element subsets of $\{1,\ldots,n\}$. A simple graph on n vertices corresponds to a coloring of E which uses two different colors: a set $\{i,j\}$ is colored q if the edge between i and j appears in a simple graph and 1 if not. For example, the graph

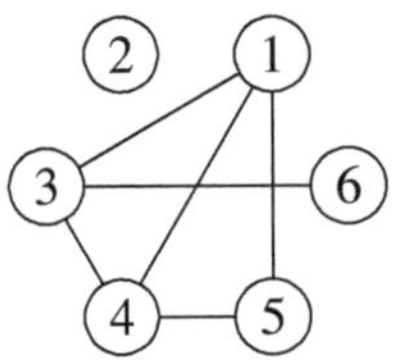

corresponds to coloring each of $\{1,3\},\{1,4\},\{1,5\},\{3,4\},\{3,6\}$, and $\{4,5\}$ with q and all other elements of E with 1. In this way, the number of edges in the graph is the number of times q is used in the coloring.

By defining $\sigma\{i,j\} = \{\sigma(i),\sigma(j)\}$ for all $\sigma \in S_n$, the symmetric group S_n acts on elements of E. Find

$$\sum_{\text{inequivalent 2 colorings } f \text{ of } E} q^{\text{the number of times color } q \text{ is used in } f}$$

when $n = 4$. Using the language of graph theory, we are finding

$$\sum_{\text{nonisomorphic simple graphs } g \text{ on 4 vertices}} q^{\text{the number edges in } g}.$$

6.3. Show that $Z_{G\times H} = Z_G Z_H$.

6.4. Let A_n be the alternating group (the subgroup of S_n containing elements with an even number of inversions), $\mathbf{Z}_n$ be the cyclic group of order n (the group generated by rotations of an n-sided regular polygon), and let D_n be the dihedral group of order $2n$ (the group generated by rotations and reflections of an n-sided regular polygon). Show that

$$\begin{aligned} Z_{S_n} &= h_n, \\ Z_{A_n} &= h_n + e_n, \\ Z_{\mathbb{Z}_n} &= \frac{1}{n}\sum_{i=1}^{n}\left(p_{n/\gcd(i,n)}\right)^{\gcd(i,n)}, \qquad \text{and} \\ Z_{D_n} &= \frac{1}{2}Z_{\mathbb{Z}_n} + \begin{cases} p_1 p_2^{(n-1)/2}/2 & \text{if } n \text{ is odd,} \\ \left(p_1^{n/2} + p_1^2 p_2^{(n-2)/2}\right)/4 & \text{if } n \text{ is even.} \end{cases} \end{aligned}$$

Solutions

6.1 By labeling the vertices of the cube in the following way

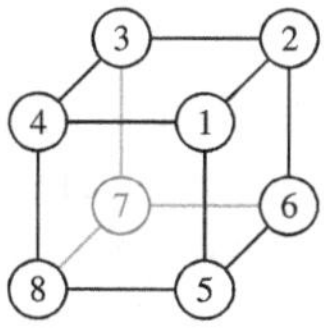

we see that the desired group is generated by the quarter turns $(1\ 2\ 6\ 5)(4\ 3\ 7\ 8)$ and $(1\ 2\ 3\ 4)(5\ 6\ 7\ 8)$. From here it can be checked that the cycle index polynomial is

$$\frac{1}{24}\left(p_1^8 + 9p_2^4 + 8p_1^2p_3^2 + 6p_4^2\right).$$

We want the coefficient of $x_1^4x_2^2x_3^2$; such a term can only come from p_1^8 or p_2^4. Therefore the answer is $(\binom{8}{4,2,2} + 9\binom{4}{2,1,1})/24 = 22$.

6.2 The set E is $\{\{1,2\},\{1,3\},\{1,4\},\{2,3\},\{2,4\},\{3,4\}\}$. Calling these elements $e_1,\ldots,e_6$, we see that the action of $(1\ 2)$ on E corresponds to sending e_1 to e_1, e_2 to e_3, e_4 to e_5, and e_6 to e_6. Therefore the action of $(1\ 2)$ corresponds to applying the permutation $(1)(2\ 4)(3\ 5)(6)$ to the subscripts of $e_1,\ldots,e_6$. Similarly, $(2\ 3)$ and $(3\ 4)$ correspond to the permutations $(1\ 2)(3)(4)(5\ 6)$ and $(1)(2\ 3)(4\ 5)(6)$ on the subscripts of $e_1,\ldots,e_6$.

Since S_4 is generated by $(1\ 2), (2\ 3)$, and $(3\ 4)$, the group G of symmetries of E under the action of S_4 is generated by $(1)(2\ 4)(3\ 5)(6)$, $(1)(2\ 3)(4\ 5)(6)$, and $(1\ 2)(3)(4)(5\ 6)$. From here we calculate the cycle index polynomial for G to be

$$\frac{1}{24}\left(p_1^6 + 9p_1^2p_2^2 + 8p_3^2 + 6p_2p_4\right) = s_{(1^6)} + s_{(2^2,1^2)} + s_{(2^3)} + s_{(3^2)} + s_{(4,2)} + s_{(6)}.$$

The desired generating function can be found by taking $x_1 = q$, $x_2 = 1$, and all other variables x_i to equal 0. When this is done, the power symmetric polynomial p_i becomes $q^i + 1$. Using this in the cycle index polynomial, our answer is

$$\frac{1}{24}\left((q+1)^6+9(q+1)^2(q^2+1)^2+8(q^2+1)^3+6(q^2+1)(q^4+1)\right).$$

When this polynomial is expanded, we find

$$1+q+2q^2+3q^3+2q^4+q^5+q^6.$$

6.3 We have

$$Z_{G\times H} = \frac{1}{|G||H|}\sum_{(g,h)\in G\times H} p_{\lambda(g)\lambda(h)} = \frac{1}{|G||H|}\sum_{g\in G}\sum_{h\in H} p_{\lambda(g)}p_{\lambda(h)} = Z_G Z_H.$$

6.4 The cycle index polynomial for the symmetric group is

$$Z_{S_n} = \frac{1}{n!}\sum_{\sigma\in S_n} p_{\lambda(\sigma)} = \frac{1}{n!}\sum_{\lambda\vdash n} p_\lambda \frac{n!}{z_\lambda}$$

by Theorem 1.10. This is equal to h_n by Theorem 2.11.

When written in one-line notation, the cycle $(1\ 2\ \cdots\ k)$ is equal to $2\ \cdots\ k\ 1$ and so this permutation has $k-1$ inversions. More generally, if $\lambda = (\lambda_1,\ldots,\lambda_\ell)$ is an integer partition of n, then the permutation $(1\ \cdots\ \lambda_1)(\lambda_1+1\ \cdots\ \lambda_1+\lambda_2)\cdots$ has $n-\ell(\lambda)$ inversions. Inversions are constants on conjugacy classes, so the number of inversions of $\sigma\in S_n$ is $(-1)^{n-\ell(\lambda(\sigma))}$. This helps us to find the cycle index polynomial for the alternating group A_n; the cycle index polynomial Z_{A_n} is

$$\frac{2}{n!}\sum_{\sigma\in A_n} p_{\lambda(\sigma)} = \frac{2}{n!}\sum_{\lambda\in S_n}\frac{1+(-1)^{n-\ell(\lambda(\sigma))}}{2}p_{\lambda(\sigma)} = \frac{1}{n!}\sum_{\lambda\vdash n}\left(1+(-1)^{n-\ell(\lambda)}\right)p_\lambda.$$

This is $h_n + e_n$ by Theorem 2.12.

The group $\mathbf{Z}_n$ is generated by $(1\ 2\ \cdots\ n)$. Since the ith power of this cycle splits into $\gcd(i,n)$ cycles of size $n/\gcd(i,n)$, the cycle index polynomial $Z_{\mathbf{Z}_n}$ is

$$\frac{1}{n}\sum_{g\in\mathbf{Z}_n} p_{\lambda(g)} = \frac{1}{n}\sum_{i=1}^{n}\left(p_{n/\gcd(i,n)}\right)^{\gcd(i,n)}.$$

The group D_n is generated by the rotation $(1\ 2\ \cdots\ n)$ and a reflection r of the n-sided regular polygon about a fixed axis. Therefore D_n is the disjoint union of $\mathbf{Z}_n$ and $r\mathbf{Z}_n$ and so

$$Z_{D_n} = \frac{1}{2n}\sum_{g\in D_n} p_{\lambda(g)} = \frac{1}{2n}\sum_{g\in\mathbf{Z}_n} p_{\lambda(g)} + \frac{1}{2n}\sum_{g\in r\mathbf{Z}_n} p_{\lambda(g)} = \frac{1}{2}Z_{\mathbf{Z}_n} + \frac{1}{2n}\sum_{g\in r\mathbf{Z}_n} p_{\lambda(g)}$$

If n is odd, the reflection r can equal

$$(1\ \ (n-1))(2\ \ (n-2))\cdots((n-1)/2\ \ (n+1)/2).$$

In this case every element in $r\mathbf{Z}_n$ has cycle type $(2^{(n-1)/2},1)$, and so

$$\frac{1}{2n}\sum_{g\in r\mathbf{Z}_n} p_{\lambda(g)} = \frac{1}{2n}\left(np_1p_2^{(n-1)/2}\right) = p_1p_2^{(n-1)/2}/2.$$

If n is even, the reflection r can equal

$$(1\ \ (n-1))(2\ \ (n-2))\cdots((n/2-1)\ \ (n/2+1)).$$

In this case, half of the elements in $r\mathbf{Z}_n$ have cycle type $(2^{(n-2)/2},1^2)$ and half of the elements have cycle type $(2^{n/2})$. These two cases arise depending on whether the reflection fixes zero or two vertices of the regular polygon. Therefore we have

$$\frac{1}{2n}\sum_{g\in r\mathbf{Z}_n} p_{\lambda(g)} = \frac{1}{2n}\left(\frac{n}{2}p_1^2p_2^{(n-2)/2}+\frac{n}{2}p_2^{n/2}\right) = \left(p_1^{n/2}+p_1^2p_2^{(n-2)/2}\right)/4.$$

Notes

Rudimentary forms of Pólya's enumeration theorem were known to Burnside in 1900. Redfield proved the first version of what is now known as Pólya's Enumeration Theorem [97]. For a modern account of Redfield's work, see [56, 57, 58]. Pólya gave the first modern formulation of Pólya's enumeration theorem [95].

The Murnaghan–Nakayama rule was proved independently by D.F. Murnaghan [91] and T. Nakayama [92, 93].

Chapter 7
Consecutive Patterns

This chapter applies the machinery we have developed in the previous chapters to find generating functions for the distribution of consecutive patterns in permutations and words.

The study of patterns in permutations and words has been a very active area of research in recent years, with explosive growth in the years since 1992. A systematic treatment of this subject is found in Kitaev's *Patterns in permutations and words* [71].

7.1 Nonoverlapping Consecutive Patterns

Given any sequence $\sigma = \sigma_1 \cdots \sigma_j$ of distinct integers, let $\mathrm{red}(\sigma)$ be the reduced permutation found by replacing the ith largest integer in σ with i. Given a permutation $\tau \in S_j$, we say that $\sigma = \sigma_1 \cdots \sigma_n \in S_n$ has a (consecutive) τ-match ending at place m if $\mathrm{red}(\sigma_{m-j+1} \cdots \sigma_m) = \tau$.

For example, if $\tau = 1\ 3\ 2$, then the permutation

$$\sigma = 5\ 1\ 8\ 6\ 12\ 7\ 2\ 3\ 11\ 4\ 10\ 9$$

has exactly 4 τ-matches. These τ-matches end at places $4, 6, 10$, and 12. There are 2 nonoverlapping τ-matches in this permutation.

Theorem 7.1. *For any permutation $\tau \in S_j$,*

$$\sum_{n=0}^{\infty} \frac{z^n}{n!} \sum_{\sigma \in S_n} x^{\tau\text{-}nlap(\sigma)} = \frac{A(z)}{(1-x) + x(1-z)A(z)},$$

where τ-$nlap(\sigma)$ is the maximum number of nonoverlapping τ-matches in $\sigma \in S_n$ and $A(z) = \sum_{n=0}^{\infty} z^n |\{\sigma \in S_n$ does not have a τ-match $\}|/n!$.

A. Mendes, J. Remmel, *Counting with Symmetric Functions*, Developments in Mathematics 43, DOI 10.1007/978-3-319-23618-6_7

Proof. For $\tau \in S_j$ and positive integers $\ell_1, \dots, \ell_m$, let $S_{(j,\ell_1,\dots,\ell_m)}$ be the set of permutations $\sigma \in S_{j+\ell_1+\cdots+\ell_m}$ such that σ has exactly $(m+1)$ τ-matches which end at places j, $j+\ell_1$, $\dots$, and $j+\ell_1+\cdots+\ell_m$. Furthermore, let J_τ be the set of all lists of the form $(j,\ell_1,\dots,\ell_m)$ such that

1. the set $S_{(j,\ell_1,\dots,\ell_m)}$ is nonempty,
2. the integers $\ell_1,\dots,\ell_m$ are all less than j, and
3. the inequality $j \le \ell_i + \ell_{i+1}$ holds for all consecutive integers ℓ_i, ℓ_{i+1} in the list.

For example, if $\tau = 432165$, then the permutation

$$\sigma = \underbrace{4\ 3\ 2\ 1\ \overbrace{8\ 7}}\overbrace{\ 6\ 5\ 13\ \underbrace{12}}\underbrace{\ 11\ 10\ 9\ 15\ 14}$$

has three τ-matches and they end at places $6, 10$, and 15. Therefore $\sigma \in S_{(6,4,5)}$. We now know that $(6,4,5) \in J_\tau$ since $S_{(6,4,5)}$ is nonempty and the integers in the list $(6,4,5)$ satisfy the appropriate inequalities.

The inequalities in the second and third conditions on the list $(j,\ell_1,\dots,\ell_m)$ in the definition of J_τ are designed so that consecutive τ-matches must overlap and no one integer is a part of more than two τ-matches. In pictures, a permutation $\sigma \in S_{(j,\ell_1,\dots,\ell_m)}$ for some $(j,\ell_1,\dots,\ell_m) \in J_\tau$ looks like this:

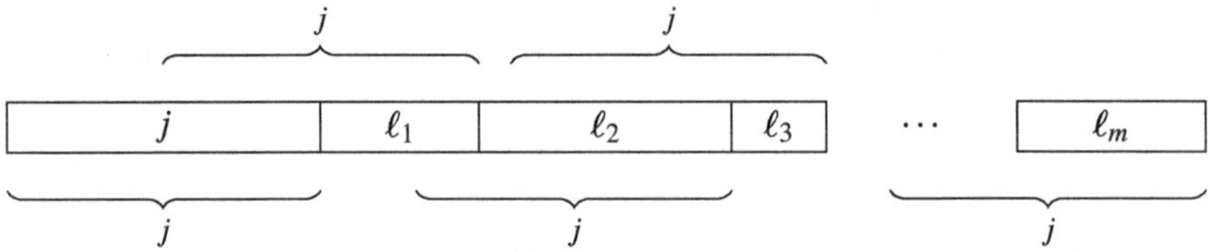

Given a permutation $\sigma \in S_{(j,\ell_1,\dots,\ell_m)}$ for some $(j,\ell_1,\dots,\ell_m) \in J_\tau$, we can determine the integers $\ell_1,\dots,\ell_m$ by finding the ending places of the τ-matches in σ.

Define a homomorphism φ by $\varphi(e_n) = (-1)^{n-1} f(n)/n!$ where $f(1) = 1$ and

$$f(n) = (1-x) \sum_{\substack{(j,\ell_1,\dots,\ell_m)\in J_\tau \\ j+\ell_1+\cdots+\ell_m=n}} (-1)^{m+1} |S_{(j,\ell_1,\dots,\ell_m)}|$$

for $n \ge 2$.

Applying φ to $n!h_n$ gives (3.3), from which we create combinatorial objects by first selecting a brick tabloid $T \in B_{\lambda,(n)}$ for some $\lambda \vdash n$ and then using the multinomial coefficient in (3.3) to assign a disjoint subset to each brick such that the union of these subsets is $\{1,\dots,n\}$.

All that remains to be used in (3.3) is the product $f(\lambda_1)f(\lambda_2)\cdots$. Since $f(1) = 1$, add no extra weight to a brick of length 1. Otherwise, for bricks of length $n \ge 2$, the function $f(n)$ tells us to

1. select a permutation $\sigma \in S_{(j,\ell_1,\dots,\ell_m)}$ for some $(j,\ell_1,\dots,\ell_m) \in J_\tau$,
2. rearrange the subset assigned to this brick of length n to create a distinct list of integers which reduce to σ,
3. place a -1 in each cell which ends a τ-match, and
4. Either keep the first -1 in each brick unchanged or change it to an x.

For example, one combinatorial object created in this way when $\tau = 132$ is

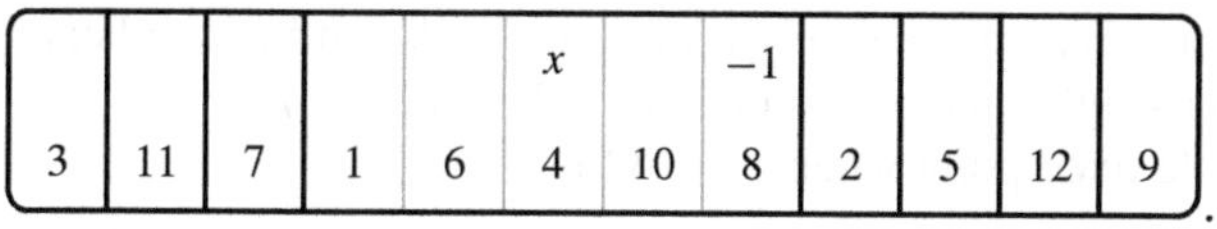

Let the weight be the product of all xs and -1s. Then $n!\varphi(h_n)$ is the weighted sum over all possible combinatorial objects.

Define an involution in by first scanning the bricks from *right to left* looking for the first instance of one of these four situations:

1. Exactly j consecutive bricks of length 1 that form a τ-match.
2. A brick of length j with a weight of -1.
3. A brick containing an element in $S_{(j,\ell_1,\ldots,\ell_m)}$ for some $(j,\ell_1,\ldots,\ell_m) \in J_\tau$ to the left of ℓ_{m+1} bricks of length 1 such that combining the bricks would create an element in $S_{(j,\ell_1,\ldots,\ell_{m+1})}$ for $(j,\ell_1,\ldots,\ell_{m+1}) \in J_\tau$.
4. A brick of length longer than j (which must have a final cell containing a -1).

If situation 1 is found first, combine the j consecutive bricks into one brick of length j and place a -1 in the terminal cell. For example, the combinatorial object shown earlier in this proof should be changed to

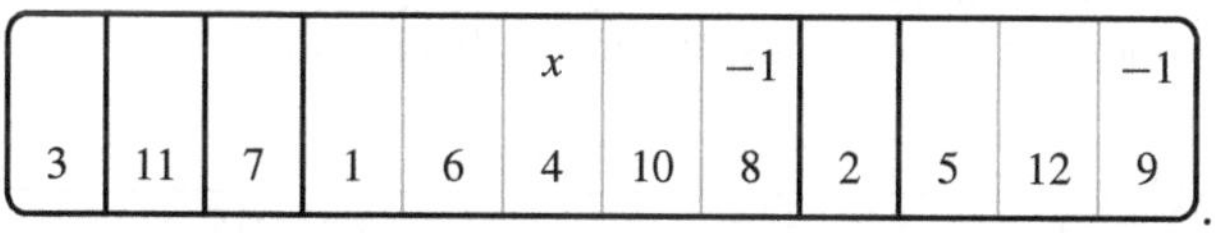

This changes every situation 1 into a situation 2, and so if a situation 2 is found first we define our involution to undo this operation.

If a situation 3 is found, combine the brick containing the element of $S_{(j,\ell_1,\ldots,\ell_m)}$ with the ℓ_{m+1} bricks to its right and place a -1 in the terminal cell. For example, this would change

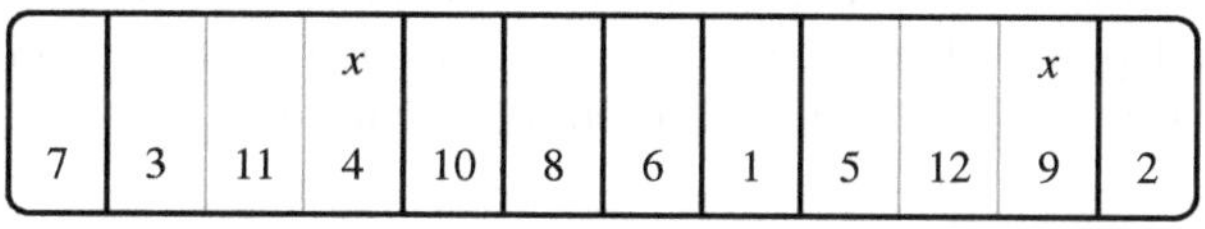

into the combinatorial object

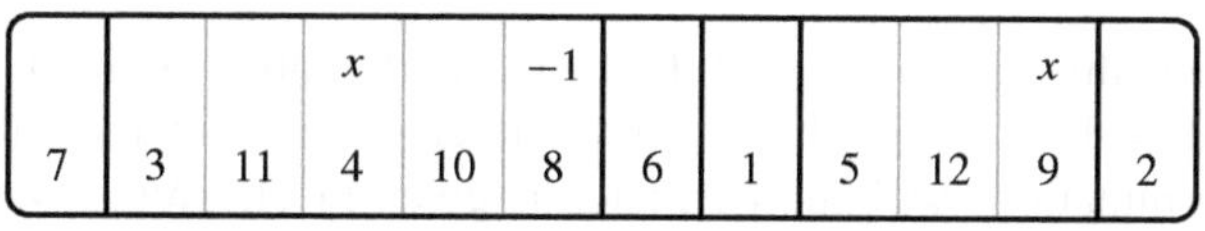

in the case of $\tau = 132$. This changes every situation 3 into a situation 4, and so if a situation 4 is found first we define our involution to undo this operation. This means we will break of the ℓ_m cells in a brick of length longer than j.

The fixed points under this involution cannot have any of the four situations listed above. This means that no j consecutive bricks of length 1 can form a τ-match, every brick of length j must have a weight of x, no brick can be lengthened using bricks of length 1 to its right, and bricks must have length 1 or j.

One possible fixed point when $\tau = 132$ is

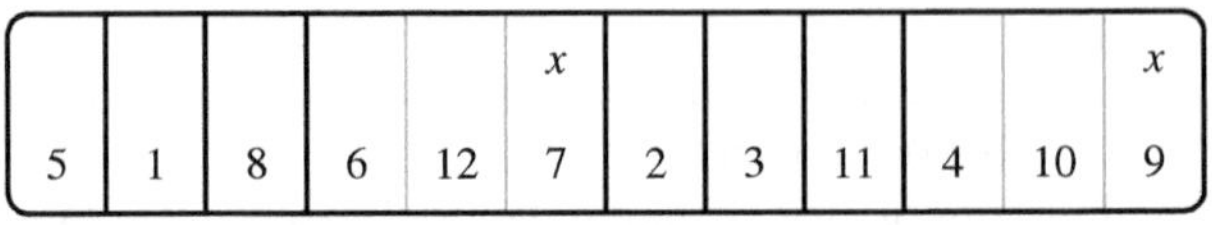

Reading right to left, we greedily assign one power of x for each nonoverlapping τ-match in a fixed point. This means $n!\varphi(h_n) = \sum_{\sigma\in S_n} x^{\text{nlap}(\sigma)}$. Applying φ to Theorem 2.5, we have

$$\sum_{n=0}^{\infty}\frac{z^n}{n!}\sum_{\sigma\in S_n} x^{\tau\text{-nlap}(\sigma)} = \frac{1}{1-z+\sum_{n=2}^{\infty}\varphi(e_n)(-z)^n}$$

$$= \frac{1}{1-z-(1-x)\sum_{n=2}^{\infty}\frac{z^n}{n!}\sum_{\substack{(j,\ell_1,\ldots,\ell_m)\in J_\tau\\ j+\ell_1+\cdots+\ell_m=n}}(-1)^{m+1}|S_{(j,\ell_1,\ldots,\ell_m)}|}. \tag{7.1}$$

Taking $x=0$ in this equation gives the generating function $A(z)$ for those permutations in S_n without any τ-matches. When $x=0$ we find

$$A(z) = \frac{1}{1-z-\sum_{n=2}^{\infty}\frac{z^n}{n!}\sum_{\substack{(j,\ell_1,\ldots,\ell_m)\in J_\tau\\ j+\ell_1+\cdots+\ell_m=n}}(-1)^{m+1}|S_{(j,\ell_1,\ldots,\ell_m)}|}, \tag{7.2}$$

which may be used to simplify (7.1), thereby giving the expression in the statement of the theorem. □

As a first example of generating functions given by Theorem 7.1, consider the generating function for the number of nonoverlapping descents, that is, the number of nonoverlapping 21-matches. There is only one permutation in S_n without a 21-match, so $A(z) = \sum_{n=0}^{\infty} z^n/n! = e^z$. Therefore we have

$$\sum_{n=0}^{\infty}\frac{z^n}{n!}\sum_{\sigma\in S_n} x^{21\text{-nlap}(\sigma)} = \frac{e^z}{(1-x)+x(1-z)e^z}.$$

Generalizing this, if $\tau = j\cdots\ 2\ 1$, then a permutation without a τ-match is a permutation without a $(j-1)$-descent. Therefore the function $A(z)$ giving the number of permutations without a τ-match is given in Theorem 3.4. Using this in Theorem 7.1, we find that for $j\geq 1$,

$$\sum_{n=0}^{\infty}\frac{z^n}{n!}\sum_{\sigma\in S_n} x^{j\cdots 21\text{-nlap}(\sigma)} = \frac{j}{jx(1-z)+(1-x)\left((1-\zeta^{j-1})e^{\zeta z}+\cdots+(1-\zeta)e^{\zeta^{j-1}z}\right)},$$

where $\zeta = e^{2\pi i/j}$ is a primitive j^{th} root of unity.

Another route to finding $A(z)$ is through (7.2). Consider the case of $\tau = 132$. The only way to have overlapping τ-matches is if exactly one integer overlaps, so the only possible lists in J_τ are of the form $(3,2,\ldots,2,2)$.

We claim that the number of $\sigma = \sigma_1 \cdots \sigma_{2n+3} \in S_{(3,2,\ldots,2,2)}$ is $(2n+1)(2n-1)\cdots 3 \cdot 1$. Clearly this is the case when $n=0$—there is only one permutation in $S_{(3)}$. For larger n there must be a τ-match starting in position $2i+1$ for all $i = 0,\ldots,n$, so the integer σ_{2i+1} must be smaller than both σ_{2i+2} and σ_{2i+3} for $i = 0,\ldots,n$. It follows that $\sigma_1 = 1$, $\sigma_3 = 2$, and σ_2 can be any one of the $2n+1$ elements in $\{3,\ldots,2n+3\}$. We have proved the claim by induction since the reduced permutation $\sigma_3 \cdots \sigma_{2n+3}$ can be any one of the $(2n-1)\cdots 3 \cdot 1$ elements in $S_{(3,2,\ldots,2)}$.

Therefore when $\tau = 132$, equation (7.2) tells us that

$$\begin{aligned} A(z) &= \frac{1}{1 - z - \sum_{n=0}^{\infty} (-1)^{n+1} (2n+1) \cdots 3 \cdot 1 \frac{z^{2n+3}}{(2n+3)!}} \\ &= \frac{1}{1 - z - \sum_{n=0}^{\infty} \left(-\frac{1}{2}\right)^{n+1} \frac{z^{2n+3}}{(2n+3)(n+1)!}} \\ &= \left(1 - \int e^{-z^2/2} dz\right)^{-1}. \end{aligned} \tag{7.3}$$

Using this $A(z)$ in Theorem 7.1 gives

$$\sum_{n=0}^{\infty} \frac{z^n}{n!} \sum_{\sigma \in S_n} x^{132\text{-nlap}(\sigma)} = \left(1 - xz + (x-1) \int e^{-z^2/2} dz\right)^{-1}.$$

In section 3.2 we adapted homomorphisms of the form $\varphi(e_n) = (-1)^{n-1} f(n)/n!$ to include a power of q to register inversions, to count common descents in two or more permutations, and to find analogous results about words. The proof of Theorem 7.1 relies on a generating function of this form and so Theorem 7.1 can be generalized in these ways as well. We record these results in our Theorems 7.2, 7.3, and 7.4.

Additionally, we can generalize Theorem 7.1 by allowing more than just one permutation τ to register a pattern match in a permutation. Let $\mathcal{T}$ be a set of permutations $\tau \in S_j$ and define a permutation $\sigma = \sigma_1 \cdots \sigma_n \in S_n$ to have a $\mathcal{T}$-match ending at place m if $\text{red}(\sigma_{m-j+1} \cdots \sigma_m) \in \mathcal{T}$.

For example, if $\mathcal{T} = \{213, 312\}$, then the permutation

$$\sigma = \underbrace{9\ 2\ \overbrace{11}}\ \overbrace{3\ \underbrace{8}}\ \underbrace{1\ 5}\ 6\ \overbrace{12\ 4\ 7}\ 10$$

has four $\mathcal{T}$-matches, ending at places 3,5,7, and 11. This choice of $\mathcal{T}$ counts the number of valleys.

Theorem 7.2. *For any subset $\mathcal{T}$ of S_j,*

$$\sum_{n=0}^{\infty} \frac{z^n}{n!} \sum_{\sigma \in S_n} x^{\mathcal{T}\text{-}nlap(\sigma)} q^{inv(\sigma)} = \frac{A_q(z)}{(1-x)+x(1-z)A_q(z)}$$

where $\mathcal{T}$-$nlap(\sigma)$ is the maximum number of nonoverlapping $\mathcal{T}$-matches in $\sigma \in S_n$ and $A_q(z) = \sum_{n=0}^{\infty} \frac{z^n}{n!} \sum_{\sigma \in S_n \text{ does not have a } \mathcal{T}\text{-match}} q^{inv(\sigma)}$.

Proof. The proof updates the proof of Theorem 7.1 for $\mathcal{T}$-matches while including the ideas in the proof of 3.7 to keep track of inversions.

Extending the definitions of J_τ and $S_{(j,\ell_1,\ldots,\ell_m)}$, we let $S_{(j,\ell_1,\ldots,\ell_m)}$ the set of permutations of length $j+\ell_1+\cdots+\ell_m$ which have $(m+1)$ $\mathcal{T}$-matches which end at places j, $j+\ell_1$, and so on. We let $J_\mathcal{T}$ the set of lists $(j,\ell_1,\ldots,\ell_m)$ defined in the same way as J_τ.

The appropriate homomorphism for this situation is $\varphi(e_n) = (-1)^{n-1} f(n)/[n]_q!$ where $f(1) = 1$ and

$$f(n) = (1-x) \sum_{\substack{(j,\ell_1,\ldots,\ell_m) \in J_\mathcal{T} \\ j+\ell_1+\cdots+\ell_m = n}} (-1)^{m+1} \sum_{\sigma \in S_{(j,\ell_1,\ldots,\ell_m)}} q^{\mathrm{inv}(\sigma)}$$

for $n \geq 2$. Then we have $[n]_q!\varphi(h_n) = \sum_{\lambda \vdash n} \begin{bmatrix} n \\ \lambda \end{bmatrix}_q |B_{\lambda,(n)}| f(\lambda_1) f(\lambda_2) \cdots$.

From this expression we build the same combinatorial objects as found in the proof of Theorem 7.1 with two differences: we consider $\mathcal{T}$-matches instead of τ-matches, and we have a power of q in each cell counting the integers to the right which are smaller. One object when $\mathcal{T} = \{213, 312\}$ is

					x		-1				
q^2	q^9	q^5	q^2	q^0	q^5	q^2	q^2	q^0	q^0	q^1	q^1
3	11	7	4	1	10	6	8	2	5	12	9

.

The involution in the proof of Theorem 7.1, provided we scan from right to left looking for $\mathcal{T}$-matches instead of τ-matches, does not rearrange the integers in the permutation. The powers of q and the number of inversions are unchanged by this operation, and so fixed points correspond to $\sum_{\sigma \in S_n} x^{\mathcal{T}\text{-nlap}(\sigma)} q^{\mathrm{inv}(\sigma)}$. The generating function in the statement of the theorem now follows from applying φ to Theorem 2.5 and simplifying in the same manner as in the proof of Theorem 7.1. □

As an example of Theorem 7.2, suppose we would like the generating function for the number of nonoverlapping valleys in S_n. A permutation has no valleys—that is, a permutation has no $\mathcal{T}$-matches when $\mathcal{T} = \{213, 312\}$—if the permutation is of the form

$$\sigma_1 < \sigma_2 < \cdots < \sigma_m = n > \sigma_{m+1} > \cdots > \sigma_n$$

for some integer m. There are $\binom{n-1}{m-1}$ choices for such a permutation, so summing over all m gives that the number of permutations without a valley is $\sum_{m=1}^{n} \binom{n-1}{m-1} = 2^{n-1}$.

The generating function for the number of permutations without any valleys is $A(z) = 1 + \sum_{n=1}^{\infty} 2^{n-1} z^n/n! = (e^{2z}+1)/2$. Taking $q = 1$ and using this $A(z)$ in Theorem 7.2 we find

$$\sum_{n=0}^{\infty} \frac{z^n}{n!} \sum_{\sigma \in S_n} x^{\text{\# nonoverlapping valleys in } \sigma} = \frac{e^{2z}+1}{2(1-x)+x(1-z)(e^{2z}+1)}. \tag{7.4}$$

A common $\mathcal{T}$-match in two permutations $\sigma, \rho \in S_n$ is an index i such that both σ and ρ have a $\mathcal{T}$-match ending at place i. If we let com-$\mathcal{T}$-nlap(σ,ρ) the number of nonoverlapping common $\mathcal{T}$-matches, then combining the methods in the proofs of Theorem 7.2 and 3.8 in a straightforward way gives us Theorem 7.3:

Theorem 7.3. *For any subset $\mathcal{T}$ of S_j,*

$$\sum_{n=0}^{\infty} \frac{z^n}{(n!)^2} \sum_{\sigma,\rho \in S_n} x^{\mathit{com}\text{-}\mathcal{T}\text{-}\mathit{nlap}(\sigma,\rho)} = \frac{A(z)}{(1-x)+x(1-z)A(z)},$$

where $A(z) = \sum_{n=0}^{\infty} |\{\sigma,\rho \text{ have no common } \mathcal{T} \text{ matches}\}| z^n/(n!)^2$.

The generalization of Theorem 7.1 for words takes a bit more care since the idea of a pattern match is slightly different from that for permutations. If $\mathcal{V}$ is a subset of $\{0,\ldots,k-1\}_j^*$, then we say that $w = w_1 \cdots w_n \in \{0,\ldots,k-1\}_n^*$ has a $\mathcal{V}$-match ending at place i if $w_{i-j+1} \cdots w_i \in \mathcal{V}$.

Unlike permutations, we do not consider reducing the word before determining if it is an element of $\mathcal{V}$. Also unlike permutations, it may be impossible for two words to have an overlapping $\mathcal{V}$-match. For example, if $\mathcal{V} = \{10,20\}$, then no two words in $\{0,1,2\}_n^*$ can have an overlapping $\mathcal{V}$-match. These type of matches are sometimes called exact $\mathcal{V}$-matches to distinguish them from the case where we allow reductions.

Theorem 7.4. *For any subset $\mathcal{V}$ of $\{0,\ldots,k-1\}_j^*$,*

$$\sum_{n=0}^{\infty} \frac{z^n}{n!} \sum_{w \in \{0,\ldots,k-1\}_n^*} x^{\mathcal{V}\text{-}\mathit{nlap}(w)} = \frac{A(z)}{(1-x)+x(1-kz)A(z)},$$

where $\mathcal{V}$-nlap(w) is the number of nonoverlapping $\mathcal{V}$-matches and

$$A(z) = \sum_{n=0}^{\infty} |\{w \in \{0,\ldots,k-1\}_n^* \textit{ does not have a } \mathcal{V}\textit{-match}\}| z^n.$$

Proof. Modifying the proof of Theorem 7.1 for words, we let $W_{(j,\ell_1,\ldots,\ell_m)}$ the set of words in $\{0,\ldots,k-1\}_{j+\ell_1+\cdots+\ell_m}^*$ which has exactly $(m+1)$ $\mathcal{V}$-matches which end at places j, $j+\ell_1$, $j+\ell_1+\ell_2$, and so on.

Let $J_{\mathcal{V}}$ be the set of all lists of the form $(j,\ell_1,\ldots,\ell_m)$ such that

1. the set $W_{(j,\ell_1,\ldots,\ell_m)}$ is nonempty,
2. the integers $\ell_1,\ldots,\ell_m$ are all less than j, and
3. the inequality $j \le \ell_i + \ell_{i+1}$ holds for all consecutive integers ℓ_i, ℓ_{i+1} in the list.

The second and third conditions here are the same as in the proof of Theorem 7.1, so the $\mathcal{V}$-matches must overlap. This means that if the set $\mathcal{V}$ does not allow overlapping $\mathcal{V}$-matches, then $J_{\mathcal{V}}$ will contain only the list (j).

The appropriate ring homomorphism is defined by $\varphi(e_n) = (-1)^{n-1} f(n)$ where $f(1) = k$ and

$$f(n) = (1-x) \sum_{\substack{(j,\ell_1,\ldots,\ell_m)\in J_{\mathcal{V}} \\ j+\ell_1+\cdots+\ell_m=n}} (-1)^{m+1} |W_{(j,\ell_1,\ldots,\ell_m)}|.$$

Applying φ to h_n gives $\sum_{\lambda\vdash n} |B_{\lambda,(n)}| f(\lambda_1) f(\lambda_2)\cdots$, from which we create objects like these (we have taken $\mathcal{V} = \{010, 020\}$ in this example):

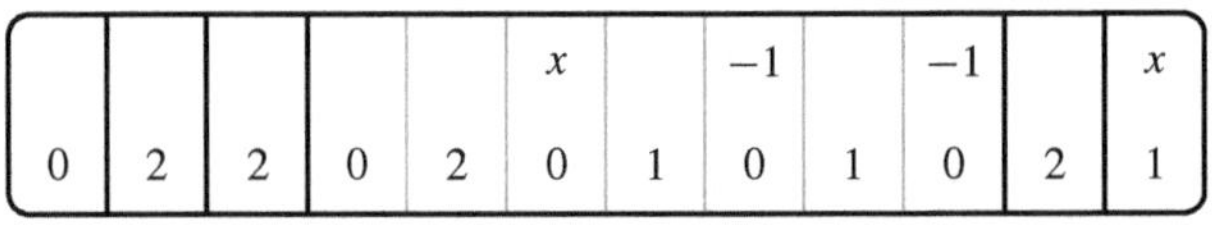

Specifically, in the same manner as before, we place a choice of $\{0,\ldots,k-1\}$ in each brick of length 1 and fill longer bricks with a choice from $W_{(j,\ell_1,\ldots,\ell_m)}$. On these longer brick we place a -1 over each place which ends a $\mathcal{V}$-match and either keep the first -1 in place or change it to an x.

Define the same sort of involution as in the proof of Theorem 7.1, scanning from right to left looking for either j bricks of length 1 to combine into one brick of length j, a brick of length j to break into j bricks of length 1, a brick of length at least j which can be combined with bricks of length 1 to its right, or a brick of length longer than length j which can have the trailing ℓ_m bricks broken off into ℓ_m bricks of length 1.

The weighted sum over all fixed points is $\sum_{w\in\{0,\ldots,k-1\}_n^*} x^{\mathcal{V}\text{-nlap}(w)}$. Applying φ to Theorem 2.5 gives

$$\sum_{n=0}^{\infty} \frac{z^n}{n!} \sum_{w\in\{0,\ldots,k-1\}_n^*} x^{\mathcal{V}\text{-nlap}(w)}$$
$$= \frac{1}{1 - mz + (1-x)\sum_{n=2}^{\infty} z^n \sum_{\substack{(j,\ell_1,\ldots,\ell_m)\in J_{\mathcal{V}} \\ j+\ell_1+\cdots+\ell_m=n}} (-1)^{m+1} |W_{(j,\ell_1,\ldots,\ell_m)}|}. \tag{7.5}$$

Taking $x = 0$ gives

$$A(z) = \frac{1}{1 - mz + \sum_{n=2}^{\infty} z^n \sum_{\substack{(j,\ell_1,\ldots,\ell_m)\in J_{\mathcal{V}} \\ j+\ell_1+\cdots+\ell_m=n}} (-1)^{m+1} |W_{(j,\ell_1,\ldots,\ell_m)}|}, \tag{7.6}$$

giving an alternative way to find $A(z)$ which can sometimes be useful in computations. Using this expression for $A(z)$ in (7.5) gives the generating function in the statement of the theorem. □

For an example of Theorem 7.4, consider counting the number of nonoverlapping $\mathcal{V} = \{0100\}$ matches in words $w \in \{0,1\}_n^*$. Every list in $J_{\mathcal{V}}$ is of the form $(4,3,\ldots,3)$

and there is only one element in $W_{(4,3,\dots,3)}$, the word $0100100100\cdots100$. Using (7.6) we find that $A(z) = \left(1-2z+\sum_{n=0}^{\infty}(-1)^{3k+4}z^{3k+4}\right)^{-1} = (1-2z+z^4/(1+z^3))^{-1}$. Putting this into Theorem 7.4 gives

$$\sum_{n=0}^{\infty} z^n \sum_{w\in\{0,1\}_n^*} x^{\#\text{ nonoverlapping 0100 matches in } w} = \frac{1+z^3}{1-2z+z^3-(1+x)z^4}.$$

7.2 Clusters

A τ-cluster of length n is a permutation $\sigma \in S_n$ with some of the (possibly overlapping) consecutive τ-matches marked in such a way that every element of σ is contained in at least one marked τ-match and any two consecutive marked τ-matches share at least one element in common. A given permutation $\sigma \in S_n$ may give rise to several τ-clusters.

For example, if $\tau = 142536$ and we indicate a marked τ-match of σ by placing an x on the element of σ that starts the marked τ-match, then

$$\overset{x}{1}\ 6\ 2\ 7\ \overset{x}{3}\ 8\ 4\ 9\ 5\ 10$$

and

$$\overset{x}{1}\ 6\ \overset{x}{2}\ 7\ \overset{x}{3}\ 8\ 4\ 9\ 5\ 10$$

are two different τ-clusters, both arising from the same underlying permutation.

Let $\mathscr{C}_{n,\tau}$ denote the set of all τ-clusters of length n. For any τ-cluster $\sigma \in \mathscr{C}_{n,\tau}$, let $mk_\tau(\sigma)$ denote the number of marked τ-matches in σ. The cluster polynomial $\mathrm{C}_{n,\tau}(x,q)$ is defined by

$$\mathrm{C}_{n,\tau}(x,q) = \sum_{\sigma\in\mathscr{C}_{n,\tau}} x^{mk_\tau(\sigma)} q^{\mathrm{inv}\sigma}.$$

Note that if $\tau \in S_j$, then $\mathrm{C}_{n,\tau}(x,q) = 0$ for $1 \le n < j$.

Theorem 7.5 is known as the cluster method of Goulden and Jackson [52, 53], adapted for permutations as described by Elizalde and Noy [36]. This section is devoted to showing how the cluster method can be proved by applying ring homomorphisms to symmetric function identities.

Theorem 7.5. *Let $\tau \in S_j$. Then*

$$\sum_{n=0}^{\infty} \frac{z^n}{[n]_q!} \sum_{\sigma\in S_n} x^{\tau\text{-}match(\sigma)} q^{inv(\sigma)} = \frac{1}{1-\left(z+\sum_{n\ge j}\mathrm{C}_{n,\tau}(x-1,q)z^n/[n]_q!\right)}, \tag{7.7}$$

where τ-match(σ) denotes the number of consecutive (possibly overlapping) τ-matches in σ.

Proof. Define ϑ by $\vartheta(e_0) = \vartheta(e_1) = 1$, $\vartheta(e_n) = 0$ for $2 \le n \le j-1$, and

$$\vartheta(e_n) = \frac{(-1)^{n-1}}{[n]_q!} C_{n,\tau}(x-1,q)$$

for $n \ge j$. Applying ϑ to $[n]_q! h_n$ and expanding in terms brick tabloids in the usual way, we find

$$[n]_q! \vartheta(h_n) = \sum_{\substack{T \in B_{\lambda,(n)} \text{ for some } \lambda \vdash n \\ \text{has bricks } b_1,\ldots,b_\ell}} \begin{bmatrix} n \\ b_1,\ldots,b_\ell \end{bmatrix}_q C_{b_1,\tau}(x-1,q) C_{b_2,\tau}(x-1,q) \cdots .$$

The right-hand side of this equation can be interpreted combinatorially. Begin by selecting a brick tabloid T. Since $C_{b_i,\tau}(x-1,q) \neq 0$ if and only if either $n = 1$ or there is a τ-cluster of size n if $n > 1$, assume that each b_i is either of size 1 or the size of some τ-cluster.

Interpret the q-binomial coefficient $\begin{bmatrix} n \\ b_1,\ldots,b_\ell \end{bmatrix}_q$ as picking sets $S_1, \ldots S_k$ where $|S_i| = b_i$ for $i = 1, \ldots, k$, placing the elements of S_i in the cells of brick b_i in increasing order, and weighting the resulting filled brick tabloid by $q^{\mathrm{inv}(\sigma)}$ where σ is the permutation that results by reading the cells of the filled brick tabloid from left to right.

Finally, use the $C_{b_1,\tau}(x-1,q) C_{b_2,\tau}(x-1,q) \cdots$ term to either take $\alpha^{(i)} = 1$ if $b_i = 1$ or to select a τ-cluster $\alpha^{(i)}$ of size b_i otherwise. With these choices, replace the elements in brick b_i by a permutation of S_i that reduces to $\alpha^{(i)}$. This will add an extra $\mathrm{inv}(\alpha^{(i)})$ inversions, accounting for the factor $q^{\mathrm{inv}(\alpha^{(i)})}$ associated to $\alpha^{(i)}$ in $C_{b_i,\tau}(x-1,q)$.

Label each cell of b_i that corresponds to a marked τ-match in $\alpha^{(i)}$ with a choice of either x or -1. The product of the labels of the cells of b_i accounts for the factor $(x-1)^{mk_\tau(\alpha^{(i)})}$ associated with $\alpha^{(i)}$ in $C_{b_i,\tau}(x-1,q)$.

An example of such an object created in this way when $\tau = 142536$ is below:

		x		-1		x		-1		x						
16	6	1	9	2	10	3	11	4	13	5	14	7	15	8	17	12

Let $\mathcal{T}$ denote the set of combinatorial objects created in this way. Let the sign of $T \in \mathcal{T}$, denoted $\mathrm{sign}(T)$, be the product of the (-1)s appearing in T and let the weight of T, denoted $w(T)$, be the product of all xs appearing in T along with a power of q registering the number of inversions found in the underlying permutation in T.

It follows from our construction that

$$[n]_q! \varphi(h_n) = \sum_{T \in \mathcal{T}} \mathrm{sign}(T) w(T).$$

Therefore, in order to complete the proof, we need to show

$$\sum_{T \in \mathcal{T}} \text{sign}(T)w(T) = \sum_{\sigma \in S_n} x^{\tau\text{-match}(\sigma)} q^{\text{inv}(\sigma)}. \tag{7.8}$$

If this identity can be verified, then the proof of the theorem is complete since the generating function in the statement of the theorem can then be found by applying the ring homomorphism ϑ to both sides of Theorem 2.5 in the usual way.

We will give two proofs of (7.8). The first and simplest proof is to replace x by $(x+1)$ in (7.8), turning the right-hand side of (7.8) into $\sum_{\sigma \in S_n} (x+1)^{\tau\text{-match}(\sigma)} q^{\text{inv}(\sigma)}$.

Let $MS_{n,\tau}$ denote set of permutations in S_n where we have marked some of the τ-matches by placing an x at the start of each marked τ-match. We let $mk_\tau(\sigma)$ denote the number of marked τ-matches in σ (these definitions extend our previous definition of $mk_\tau(\sigma)$ to include any permutation σ, not just clusters). It follows that

$$\sum_{\sigma \in S_n} (x+1)^{\tau\text{-match}(\sigma)} q^{\text{inv}(\sigma)} = \sum_{\beta \in MS_{n,\tau}} x^{mk_\tau(\sigma)} q^{\text{inv}(\sigma)}. \tag{7.9}$$

The effect of replacing x by $(1+x)$ on the left-hand side of (7.8) is the signed sum of the weights of $T \in \mathcal{T}$ where the choice of x or -1 labels has been replaced with only the choice of x.

There is a natural one-to-one correspondence between such $T \in \mathcal{T}$ and elements of $MS_{n,\tau}$: simply send T to the element of $MS_{n,\tau}$ found by removing the brick structure. We only need to show that we can recover the brick structure on T from the labels.

Given a consecutive subsequence of a permutation β, say $\beta_{i+1},\ldots,\beta_{i+k}$, let $\text{red}(\beta_i\beta_{i+1}\ldots\beta_{i+k})$ denote the element of $MS_{j,\tau}$ that results by replacing the underlying permutation with its reduction and marking all corresponding τ-matches that are contained in $\beta_{i+1}\ldots\beta_{i+k}$. For example, if $\tau = 123$, then

$$\text{red}(\overset{x}{1}\ \overset{x}{2}\ 3\ 5\ \overset{x}{4}\ 7\ 9) = \overset{x}{1}\ \overset{x}{2}\ 3\ 5\ \overset{x}{4}\ 6\ 7.$$

We say that a consecutive sequence $\beta_i\beta_{i+1}\ldots\beta_{i+k}$ of β is a τ-subcluster if $\text{red}(\beta_i\beta_{i+1}\ldots\beta_{i+k})$ is a τ-cluster. We say that it is a maximal τ-subcluster if it is a τ-subcluster of β and it is not contained in a strictly larger τ-subcluster of β. For example, the maximal 123-subclusters of β given above are $\overset{x}{1}\ \overset{x}{2}\ 3\ 5$ and $\overset{x}{4}\ 7\ 9$. If we start with $T \in \mathcal{T}$, the bricks that are not of size 1 cover the maximal τ-subclusters of β, meaning that we can recover T from β. This proves (7.9) and therefore completes our first proof of the theorem.

Our second proof also shows that (7.9) follows from (7.8). Given that most of the generating functions in this book come from proofs using involutions, it is natural to ask if this passage from (7.8) to (7.9) can as well be realized by a series of involutions. Our second proof will show that this is indeed the case. Unfortunately, this second proof is more involved than the first.

Given a permutation $\beta = \beta_1\ldots\beta_n \in S_n$, let $\overline{\beta}$ denote the element of $MS_{n,\tau}$ where the first element of every τ-match in β is marked. We shall call $\overline{\beta}$ the fully τ-marked version of β.

Let $\mathcal{E}_{n,\tau,\beta}$ denote the set of $T \in \mathcal{T}$ with underlying permutation β. If β_i is not an element of maximal τ-subcluster in $\overline{\beta}$, then β_i must be covered by a brick of size 1. Thus if β has no τ-matches, then there is only one element in $\mathcal{E}_{n,\tau,\beta}$, namely, the one consisting entirely of bricks of size 1.

Fix β such that β has at least one τ-match. If $T \in \mathcal{E}_{n,\tau,\beta}$, we say that a maximal τ-subcluster $\beta_{i+1}\ldots\beta_{i+k}$ in $\overline{\beta}$ is fully τ-marked if the elements of $\beta_{i+1}\ldots\beta_{i+k}$ are covered by a single brick of size k in T and the start of any τ-match contained in $\beta_{i+1}\ldots\beta_{i+k}$ is marked with an x.

As described above, there is only one element of $T^* \in \mathcal{E}_{n,\tau,\beta}$ in which every maximal τ-subcluster $\beta_{i+1}\ldots\beta_{i+k}$ in $\overline{\beta}$ is fully τ-marked. That is, T^* has the brick structure associated with $\overline{\beta}$ and the start of any τ-match in β is marked with an x. In such a case, the weight of T^* is $x^{\tau\text{-match}(\beta)}q^{\mathrm{inv}(\beta)}$.

Our goal in this second proof of (7.9) is to define a series of involutions which will cancel out all the elements of $\mathcal{E}_{n,\tau,\beta}$ which are not equal to T^*. That is, suppose that we are given a $T \in \mathcal{E}_{n,\tau,\beta} \setminus \{T^*\}$ and $\beta_{i+1}\beta_{i+2}\ldots\beta_{i+k}$ is the left-most maximal τ-subcluster of $\overline{\beta}$ which is not fully τ-marked in T. Thus the brick structure on the elements $\beta_1,\ldots,\beta_i$ is completely determined.

Our first involution I_1 looks at the element β_{i+1}. Since β_{i+1} is the start of maximal τ-subcluster of $\overline{\beta}$, $\beta_{i+1}\beta_{i+2}\ldots\beta_{i+j}$ is a τ-match. If in T, if β_{i+1} is covered with a brick of size greater than 1, and if the label on cell $i+1$ is x, then T is a fixed point of I_1. Otherwise, we have two cases to consider.

Case A1. The number β_{i+1} is covered with a brick of size 1 in T. In this case, if each of $\beta_{i+1}\beta_{i+2}\ldots\beta_{i+j}$ is covered with bricks of size 1 in T, then let $I_1(T) = T'$ where T' results from replacing the j bricks of size 1 on $\beta_{i+1}\beta_{i+2}\ldots\beta_{i+j}$ by a single brick of size j and the label on cell $i+1$ is changed to -1.

If $\beta_{i+1}\beta_{i+2}\ldots\beta_{i+j}$ are not all covered with bricks of size 1 in T, there is an $s > 1$ such that $\beta_{i+1},\ldots,\beta_{i+s-1}$ are covered by bricks of size 1 and β_{i+s} is covered by some brick b of size u where $u \geq j$. In this case, $I_1(T) = T'$ where T' is found by replacing the $s-1$ bricks of size 1 on $\beta_{i+1}\beta_{i+2}\ldots\beta_{i+s-1}$ plus the brick b by a single brick b^* of size $s-1+u$ and replacing the label on cell $i+1$ by -1.

Case A2. The number β_{i+1} is covered with a brick b size u where $u > 1$ in T and $L(i+1) = -1$. In this case if $u = j$, meaning that $\beta_{i+1}\beta_{i+2}\ldots\beta_{i+j}$ are covered by a single brick b of size j, then let $I_1(T) = T'$ where T' is found by replacing brick b by j bricks of size 1 and replacing the label -1 of cell $i+1$ by 1.

If $u > j$, then by our conditions, one of the cells $i+2,\ldots,i+j$ must be labeled with either -1 or x since consecutive τ-matches in b which are labeled with either -1 or x must have at least one element in common. In this case, let s be the least number greater than 1 such that cell $i+s$ is labeled with either -1 or x. Let $I_1(T) = T'$ where T' is found by replacing b by $s-1$ bricks of size 1 followed by a single brick of size $u-s+1$ and replacing the label -1 on cell $i+1$ by 1.

By construction, I_1 is a sign reversing and weight preserving involution. Thus I_1 shows that for fixed β with τ-match$(\beta) \geq 1$,

$$\sum_{T \in \mathcal{E}_{n,\tau,\beta}} w(T) = \sum_{T \in \mathcal{E}_{n,\tau,\beta}, I_1(T)=T} w(T).$$

Let $\mathcal{E}1_{n,\tau,\beta}$ denote set of all fixed points of I_1, that is, $\mathcal{E}1_{n,\tau,\beta}$ consists of all those $T \in \mathcal{E}_{n,\tau,\beta} \setminus \{T^*\}$ such that the left most τ-subcluster $\beta_{i+1} \ldots \beta_{i+k}$ of $\overline{\beta}$ which is not fully τ-marked is such that β_{i+1} is covered with a brick of size greater than 1 and cell $i+1$ is labeled with an x.

Our second involution considers the element β_{i+k}. Since $\beta_{i+1} \ldots \beta_{i+k}$ is a maximal τ-subcluster of $\overline{\beta}$, we know that $\beta_{i+k-j+1}\beta_{i+k-j+2} \cdots \beta_{i+k}$ is a τ-match. If the elements of $\beta_{i+k-j+1}\beta_{i+k-j+2} \cdots \beta_{i+k}$ are covered by a single brick and the label on cell $i+k-j+1$ is x, then T is a fixed point of I_2. Otherwise, we have two cases to consider.

Case B1. The number β_{i+k} is covered with a brick of size 1 in T. In this case, if each of $\beta_{i+k-j+1}\beta_{i+k-j+2} \cdots \beta_{i+k}$ is covered by a brick of size 1 in T, then we let $I_2(T) = T'$ where T' is found by replacing the j bricks of size 1 on $\beta_{i+k-j+1}\beta_{i++k-j+2} \cdots \beta_{i+k}$ by a single brick of size j and replacing the label on cell $i+k-j+1$ by -1.

If $\beta_{i+k-j+1}\beta_{i+k-j+2} \cdots \beta_{i+k}$ are not all covered with bricks of size 1 in T, there is an $s > 1$ such that $\beta_{i+k-j+1}, \ldots, \beta_{i+k-j+s}$ are contained in a single brick of size $u > 1$ and $\beta_{i+k-j+s+1}, \ldots, \beta_{i+k}$ are covered by bricks of size 1 in T. In this case, let $I_2(B, \beta, L) = T'$ where T' is found by replacing the brick b and the following bricks of size 1 by a single brick b^* and replacing the label on cell $i+k-j+1$ by -1.

Case B2. The number β_{i+k} is covered with a brick b of size u where $u > 1$ in T and $L(i+k-j+1) = -1$. In this case, if $u = j$ so that $\beta_{i+k-j+1}\beta_{i+k-j+2} \cdots \beta_{i+k}$ are covered by a single brick b of size j, then let $I_2(T) = T'$ where T' is found by replacing brick b by j bricks of size 1 and replacing the label -1 of cell $i+k-j+1$ by 1.

If $u > j$, then by our conditions, the second to last τ-match in b which is labeled with x or -1 must end in some cell $i+s$ where $i+k-j+1 \leq s < i+k$. Then let $I_2(T) = T'$ where T' is found by replacing b by a brick that starts at the first cell of b and ends at cell $i+s$ followed by bricks of size 1 and replacing the label -1 on cell $i+k-j+1$ by 1.

By construction, I_2 is a sign reversing weight preserving involution. Thus I_2 shows that for fixed β with at least one τ-match,

$$\sum_{T \in \mathcal{E}1_{n,\tau,\beta}} w(T) = \sum_{T \in \mathcal{E}1_{n,\tau,\beta}, I_2(T)=T} w(T).$$

Let $\mathcal{E}2_{n,\tau,\beta}$ denote set of all fixed points of I_2. That is, $\mathcal{E}2_{n,\tau,\beta}$ consists of all $T \in \mathcal{E}_{n,\tau,\beta} \setminus \{T^*\}$ such that in the left-most τ-subcluster $\beta_{i+1} \ldots \beta_{i+k}$ of $\overline{\beta}$ which is not fully τ-marked in T, β_{i+1} is covered with a brick of size greater than 1 and cell $i+1$ is labeled with an x, the elements $\beta_{i+k-j+1}\cdots\beta_{i+k}$ are contained in a single brick, and the label on cell $i+k-j+1$ is x.

Given $T \in \mathcal{E}2_{n,\tau,\beta}$, we say that maximal τ-subcluster $\beta_{i+1}\ldots\beta_{i+k}$ of $\overline{\beta}$ is x-connected in T if when we remove the -1 labels, $\mathrm{red}(\beta_{i+1}\ldots\beta_{i+1})$ reduces to a τ-cluster. Thus $\beta_{i+1}\ldots\beta_{i+k}$ is x-connected in T if every element in $\beta_{i+1}\ldots\beta_{i+k}$ is an element of τ-match whose first element is labeled with x in L and any two consecutive τ-matches whose first elements are labeled with x in L have at least one element in common.

If the left-most maximal τ-subcluster of $\overline{\beta}$ which is not fully τ-marked in T is x-connected in T, then T will be a fixed point of our third involution I_3. Otherwise, suppose that we are given $T \in \mathcal{E}2_{n,\tau,\beta}$ and $\beta_{i+1}\ldots\beta_{i+k}$ is the left-most maximal τ-subcluster of $\overline{\beta}$ which is not x-connected in T. Then we let

$\hat{\beta}_{i+s} = \beta_{i+s}\beta_{i+s+1}\cdots\beta_{i+s+j-1}$ and $\hat{\beta}_{i+t} = \beta_{i+t}\beta_{i+t+1}\cdots\beta_{i+t+j-1}$ be the left-most pair of consecutive τ-matches in $\beta_{i+1}\ldots\beta_{i+k}$ whose first cells are marked with an x and which have no elements in common. Since we are assuming that the first and last τ-matches in $\beta_{i+1}\ldots\beta_{i+k}$ are marked with an x in T, such a pair must exist.

In such a case, we must have $i+s+j-1 < i+t$. Let $u_0 = i+s < u_1 < u_2 < \cdots < u_m < i+t$ be the start of all the τ-matches in $\beta_{i+1}\ldots\beta_{i+k}$ whose first elements lie between $i+s$ and $i+t$. Let $\hat{\beta}_{u_s}$ denote the τ-match which starts at u_s for $1 \le s \le m$. Since $\beta_{i+1}\ldots\beta_{i+k}$ is a maximal τ-subcluster of $\overline{\beta}$, we know that $\hat{\beta}_{u_m}$ must have at least one element in common with $\hat{\beta}_{i+t}$.

However there may be more than one of the $\hat{\beta}_{u_s}$ which have at least one element in common with $\hat{\beta}_{i+t}$. Hence we let ℓ the smallest s such that $\hat{\beta}_{u_s}$ has at least one element in common with $\hat{\beta}_{i+t}$. We then have five cases to consider depending on the configuration for the τ-match $\hat{\beta}_{u_m} = \beta_{u_m}\beta_{u_m+1}\cdots\beta_{u_m+j-1}$.

Case C1. The number β_{u_m} is covered by a brick of size 1. In this case, because $\beta_{i+1}\ldots\beta_{i+k}$ is a maximal τ-subcluster of $\overline{\beta}$, we know that $\hat{\beta}_{u_m}$ and $\hat{\beta}_{i+t}$ must have at least one element in common. It follows that all the cells that lie between u_m and $i+t$ must be covered by bricks of size 1 and that β_{i+t} must start with a brick b of size greater than 1. In this case, we let $I_3(T) = T'$ where T' is found by replacing the bricks of size 1 on $\beta_{u_m}\beta_{u_m+1}\cdots\beta_{i+t-1}$ plus the brick b by a single brick b^* and changing the label on cell u_m from 1 to -1.

Case C2. The number β_{u_m} starts a brick b of size greater than 1 and is labeled with -1. In this case, we let $I_3(T) = T'$ where T' is found by replacing brick b by bricks of size 1 on the elements $\beta_{u_m}\beta_{u_m+1}\cdots\beta_{i+t-1}$ followed by a brick b^* which covers the remaining cells of b and changing the label on cell u_m from -1 to 1.

As an example, when $\tau = 1324$, the function I_3 sends this combinatorial object:

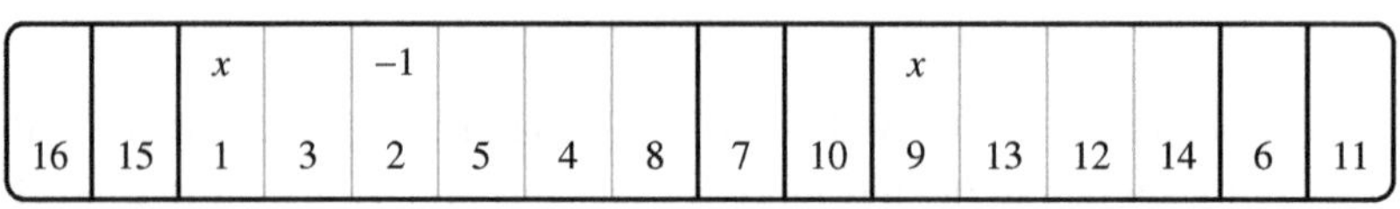

to the combinatorial object pictured here:

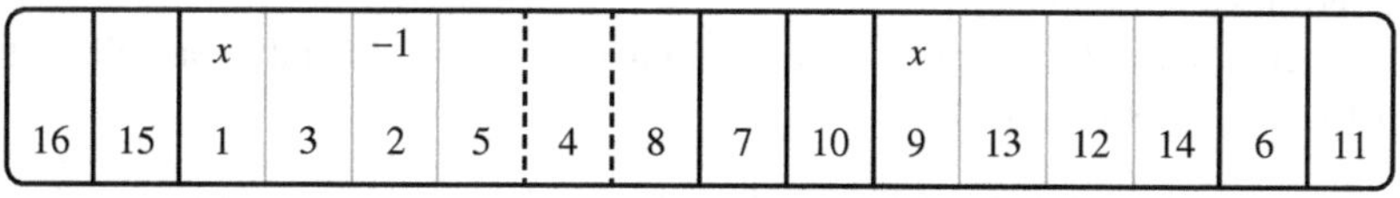

If T is in case C1, then $I_3(T)$ will be in case C2. Similarly, if T is in case C2, then $I_3(T)$ will be in case C1. When restricted to these two cases, I_3 is an involution.

Case C3. The number β_{u_m} is contained in a brick b of size greater than 1 which ends before cell $i+t$. In this case, a brick b' must start at cell $i+t$. It is possible that there may be some bricks of size 1 in between b and b'. Moreover, if $u_\ell < u_m$, then all the cells from u_ℓ up to u_m must be labeled with 1 since there can be no τ-match starting as such cells which are contained in b as all such τ-matches have an element in common with $\hat{\beta}_{i+t}$.

Let $I_3(T) = T'$ where T' is found by replacing b, b', and any brick of size 1 between b and b' by a single brick b^* and changing the label on cell u_m from 1 to -1. Note that if $u_\ell < u_m$, then all the cells from u_ℓ up to u_m must be labeled with 1 in T'.

Case C4. The number β_{u_m} is contained in a brick b which also contains β_{i+t} and is labeled with -1 in T and if $u_\ell < u_m$, then u_ℓ is also in brick b and all the cells between $u_\ell - 1$ and u_m are labeled with 1 in T. Consider the right-most cell before u_m in b which has a label of either -1 or x. This cell must be of the form u_r where $0 \leq r < \ell$ and $\hat{\beta}_{u_r}$ and $\hat{\beta}_{u_m}$ have at least one element in common.

Let $I_3(T) = T'$ where T' is found by replacing the brick b with a brick b^* which starts at the first cell of b and ends at the end of the τ-match $\hat{\beta}_{u_r}$, a brick b^{**} which starts at cell $i+t$ and contains all the cells to the left of $i+t$ in b, plus a sequence of bricks of size 1 between b^* and b^{**} and changing the label on cell u_m from -1 to 1. Note that if $u_\ell < u_m$, then all the cells from u_ℓ up to u_m must be labeled with 1 in T'.

As an example, when $\tau = 142536$, the function I_3 sends this combinatorial object:

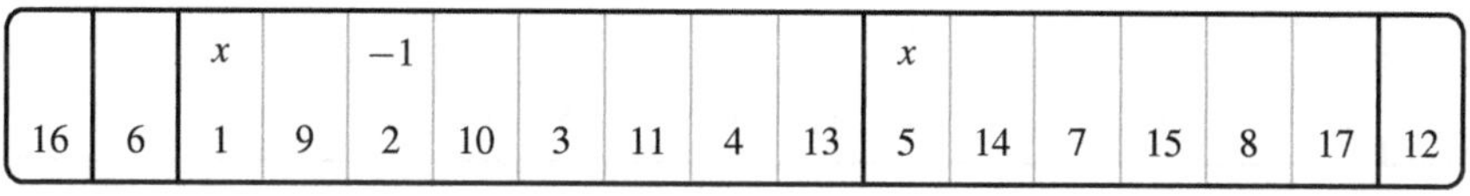

to the combinatorial object pictured here:

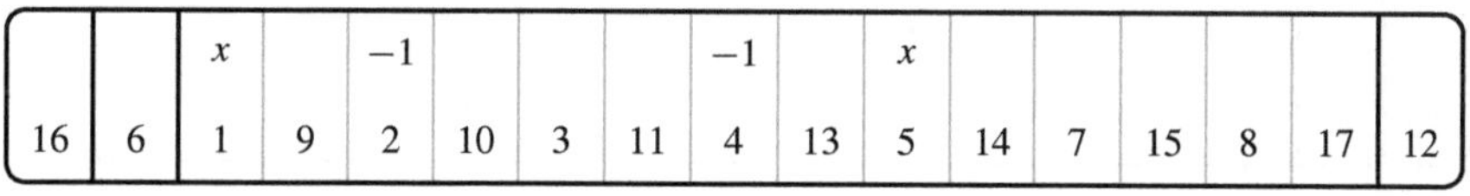

If T is in case C3, then $I_3(T)$ will be in case C4. Similarly, if T is in case C4, then $I_3(T)$ will be in case C3. When restricted to these two cases, I_3 is an involution.

Case C5. Suppose we are not in cases C1–C4. In this case, it must that u_m is in the same brick as $i+t$, $u_\ell < u_m$, and one of $u_\ell, u_{\ell+1}, \ldots, u_{m-1}$ must be labeled with -1 in T. It follows that if the label on u_m can be either 1 or -1 because even if the label on u_m is 1, the labels on $u_\ell, u_{\ell+1}, \ldots, u_{m-1}$ ensure that T meets the conditions to be in $\mathcal{E}2_{n,\tau,\beta}$.

Let $I_3(T) = T'$ where all we do to find T' is change the label on cell u_m from 1 to -1 or from -1 to 1.

As an example, when $\tau = 142536$, the function I_3 sends this combinatorial object

		x				-1				x						
16	6	1	9	2	10	3	11	4	13	5	14	7	15	8	17	12

to the combinatorial object pictured here:

		x				-1		-1		x						
16	6	1	9	2	10	3	11	4	13	5	14	7	15	8	17	12

The map I_3 is a sign reversing weight preserving involution when restricted to case C5.

This completes the description of the involution I_3. At this point we can now consider those combinatorial objects fixed by I_3. Let $\mathcal{E}3_{n,\tau,\beta}$ denote set of all fixed points of I_3, That is, $\mathcal{E}3_{n,\tau,\beta}$ consists of all those $T \in \mathcal{E}_{n,\tau,\beta} - \{T^*\}$ such that in the left-most τ-subcluster $\beta_{i+1}\ldots\beta_{i+k}$ of $\overline{\beta}$ which is not fully τ-marked, the start of the first τ-match is labeled with x, the start of the final τ-match is labeled with x, and $\beta_{i+1}\ldots\beta_{i+k}$ is x-connected which means that the elements $\beta_{i+1}\ldots\beta_{i+k}$ must be covered by a single brick b of size k.

This brings us to our final involution I_4. Let $\hat{\beta}_{i+p} = \beta_{i+p}\beta_{i+p+1}\cdots\beta_{i+p+j-1}$ be the left-most τ-match in $\beta_{i+1}\ldots\beta_{i+k}$ which is not marked with an x. In this case, let $I_4(T) = T'$ where the only change made to T is that we change the label on cell $i+p$ from 1 to -1 if the label on cell $i+p$ was 1 or we change the label on cell $i+p$ from -1 to 1 if the label on cell $i+p$ was -1.

As an example, when $\tau = 142536$, the function I_3 sends this combinatorial object

		x		-1		x		-1		x						
16	6	1	9	2	10	3	11	4	13	5	14	7	15	8	17	12

to the combinatorial object pictured here:

		x				x		-1		x						
16	6	1	9	2	10	3	11	4	13	5	14	7	15	8	17	12

The map I_4 is a sign reversing and weight preserving involution.

There are no fixed points under all of the successive involutions I_1, I_2, I_3, and I_4, showing that the signed, weighted sum over all elements in $\mathcal{E}_{n,\tau,\beta} \setminus \{T^*\}$ is 0. This is exactly what is needed to finish our second proof that (7.9) follows from (7.8). □

7.3 The Minimal Overlapping Property

The basis $p_{\nu,n}$ can be used to find a generating function for (possibly overlapping) $\mathcal{T}$-matches in permutations for sets $\mathcal{T}$ with the "minimal overlapping property."

We say that a subset $\mathcal{T}$ of S_j has the minimal overlapping property if the smallest integer n such that a permutation in S_n with two $\mathcal{T}$-matches exists is $n = 2j-1$. Put another way, every pair of overlapping $\mathcal{T}$-matches must share exactly one integer. For example, $\mathcal{T} = \{213, 312\}$ has the minimal overlapping property since no permutation in S_4 has two $\mathcal{T}$-matches (the permutation $21534 \in S_5$ has two $\mathcal{T}$-matches).

Theorem 7.6. *If $\mathcal{T}$ is a subset of S_{j+1} with the minimal overlapping property, then*

$$\sum_{n=1}^{\infty} \frac{z^n}{n!} \sum_{\sigma \in S_n} x^{\#\text{ of } \mathcal{T}\text{-matches in } \sigma} = \sum_{m=0}^{j-1} \frac{(1-x)^{m/j} A_m\left((1-x)^{1/j} z\right)}{1-x-xA_0\left((1-x)^{1/j} z\right)}$$

where $A(z) = \sum_{n=1}^{\infty} a_n z^n$ for $a_n = |\{\sigma \in S_n$ does not have a $\mathcal{T}$-match $\}|/n!$ and

$$A_m(z) = \sum_{n=1}^{\infty} a_{nj-m} z^{nj-m}$$

for $m = 0, \ldots, j-1$.

Proof. Let φ be the ring homomorphism defined by $\varphi(e_n) = (-1)^{n-1} x (1-x)^{n-1} a_{nj}$ and let ν be the function defined by $\nu(n) = a_{nj-m}/(x a_{nj})$. Then $(nj-m)!\varphi(p_{\nu,n})$ is equal to

$$(nj-m)! \sum_{\lambda \vdash n} w(B_{\lambda,(n)}) x^{\lambda_1} (1-x)^{\lambda_1 - 1} a_{\lambda_1 j} x^{\lambda_2} (1-x)^{\lambda_2 - 1} a_{\lambda_2 j} \cdots,$$

which in turn may be written as

$$\sum_{\substack{T \in B_{\lambda,(n)} \text{ for } \lambda \vdash n \\ \text{has bricks } b_1, \ldots, b_\ell}} \binom{nj-m}{jb_1, \ldots, jb_\ell - m} x^{\ell-1} (1-x)^{n-\ell} (jb_1)! a_{jb_1} \cdots (jb_\ell - m)! a_{jb_\ell - m}.$$

This expression tells us to choose a brick tabloid $T \in B_{\lambda \vdash n}$ for some $\lambda \vdash n$ with bricks of lengths $b_1, \ldots, b_\ell$. Scale each brick by a factor of j. Use the multinomial coefficient to fill these bricks with disjoint sequences of integers with union $1, \ldots, nj-m$, leaving the last m cells in the final brick empty. The terms of the form $(jb_i)! a_{jb_i}$ permit us to rearrange the integers in each brick such that no brick contains a $\mathcal{T}$-match (although $\mathcal{T}$-matches may straddle consecutive bricks). Finally, use the $x^{\ell-1}(1-x)^{n-\ell}$ term to place a choice of either 1 or $-x$ in every j^{th} cell not at the end of a brick and place an x in the last cell of each nonterminal brick in T.

One such object when $n = 6$, $m = 1$, and $\mathcal{T} = \{213, 312\}$ (and so $j = 2$) is

	x		1		$-x$		x		$-x$		
5	1	2	3	4	11	9	8	6	10	7	

Scan the bricks from left to right looking for the first $-x$ or two consecutive bricks which can be combined without creating a brick containing a $\mathcal{T}$-match. If a $-x$ is found, break the brick into two bricks and reverse the sign on the $-x$. If consecutive bricks can be combined, do so, changing the x in the middle to a $-x$.

A fixed point under this involution cannot have $-x$s and every x must appear in a brick which cannot be combined with the brick to its immediate right without introducing a brick with a $\mathcal{T}$-match. For example, one such fixed point is

	x		1		1		x		1		
5	1	2	3	4	11	9	8	6	10	7	

The hypothesis that $\mathcal{T}$ possesses the minimal overlapping property implies that there is exactly one $\mathcal{T}$-match for every x in a fixed point, since the difference between the places of $\mathcal{T}$-matches must be at least j. Therefore $(nj-m)!\varphi(p_{\nu,n})$ is equal to $\sum_{\sigma \in S_{nj-m}} x^{\#\text{ of }\mathcal{T}\text{-matches in }\sigma}$. Summing over all values of m,

$$\begin{aligned}\sum_{n=1}^{\infty} \frac{z^n}{n!} \sum_{\sigma \in S_n} x^{\#\text{ of }\mathcal{T}\text{-matches in }\sigma} &= \sum_{m=0}^{j-1} \sum_{n=1}^{\infty} \frac{z^{nj-m}}{(nj-m)!} \sum_{\sigma \in S_{nj-m}} x^{\#\text{ of }\mathcal{T}\text{-matches in }\sigma} \\ &= \sum_{m=0}^{j-1} \frac{1}{z^m} \varphi\left(\sum_{n=1}^{\infty} p_{\nu,n} z^{nj} \right). \\ &= \sum_{m=0}^{j-1} \frac{1}{z^m} \frac{\sum_{n=1}^{\infty} (-1)^{n-1} \nu(n) \varphi(e_n) z^{nj}}{1 + \sum_{n=1}^{\infty} \varphi(e_n)(-z^j)^n}.\end{aligned}$$

The definitions of $\varphi(e_n)$ and $\nu(n)$ turn the above generating function into the statement of the theorem. □

As a double check of Theorem 7.6, consider $\mathcal{T} = \{2\ 1\}$ so that each $\mathcal{T}$-match is actually a descent. Here $j = 1$ and $A(z) = e^z - 1$ is the exponential generating function starting at $n = 1$ for the permutations in S_n without any descents. Theorem 7.6 gives

$$1 + \sum_{n=1}^{\infty} \frac{z^n}{n!} \sum_{\sigma \in S_n} x^{\text{des}(\sigma)} = 1 + \frac{e^{(1-x)z} - 1}{1 - x - x\left(e^{(1-x)z} - 1\right)} = \frac{x-1}{x - e^{(x-1)z}},$$

which matches Corollary 3.2 as it should.

For a new example of Theorem 7.6, suppose $\mathcal{T} = \{1\ 3\ 2\}$ (so that $j = 2$) and let $A(z)$ be the generating function in (7.3), giving the number of permutations in S_n without any 1 3 2 matches. This way, $A_0(z) = (A(z) + A(-z))/2 - 1$ and $A_1(z) = (A(z) - A(-z))/2$. Using Theorem 7.6 and cleaning up the result gives

$$\sum_{n=0}^{\infty}\frac{z^n}{n!}\sum_{\sigma\in S_n}x^{\#\text{ of 132-matches in }\sigma}=\left(1-\int e^{(x-1)z^2/2}\,dz\right)^{-1}. \tag{7.10}$$

For a third example of Theorem 7.6, let us find the generating function for the number of valleys in permutations in S_n. Take $\mathcal{T}=\{213,312\}$ (so that $j=2$). The exponential generating function giving the permutations with no valleys is $A(z)=\sum_{n=1}^{\infty}2^{n-1}z^n/n!=(e^{2z}-1)/2$, which was found in the discussion immediately preceding (7.4). Using $A_0(z)=(A(z)+A(-z))^2/2=\sinh^2 z$ and $A_1(z)=(A(z)-A(-z))^2/2=\sinh 2z/2$ in Theorem 7.6 gives

$$\sum_{n=0}^{\infty}\frac{z^n}{n!}\sum_{\sigma\in S_n}x^{\mathrm{val}(\sigma)}=\frac{\sinh^2\left(z\sqrt{1-x}\right)+\sqrt{1-x}\sinh\left(2z\sqrt{1-x}\right)/2}{1-x-x\sinh^2\left(z\sqrt{1-x}\right)}.$$

Using the identity $\sinh iz=i\sin z$, this simplifies to the generating function in Corollary 4.5, as it should. We have now found the generating function for the number of valleys in a permutation in two different ways, which is good since no fact in mathematics should be considered true unless two different proofs are given (a joke).

Just as in the case of Theorem 7.1, Theorem 7.6 can be q-analogued to keep track of inversions, extended to count common matches in tuples of permutations and adapted to the case of words. These extensions are so similar to those given in Theorems 7.2, 7.3, and 7.4 that they are left to the reader.

Let τ be a permutation which satisfies the minimal overlapping property. Theorem 7.6 used permutations without any pattern matches in order to find a generating function for the number of τ-matches in permutations. An alternative method to finding this generating function can be found by using permutations with the maximum possible number of τ-matches. This is the approach we used in the proof of Theorem 7.1 when finding a generating function for nonoverlapping τ-matches.

If $\tau\in S_j$ has the minimal overlapping property, then the shortest permutation σ such that σ has n τ-matches must have length $n(j-1)+1$. Let $MP_{\tau,n(j-1)+1}$ be the subset of $S_{n(j-1)+1}$ containing those permutations with n τ-matches. We shall refer to the permutations in $MP_{n,n(j-1)+1}$ as maximum packings for τ. Let $mp_{\tau,n(j-1)+1}=|MP_{\tau,n(j-1)+1}|$ and

$$mp_{\tau,n(j-1)+1}(q)=\sum_{\sigma\in MP_{\tau,n(j-1)+1}}q^{\mathrm{inv}(\sigma)}.$$

As we see in our next theorem, a simple formula for $mp_{\tau,n(j-1)+1}(q)$ exists in the special case when τ begins or ends with a 1.

Theorem 7.7. *Suppose that $\tau=\tau_1\cdots\tau_j\in S_j$ has the minimal overlapping property where $\tau_1=1$ and $\tau_j=s$. Then, for all $n\geq 1$,*

$$mp_{\tau,(n+1)(j-1)+1}(q)=q^{inv(\tau)}\begin{bmatrix}(n+1)(j-1)+1-s\\ j-s\end{bmatrix}_q mp_{\tau,n(j-1)+1}(q).$$

Unwinding this recursion gives

$$mp_{\tau,(n+1)(j-1)+1}(q) = q^{(n+1)inv(\tau)} \prod_{i=1}^{n+1} \begin{bmatrix} i(j-1)+1-s \\ j-s \end{bmatrix}_q.$$

Proof. Suppose $\sigma = \sigma_1 \cdots \sigma_{(n+1)(j-1)+1}$ is a maximum packing for τ. Since $\tau_1 = 1$, the integer σ_1 is less than $\sigma_2, \ldots, \sigma_j$, the integer σ_j is less than $\sigma_{j+1}, \ldots, \sigma_{j+(j-1)}$, the integer σ_{2j-1} is less than $\sigma_{2j}, \ldots, \sigma_{2j-1+(j-1)}$, and so on.

It follows that $\sigma_1 = 1$ and that σ_j must be less than $\sigma_{j+1}, \ldots, \sigma_{(n+1)(j-1)+1}$. We claim that σ_j must be s. We know that $\sigma_j \geq s$ since $\sigma_1 \cdots \sigma_j$ being a τ-match means that there must be $s-1$ elements of $\sigma_1, \ldots, \sigma_{j-1}$ which are less than σ_j. However if $\sigma_j > s$, then $1, \ldots, \sigma_j - 1$ must be among $\sigma_1, \ldots, \sigma_{j-1}$, which violates the fact that σ_j is the sth smallest element among $\sigma_1, \ldots, \sigma_j$.

Therefore $1, \ldots, s$ must be among $\sigma_1, \ldots, \sigma_j$, meaning that the positions of $1, \ldots, s$ in $\sigma_1, \ldots, \sigma_j$ must be the same as the positions of $1, \ldots, s$ in τ. There are $\binom{(n+1)(j-1)+1-s}{j-s}$ ways to choose the remaining $j-s$ elements in $\sigma_1, \ldots, \sigma_j$. Once these are chosen, then their positions are completely determined by τ. Moreover, $\text{red}(\sigma_j \cdots \sigma_{(n+1)(j-1)+1})$ must be an element of $MP_{\tau,n(j-1)+1}$. It follows that

$$mp_{\tau,(n+1)(j-1)+1} = \binom{(n+1)(j-1)+1-s}{j-s} mp_{\tau,n(j-1)+1}.$$

As for the powers of q, we can count the inversions in σ by

1. counting the inversions in $\sigma_1 \ldots \sigma_j$, which contribute a factor of $q^{\text{inv}\tau}$ to $q^{\text{inv}\sigma}$,
2. counting inversions among $\{\sigma_1, \ldots, \sigma_j\} \setminus \{1, \ldots, s\}$ and $\sigma_{j+1} \ldots \sigma_{(n+1)(j-1)+1}$, which contribute a factor of $\begin{bmatrix} (n+1)(j-1)+1-s \\ j-s \end{bmatrix}_q$ to $mp_{\tau,(n+1)(j-1)+1}(q)$ as we vary over all choices $\{\sigma_1, \ldots, \sigma_j\} \setminus \{1, \ldots, s\}$, and
3. counting the inversions in $\sigma_j \ldots \sigma_{(n+1)(j-1)+1}$, contributing $mp_{\tau,n(j-1)+1}(q)$.

Therefore

$$mp_{\tau,(n+1)(j-1)+1}(q) = q^{\text{inv}(\tau)} \begin{bmatrix} (n+1)(j-1)+1-s \\ j-s \end{bmatrix}_q mp_{\tau,n(j-1)+1}(q).$$

Iterating this recursion proves the second identity in the statement of the theorem. □

Our interest in $mp_{\tau,n(j-1)+1}(q)$ comes from applications of the next theorem, where we use these polynomials in q to find the generating function for the distribution of τ-matches in permutations for various choices of τ.

Theorem 7.8. *If $\tau \in S_j$ has the minimal overlapping property, then*

$$\sum_{n=0}^{\infty} \frac{z^n}{[n]_q!} \sum_{\sigma \in S_n} x^{\text{the number of } \tau\text{-matches in } \sigma} q^{inv(\sigma)}$$

$$= \frac{1}{1 - \left(z + \sum_{n=1}^{\infty} \frac{z^{n(j-1)+1}}{[n(j-1)+1]_q!} (x-1)^n mp_{\tau,n(j-1)+1}(q)\right)}.$$

We will give two proofs of this theorem. The first proof uses the technique of using a ring homomorphism on the ring of symmetric functions. As usual, it can be modified to prove a number of similar results. The second proof is shorter but less direct.

Proof (using a ring homomorphism). Define a ring homomorphism φ by $\varphi(e_n) = (-1)^{n-1} f(n)/[n]_q!$ where

$$f(n) = (x-1)^{(n-1)/(j-1)} mp_{\tau,n}(q)$$

if $n = s(j-1)+1$ for some $s \geq 0$ and $f(n) = 0$ otherwise. Applying φ to $[n]_q!h_n$ gives

$$[n]_q!\varphi(h_n) = \sum_{\substack{T \in B_{\lambda,(n)} \text{ for some } \lambda \vdash n \\ \text{has bricks } b_1,\ldots,b_\ell}} \begin{bmatrix} n \\ b_1,\ldots,b_{\ell(\lambda)} \end{bmatrix}_q f(b_1)f(b_2)\cdots. \tag{7.11}$$

From here we create combinatorial objects by first selecting a brick tabloid $T \in B_{\lambda,(n)}$ for some $\lambda \vdash n$ and then using the q-multinomial coefficient to assign a disjoint subset to each brick such that the union of these subsets is $\{1,\ldots,n\}$. With these assignments of disjoint subsets comes a power of q registering inversions within disjoint subsets.

If $b_i = s(j-1)+1$ for $s \geq 1$, then let $1 \leq a_1^i < \cdots < a_{s(j-1)+1}^i \leq n$ be the subset of $\{1,\ldots,n\}$ assigned to the brick b_i. We interpret the term

$$(x-1)^{(b_i-1)/(j-1)} mp_{\tau,b_i}(q) = (x-1)^s mp_{\tau,s(j-1)+1}(q)$$

as the number of ways of filling b_i with a permutation $\beta_i \in MP_{\tau,s(j-1)+1}$ and then labeling each cell in b_i which is the start of τ-match in β_i with either x or -1. In this case, we weight β_i with $q^{\mathrm{inv}(\beta_i)}$. Finally, we replace the numbers by $1,\ldots,s(j-1)+1$ that occur in β_i by $a_1,\ldots,a_{s(j-1)+1}$, respectively. Doing this for each brick will result in a filling of the cells of T with a permutation $\sigma \in S_n$. It follows that

$$\mathrm{inv}(\alpha) + \sum_{b_i>1} \mathrm{inv}(\beta_i) = \mathrm{inv}(\sigma)$$

since $\mathrm{inv}(\alpha)$ accounts for the inversions that come from pairs of elements that lie in two different bricks and $\sum_{b_i>1} \mathrm{inv}(\beta_i)$ accounts for the inversions that come from pairs of σ that lie in the same brick.

For example, here are four different possible combinatorial objects—referred to later in the proof as T_1, T_2, T_3, and T_4—created in this way when $\tau = 213$ and $n = 17$:

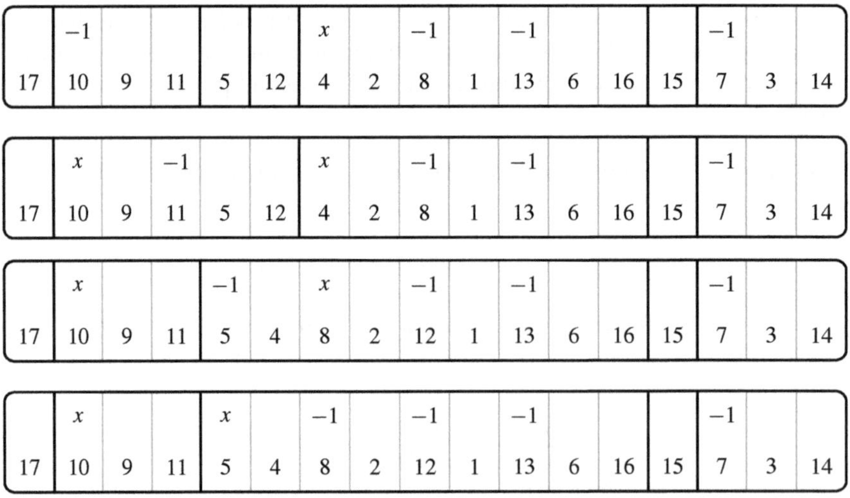

There is a weight coming from -1 and x terms in each of these four examples; they are x, x^2, x^2, and x^2, respectively. All four of the these combinatorial objects have 72 inversions, and so there is a weight of q^{72} associated with all of them in addition to the weight coming from the -1 and x terms. The signed weighted sum over all combinatorial objects is equal to (7.11).

Define a sign reversing weight preserving involution I on the collection of such combinatorial objects T by scanning the cells from left to right looking for the first time we are in one of the following cases:

Case 1. There is a brick b_i of size j whose first cell is labeled with -1. In this case, let $I(T)$ be the object T with this brick of size j replaced with j bricks of size 1 and the -1 sign removed.

Case 2. There are j consecutive bricks of size 1 in T, namely $b_i, b_{i+1}, \ldots, b_{i+j-1}$, such that the letters in these cells form a τ-match. In this case, $I(T)$ is T with the bricks $b_i, b_{i+1}, \ldots, b_{i+j-1}$ combined into a single brick b of size j and the first cell of b is labeled with -1.

Case 3. There is a brick b_i of size $(s+1)(j-1)+1$ where $s \geq 1$ such that all the labels on b_i are xs except for the cell that is j cells from the right which is labeled with -1. In this case, $I(T)$ is found by replacing the brick b_i by a brick of size $s(j-1)+1$ followed by $j-1$ bricks of size 1 and removing the -1 label that was in b_i.

Case 4. There are j consecutive bricks in T, namely $b_i, b_{i+1}, \ldots, b_{i+j-1}$ such that $b_i = s(j-1)+1 > 1$ and $b_{i+1}, \ldots, b_{i+j-1}$ are of size 1, all the labels on b_i are xs, and the letters in these bricks form a maximum packing for τ of size $(s+1)(j-1)+1$. In this case, $I(T)$ is found by replacing the bricks $b_i, b_{i+1}, \ldots, b_{i+j-1}$ by a single brick b of size $(s+1)(j-1)+1$ and by labeling the last cell of b_i with a -1.

Case 5. There is a brick b_i of size $(s+1)(j-1)+1$ where $s \geq 1$ such that the first cell of b_i is labeled with -1. In this case, $I(T)$ is found by replacing the brick b_i

by $j-1$ bricks of size 1 followed by a brick of size $s(j-1)+1$ and by removing the -1 label that was on the first cell of b_i.

Case 6. There are j consecutive bricks in T, namely $b_i, b_{i+1}, \ldots, b_{i+j-1}$, such that $b_i, \ldots, b_{i+j-2}$ are of size 1, $b_{i+j-1} = s(j-1)+1 > 1$, and the letters in these bricks form a maximum packing for τ of size $(s+1)(j-1)+1$. In this case, $I(T)$ is found by replacing the bricks $b_i, b_{i+1}, \ldots, b_{i+j-1}$ by a single brick b of size $(s+1)(j-1)+1$ and by labeling the first cell of b with a -1.

Case 7. There is a brick b_i of size $s(j-1)+1$ where $s \geq 3$ such that the first cell is labeled with an x and there is a cell which has a label -1 which is not the jth cell from the right. Let t be the left-most cell of b_i which is labeled with -1. In this case, $I(T)$ is found by replacing the brick b_i with j consecutive bricks $c_1, c_2, \ldots, c_{j-1}, c_j$ where c_1 contains all the cells of b_i up to and including cell t, $c_2, \ldots, c_{j-1}$ are bricks of size 1, and c_j contains the remaining cells of b_i. Remove the -1 label from cell t.

Case 8. There are j consecutive bricks in T, namely $b_i, b_{i+1}, \ldots, b_{i+j-1}$, such that $b_i = c(s-1)+1 > 1$ and has no -1 labels, $b_{i+1}, \ldots, b_{i+j-2}$ are bricks of size 1, $b_{i+j-1} = d(j-1)+1 > 1$, and the letters in these three bricks form a maximum packing for τ of size $(c+d+1)(j-1)+1$. In this case, $I(T)$ is found by replacing the j bricks $b_i, b_{i+1}, \ldots, b_{i+j-1}$ by a single brick b and adding a label -1 on the last cell of b_i.

Case 9. If none of the previous 8 cases apply, set $I(T) = T$.

For example, consider images of the combinatorial objects T_1, T_2, T_3, and T_4 pictured earlier in this proof. Case 1 applies to T_1, and so $I(T_1)$ is equal to

						x		-1		-1				-1		
17	10	9	11	5	12	4	2	8	1	13	6	16	15	7	3	14

The object $I(T_1)$ now falls under the jurisdiction of case 2. Applying I yet again will yield T_1, showing that I is indeed an involution in this example.

Case 3 applies to T_2. The involution I replaces b_2 in T_2 with a brick of size 3 followed by two bricks of size 1 and removing the -1 label from cell 4, as shown:

	x					x		-1		-1				-1		
17	10	9	11	5	12	4	2	8	1	13	6	16	15	7	3	14

Now $I(T_2)$ is in case 4, and it may be verified that $I^2(T_2) = T_2$.

The object T_3 is in case 5 where the needed b_i is the third brick. Thus we obtain $I(T_3)$ by replacing b_3 by two bricks of size 1 followed by a brick of size 7 and removing the -1 label on cell 5:

	x					x		-1		-1				-1		
17	10	9	11	5	4	8	2	12	1	13	6	16	15	7	3	14

The result $I(T_3)$ is in case 6, and so $I^2(T_3) = T_3$.

Finally, T_4 is in case 7, with $t = 9$ so that we replace the fifth brick by three consecutive bricks of sizes 3, 1, and 3, reading from left to right, and remove the -1 label for cell 9:

	x			x				-1		-1				-1		
17	10	9	11	5	4	8	2	12	1	13	6	16	15	7	3	14

The result $I(T_3)$ is in case 8, and it may be checked that $I^2(T_4) = T_4$.

To prove that I is an involution we proceed by a case-by-case analysis. Let T be a combinatorial object with $I(T) \neq T$. In all cases, $I(T)$ is defined by changing the brick structure on some cells $s, s+1, \cdots, s+j-1$ where $\sigma_s \sigma_{s+1} \ldots \sigma_{s+j-1}$ is a τ-match in σ.

Case 1 and Case 2. Suppose $I(T)$ was defined using case 1 and that the brick of size j that was used in the definition of $I(T)$ is b_i and b_i covers cells $t, t+1, \cdots, t+j-1$. Then in $I(T)$, we have the possibility to recombine the bricks of size 1 that now cover cells $t, t+1, \cdots, t+j-1$.

Thus if $I^2(T) \neq T$, then it must be that we took some action which involved a τ-match $\sigma_s \sigma_{s+1} \ldots \sigma_{s+j-1}$ where $s < t$. Now it cannot be that $s+j-1 < t$ since otherwise we could have taken the same action by changing the brick structure on cells $s, s+1, \ldots, s+j-1$ in T which would violate the fact that we always take an action on the left-most possible cells that we can when defining $I(T)$.

Because σ has the minimal overlapping property, the only other possibility is that $t = s+j-1$. Now if $s > 1$, then the minimal overlapping property for τ implies that $\sigma_{s-1}\sigma_s \ldots \sigma_{s+j-2}$ is not a τ-match and hence the cells $s-1, s, \ldots, s+j-2$ cannot lie in a single brick since the last j cells in any brick b of size greater than 1 must correspond to a τ-match in σ. Thus it must be that in T, cells $s+1, \cdots, s+j-2$ must be covered by bricks of size 1. If cell s is also covered by a brick of size 1, then we could apply case 6 to T using the $j-1$ bricks of size 1 covering cells $s, \cdots, s+j-2$ plus b_i which would contradict the fact that for T, we are in case 1 using brick b_i. If cell s is part of brick b of size > 1, then we could apply case 8 to T using b plus the $j-2$ bricks of size 1 covering cells $s+1, \ldots, s+j-2$ plus b_i which again would contradict the fact that for T, we are in case 1 using brick b_i. If $s = 1$, then cells $s, \ldots, s+j-2$ must be covered by bricks of size 1 so that we could apply case 6 to T using the $j-1$ bricks of size 1 covering cells $s, \ldots, s+j-2$ plus b_i which would contradict the fact that for T, we are in case 1 using brick b_i.

Thus the left-most τ-match that we can use to define the image of I for $I(T)$ is the τ-match that lies in the j bricks of size 1 covering cells $t, t+1, \cdots, t+j-1$ in which case we know that $I^2(T) = T$. An entirely similar analysis will show that if $I(T)$ is defined using case 2, then $I^2(T) = T$.

Case 3 and Case 4. Suppose $I(T)$ was defined using case 3 using a brick b_i of size $a(j-1)+1$ where $a \geq 2$ and b_i covers cells $t, t+1, \ldots, t+a(j-1)$. Then in

$I(T)$, there is a single brick b covering cells $t, t+1, \cdots, t+(a-1)(j-1)$ followed by $j-1$ bricks of size 1 covering cells $t+(a-1)(j-1)+1, \ldots, t+a(j-1)$ and all the labels on b are xs.

If $I^2(T) \neq T$, then it must be the case that we took some action which involved a τ-match $\sigma_s\sigma_{s+1}\ldots\sigma_{s+j-1}$ where $s < t$. But then we could have taken some action by changing the brick structure on cells $s, s+1, \cdots, s+j-1$ in T which would violate the fact that we always take an action on the left-most possible cells that we can when defining $I(T)$. Thus it must be the case that the left-most action that we can take to define I on $I(T)$ is to combine b with the $j-1$ bricks of size 1 that follow b and hence $I^2(T) = T$. A similar analysis will show that if $I(T)$ was defined using case 4, then $I^2(T) = T$.

Case 5 and Case 6. Suppose $I(T)$ was defined using case 5 using brick $b_i = a(j-1)+1$ where $a \geq 2$. The analysis in this case is essentially the same as the analysis of Case 1. That is, suppose that b_i covers cells $t, t+1, \cdots, t+a(j-1)$. We are assuming that cell t is labeled with -1.

The first $j-1$ cells of b_i in $I(T)$ will be covered with bricks of size 1 and the remaining cells of b_i with a single brick b. Thus if $I^2(T) \neq T$, then it must be the case that we took some action which involved a τ-match $\sigma_s\sigma_{s+1}\ldots\sigma_{s+j-1}$ where $s < t$. Now it cannot be that $s+j-1 < t$ since otherwise we could have taken the same action by changing the brick structure on cells $s, s+1, \ldots, s+j-1$ in T which would violate the fact that we always take an action on the left-most possible cells that we can when defining $I(T)$.

Because τ has the minimal overlapping property, the only other possibility is that $t = s+j-1$. Now if $s > 1$, then the minimal overlapping property for τ implies that $\sigma_{s-1}\sigma_s\ldots\sigma_{s+j-2}$ is not a σ-match and hence the cells $s-1, s, \cdots, s+j-2$ cannot lie in a single brick since the last j cells in any brick b of size greater than 1 must correspond to a τ-match in σ. Thus it must be that in T, cells $s+1, \cdots, s+j-2$ must be covered by bricks of size 1. If cell s is also covered by a brick of size 1, then we could apply case 6 to T using the $j-1$ bricks of size 1 covering cells $s, \ldots, s+j-2$ plus b_i which would contradict the fact that for T, we are in case 5 using brick b_i.

If cell s is part of brick b of size > 1, then we could apply case 8 to T using b plus the $j-2$ bricks of size 1 covering cells $s+1, \cdots, s+j-2$ plus b_i which again would contradict the fact that for T, we are in case 5 using brick b_i. If $s = 1$, then cells $s, \cdots, s+j-2$ must be covered by bricks of size 1 so that we could apply case 6 to T using the $j-1$ bricks of size 1 covering cells $s, \ldots, s+j-2$ plus b_i which would contradict the fact that for T, we are in case 5 using brick b_i.

Thus the left-most τ-match that we can use to define the image of I for $I(T)$ is the τ-match that lies $j-1$ bricks of size 1 covering cells $t, t+1, \cdots, t+j-1$ plus the brick b in which case we know that $I^2(T) = T$. An entirely similar analysis will show that if $I(T)$ is defined using case 6, then $I^2(T) = T$.

Case 7 and Case 8. Suppose $I(T)$ was defined using case 7 using a brick b_i of size $a(j-1)+1$ where $a \geq 3$. Suppose that b_i covers cells $t, t+1, \ldots, t+a(j-1)$. We are assuming that cell t has label x and that the left-most cell of b_i which is labeled with -1 occurs on cell $t+b(j-1)$ where $1 \leq b < a-1$.

Then in $I(T)$ there is a single brick b^* covering cells $t, t+1, \ldots, t+b(j-1)$ followed by $j-2$ bricks of size 1 covering cells $t+b(j-1)+1, \cdots, t+b(j-1)+j-2$ followed by a brick b^{**} covering the remaining cells of b_i. Moreover all the labels on b^* are xs. In this case, if $I^2(T) \neq T$, then it must be the case that we took some action which involved a τ-match $\sigma_s \sigma_{s+1} \ldots \sigma_{s+j-1}$ where $s < t$. But then we could have taken some action by changing the brick structure on cells $s, s+1, \ldots, s+j-1$ in T which would violate the fact that we always take an action on the left-most possible cells that we can when defining $I(T)$.

Thus it must be the case that the left-most action that we can take to define is to recombine b^* plus the following $j-2$ bricks of size 1 plus b^{**} into a single brick so that $I^2(T) = T$. An entirely similar analysis will show that if $I(T)$ is defined using Case 8, then $I^2(T) = T$.

We have now shown that I is a sign reversing weight preserving involution. It is time to move on to describing the fixed points of the involution I.

Let T be a fixed point with bricks $b_1, \ldots, b_\ell$. There cannot be any -1 labels on any of the bricks in B since otherwise we could apply cases 1, 3, 5, or 7. The x weight of T is x^c where c is the number of τ-matches in σ that lie entirely within some brick b_i in B.

We claim that any τ-match in σ must lie entirely within some brick. That is, suppose that $\sigma = \sigma_1 \ldots \sigma_n$ and $\sigma_s \sigma_{s+1} \ldots \sigma_{s+j-1}$ a τ-match that does not lie in a single brick. Because τ has the minimal overlapping property, there are only four possibilities, namely,

a. cells $s, s+1, \cdots, s+j-1$ are covered by bricks of size 1,
b. cell s is part of brick b_i of size > 1 and cells $s+1, \ldots, s+j-1$ are covered by bricks of size 1,
c. cell $s+j-1$ is part of brick b_i of size > 1 and cells $s, \ldots, s+j-2$ are covered by bricks of size 1, and
d. cell s is part of a brick b_i of size > 1, $b_{i+1}, \ldots, b_{i+j-2}$ are bricks of size 1 covering cells $s+1, \cdots, s+j-2$, and cell $s+j-1$ is part of brick b_{i+j-1} which is of size > 1.

In case (a), we could apply case 2 of the definition of I to cells $s, s+1, \ldots, s+j-1$. In case (b), we could apply case 4 of the definition of I to cells of b_i plus cells $s+1, \cdots, s+j-1$. In case c., we could apply case 6 of the definition of I to cells $s, \ldots, s+j-2$ plus the cells of b_i. In case d., we can apply case 8 of the definition of I to the cells of contained in the bricks $b_i, \cdots, b_{i+j-1}$.

Thus in each of the cases (a)–(d), $I(T) \neq T$. This contradicts our choice of T. Thus we have shown that if $I(T) = T$, then the x weight of T is equal to $x^{\tau\text{-mch}(\sigma)}$.

Finally note that if $\sigma = \sigma_1 \cdots \sigma_n \in S_n$, then we can construct a fixed point of I by placing bricks which cover the maximal length maximum packings in σ, covering the remaining cells by bricks of size 1, and labeling the start of each τ-match in σ

by x. It thus follows that $[n]_q!\varphi(h_n)$ is equal to the sum of weights over all fixed points, that is, we have shown that

$$[n]_q!\varphi(h_n) = \sum_{\sigma\in S_n} x^{\text{the number of } \tau\text{-matches in } \sigma} q^{\text{inv}(\sigma)}.$$

Applying φ to Theorem 2.5 gives the generating function in the statement of the theorem. □

Proof (not using a ring homomorphism). The idea of this second proof is to replace x with $x+1$ in the statement of the theorem. Let $MS_{\tau,n}$ denote the set of all permutations $\sigma \in S_n$ where some of the τ-matches of σ are marked by marking the first element of τ-match with a $*$. For example, if $\tau = 132$ and $\sigma = 2546175$, then there are two τ-matches in σ, namely, 245 and 175. Thus σ gives rise to the following four elements $MS_{132,7}$:

$$2546175,\ 2^*546175,\ 25461^*75,\ \text{and}\ 2^*5461^*75.$$

If $\sigma \in MS_{\tau,n}$, we let $*(\sigma)$ denote the number of $*$s that occur in σ. When we replace x by $x+1$ in the generating function in the statement of the theorem, we find

$$1+\sum_{n\geq 1}\frac{z^n}{[n]_q!}\sum_{\sigma\in MS_{\tau,n}} x^{*(\sigma)} = \frac{1}{1-(z+\sum_{n\geq 1}\frac{z^{n(j-1)+1}}{[n(j-1)+1]_q!}x^n mp_{\tau,n(j-1)+1}(q))}.$$

Our goal is to prove this identity is true.

Extracting the coefficient of $z^n/[n]_q!$ on the right-hand side of this identity by expanding the right-hand side as a geometric series, we find that $\sum_{\sigma\in MS_{\tau,n}} x^{*(\sigma)}$ is equal to

$$[n]_q!\sum_{k\geq 1}\sum_{\substack{b_1,\cdots,b_k\in\{1\}\cup\{n(j-1)+1:n\geq 1\}\\ b_1+b_2+\cdots+b_k=n}}\prod_{i=1}^{k}\frac{(\chi(b_i=1)+\chi(b_i\geq 1)x^{\frac{b_i-1}{j-1}}mp_{\tau,b_i}(q))}{[b_i]_q!}$$

$$=\sum_{k\geq 1}\sum_{\substack{b_1,\cdots,b_k\in\{1\}\cup\{n(j-1)+1:n\geq 1\}\\ b_1+b_2+\cdots+b_k=n}}\begin{bmatrix} n \\ b_1,\cdots,b_k\end{bmatrix}\prod_{i=1}^{k}(\chi(b_i=1)+\chi(b_i>1)x^{\frac{b_i-1}{j-1}}mp_{\tau,b_i}(q)),$$

where for any statement A, $\chi(A)$ is 1 if A is true and 0 if A is false.

With the terms of the form

$$\begin{bmatrix} n \\ b_1,\cdots,b_k\end{bmatrix}\prod_{i=1}^{k}(\chi(b_i=1)+\chi(b_i>1)x^{\frac{b_i-1}{j-1}}mp_{\tau,b_i}(q))$$

in this expression we can associate an ordered triple of the form (B,σ,L) such that

1. B is a list of integers of the form $(b_1,\ldots,b_k)$ which sum to n. With these integers we can associate a brick tabloid with bricks $b_1,\ldots,b_k$.

2. σ is an element of S_n such that in each brick $b_i > i$, the elements of σ reduces to a maximum packing of MP_{τ,b_i}, and
3. L is the labeling of the cells of B such that the start of each cell which starts a τ-match that lies entirely with a brick b_i is labeled with an x and all other cells are labeled with 1.

We define the weight of such a triple to be $q^{\mathrm{inv}\sigma}$ times the product of the labels in each cell.

Given such a triple (B,σ,L), we can obtain an element of $MS_{\tau,n}$ by removing the brick structure and putting a $*$ on each element of σ whose cell was labeled with an x. On the other hand, given an element of $\sigma \in MS_{\tau,n}$, we can reconstruct the brick structure as follows. Let us say that two marked τ-matches in σ are linked if they have an element in common. Since τ is a minimal overlapping permutation, two τ-matches in σ are linked if and only if the last element of the first τ-match is equal to the first element of the second τ-match.

We call two marked τ-matches M and N of σ marked τ-match connected in σ if there is a sequence of marked τ-matches of σ, $M = M_1, M_2, \cdots, M_k = N$ such that for all i, M_i and M_{i+1} are linked. The maximal marked τ-match connected components of σ under this relation are just consecutive sequences on σ which reduce to a maximum packing of τ. Thus for each maximal marked τ-match connected component, we simply cover the elements with a single brick b and cover all elements which are not part of a maximal marked τ-match component by a brick of size 1 and label each cell that corresponds to element with a $*$ on it with x and label all remaining cells with 1, we can recover (B,σ,L). This proves the desired equality. □

Combining the results in Theorems 7.7 and 7.8 that if $j \geq 3$ and $\tau = \tau_1 \cdots \tau_j \in S_j$ has the minimal overlapping property with $\tau_1 = 1$ and $\tau_j = s$, then

$$\sum_{n\geq 0} \frac{z^n}{[n]_q!} \sum_{\sigma\in S_n} x^{\text{the number of } \tau\text{-matches}(\sigma)} q^{\mathrm{inv}(\sigma)}$$

$$= \frac{1}{1-\left(z+\sum_{n\geq 1} \frac{(x-1)^n z^{n(j-1)+1}}{[n(j-1)+1]_q!} \left(q^{\mathrm{inv}(\tau)}\right)^{n+1} \prod_{i=1}^{n+1} \left[\begin{smallmatrix} i(j-1)+1-s \\ j-s \end{smallmatrix}\right]\right)}.$$

For example, 132 has the minimal overlapping property and it can be shown that

$$mp_{132,2n+1}(q) = q^n \prod_{i=1}^{n} [2i-1]_q.$$

Applying our results to the permutation 132 gives us a q-analogue of (7.10):

$$\sum_{n\geq 0} \frac{z^n}{[n]_q!} \sum_{\sigma\in S_n} x^{\#\text{ of 132-matches}(\sigma)} q^{\mathrm{inv}(\sigma)} = \frac{1}{1-\left(z+\sum_{n\geq 1} \frac{(x-1)^n z^{2n+1}}{[2n+1]_q!} q^n \prod_{i=1}^{n} [2i-1]_q\right)}$$

$$= \frac{1}{1-\sum_{n\geq 0} \frac{q^n (x-1)^n z^{2n+1}}{[2n+1]_q} \prod_{i=1}^{n} [2i]_q}.$$

As another example, the permutation 1342 also has the minimal overlapping property. Elizalde and Noy [35] showed that

$$\sum_{n\geq 0}\frac{z^n}{n!}\sum_{\sigma\in S_n} x^{\#\text{ of 1342-matches}(\sigma)} = \frac{1}{1-\int_0^z e^{(x-1)t^3/6}\,dt} = \frac{1}{1-\sum_{n\geq 0}\frac{(x-1)^n z^{3n+1}}{6^n(n!)(3n+1)}}.$$

Since

$$mp_{1342,2n+1} = (q^2)^n\prod_{i=1}^{n}\begin{bmatrix}3n+1-2\\2\end{bmatrix} = q^{2n}\prod_{i=1}^{n}\frac{[3n-1]_q[3n-2]_q}{[2]_q},$$

we can use Theorem 7.8 to find a q-analogue:

$$\begin{aligned}\sum_{n\geq 0}\frac{z^n}{n!}\sum_{\sigma\in S_n} & x^{\#\text{ of 1342-matches}(\sigma)}q^{\text{inv}(\sigma)}\\ &= \frac{1}{1-(z+\sum_{n\geq 1}\frac{(x-1)^n z^{3n+1}}{[3n+1]_q!}q^{2n}\frac{1}{[2]_q^n}\prod_{i=1}^{n}[3i-1]_q[3i-2]_q)}\\ &= \frac{1}{1-\sum_{n\geq 0}\frac{q^{2n}(x-1)^n z^{3n+1}}{[3n+1]_q[2]_q^n\prod_{i=1}^{n}[3i]_q}}.\end{aligned}$$

As a third example, suppose τ is a permutation of the form $\tau = 12\cdots a\sigma(a+1)$ where σ is a permutation of $\{a+2,\cdots,k+1\}$, then τ has the minimal overlapping property and $\text{inv}(\tau) = (k-a)+\text{inv}(\sigma)$. Using Theorems 7.7 and 7.8, we have

$$\begin{aligned}\sum_{n\geq 0}\frac{z^n}{[n]_q!}\sum_{\sigma\in S_n} & x^{\#\text{ of }\tau\text{-matches}(\sigma)}q^{\text{inv}(\sigma)}\\ &= \frac{1}{1-(z+\sum_{i\geq 0}\frac{(x-1)^{i+1}z^{ik+1}}{[ik+1]_q!}(q^{k-a+\text{inv}(\sigma)})^{(i+1)}\prod_{j=2}^{i}\begin{bmatrix}jk-a\\k-a\end{bmatrix}}. \qquad (7.12)\end{aligned}$$

We pause to observe two other consequences of Theorem 7.8. The permutations α and β are called c-Wilf equivalent if the number of permutations in S_n without any consecutive α matches is equal to the number of permutations in S_n without any consecutive β matches for all $n \geq 1$. We call these two permutations strongly c-Wilf equivalent if

$$\sum_{n=0}^{\infty}\frac{z^n}{n!}\sum_{\sigma\in S_n} x^{\#\text{ of }\alpha\text{-matches in }\sigma} = \sum_{n=0}^{\infty}\frac{z^n}{n!}\sum_{\sigma\in S_n} x^{\#\text{ of }\beta\text{-matches in }\sigma}.$$

Clearly, if α and β are strongly c-Wilf equivalent, then α and β are also c-Wilf equivalent. The surprise is that Theorem 7.8 implies the converse in the case of minimal overlapping permutations. To see this, suppose that α and β are c-Wilf equivalent and take $x = 0$ in Theorem 7.8 to see that

$$\frac{1}{1-(z+\sum_{n\geq 1}\frac{z^{n(j-1)+1}}{(n(j-1)+1)!}(-1)^n mp_{\alpha,n(j-1)+1})}$$
$$= \frac{1}{1-(z+\sum_{n\geq 1}\frac{z^{n(j-1)+1}}{(n(j-1)+1)!}(-1)^n mp_{\beta,n(j-1)+1})}$$

from which it follows that $mp_{\beta,n(j-1)+1} = mp_{\alpha,n(j-1)+1}$ for all $n \geq 1$. Then Theorem 7.8 implies that the two generating functions for consecutive α and β matches are the same.

A second consequence of Theorem 7.8 is a proof of a conjecture of Elizalde [34] that if $\alpha = \alpha_1 \cdots \alpha_j$ and $\beta = \beta_1 \cdots \beta_j$ are permutations in S_j with the minimal overlapping property and $\alpha_1 = \beta_1$ and $\alpha_j = \beta_j$, then α and β are strongly c-Wilf equivalent. This was proved independently by Duane and Remmel [32] using Theorem 7.9 below and by Dotsenko and Khoroshkin [29].

Theorem 7.9. *Suppose $\alpha = \alpha_1 \cdots \alpha_j$ and $\beta = \beta_1 \cdots \beta_j$ are minimal overlapping permutations in S_j and $\alpha_1 = \beta_1$ and $\alpha_j = \beta_j$. Then for all $n \geq 1$,*

$$mp_{\alpha,n(j-1)+1} = mp_{\beta,n(j-1)+1}.$$

If in addition $q^{inv(\alpha)} = q^{inv(\beta)}$, then

$$mp_{\alpha,n(j-1)+1}(q) = mp_{\beta,n(j-1)+1}(q).$$

Proof. Let us try constructing all maximum packings $\sigma = \sigma_1 \ldots \sigma_{n(j-1)+1}$ of size $n(j-1)+1$ for α or β. One way to do this is to partition $\{1,\ldots,n(j-1)+1\}$ into sets $T_1,\ldots,T_n$ where $|T_1| = j$ and $|T_i| = j-1$ for $i \geq 2$ and use the elements of T_1 for $\sigma_1 \ldots \sigma_j$, use the elements of T_2 for $\sigma_{j+1} \ldots \sigma_{2j-1}$, use the elements of T_3 for $\sigma_{2j} \cdots \sigma_{3j-2}$, and so on.

This may not work for all choices of $T_1,\ldots T_n$. For instance, if $\alpha = 132$ and we pick $T_1 = \{4,5,6\}$ and $T_2 = \{1,2\}$, then there will be no way to use T_1 for the elements $\sigma_1\sigma_2\sigma_3$ and use T_2 for the elements $\sigma_4\sigma_5$ to produce a maximum packing for 132 because we must let $\sigma_1 = 3$, $\sigma_2 = 5$, and $\sigma_3 = 4$. But then σ_3 will be greater than σ_4 and σ_5 so that this choice will not allow us to construct a maximum packing for 132.

Our claim is that for any choice of $T_1,\ldots,T_n$, either we cannot construct a maximum packing for either α or β in this way or we can construct a maximum packing for both α and β in this way.

For example, if $\alpha = 24153$, then we first choose $T_1,\ldots,T_n$; suppose we pick $T_1 = \{2,5,6,7,10\}, T_2 = \{1,8,9,15\}, T_3 = \{3,11,13,14\}$, and $T_4 = \{4,12,16,17\}$. This means that when creating the permutation σ, these sets of integers must appear in the blank spaces as shown below:

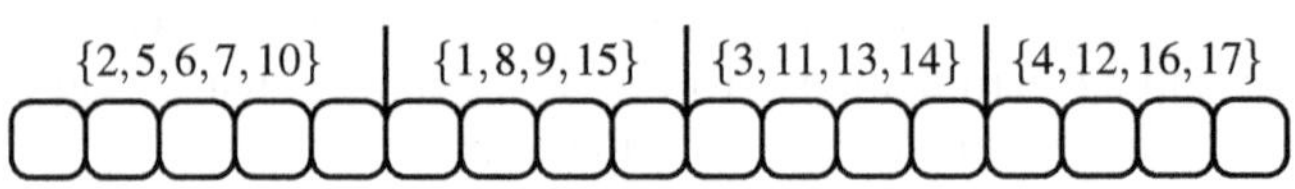

Given the choice for T_1, there is only one way to fill in the first 5 integers in σ as to form an α match; we must have this:

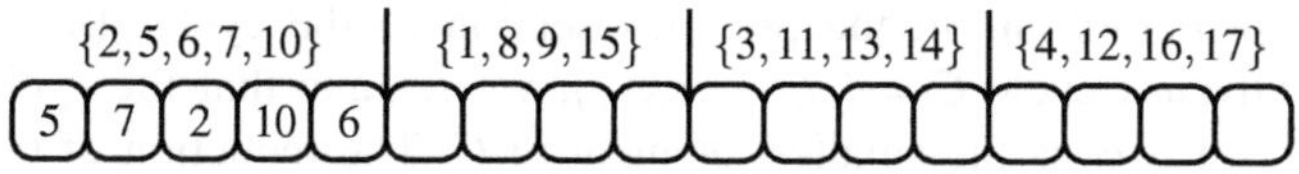

In this process, $\sigma_5 = 6$ must be the third largest element of T_1 since $\alpha_5 = 3$. In order to continue, it must be that 6 is the second largest element in $\{6\} \cup T_2$ if $\sigma_5 \cdots \sigma_9$ is an α-match. Since 6 is the second largest element in $\{6\} \cup T_2$, the positions of the elements of T_2 are then forced by the requirement that $\sigma_5 \cdots \sigma_9$ be an α-match which is pictured below:

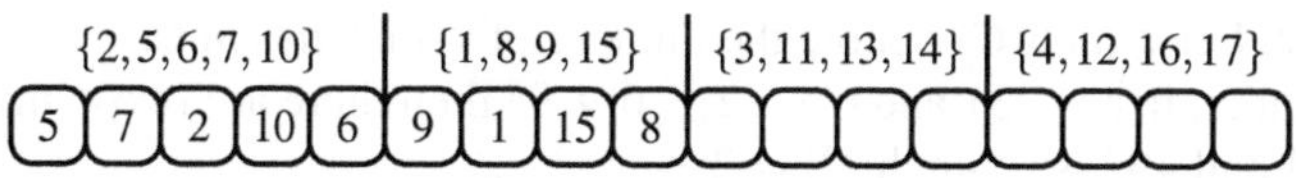

In particular, σ_9 must be the third largest element of $\{6\} \cup T_2$ so that $\sigma_9 = 8$. To continue, it must be that 8 is the second largest element of $\{8\} \cup T_3$ if $\sigma_9 \cdots \sigma_{13}$ is an α-match and σ_{13} must be the third largest element of $\{8\} \cup T_3$. In this case, 8 is the second largest element of $\{8\} \cup T_3$ and we are forced to have $\sigma_{13} = 11$ since 11 is the third largest element of $\{8\} \cup T_3$:

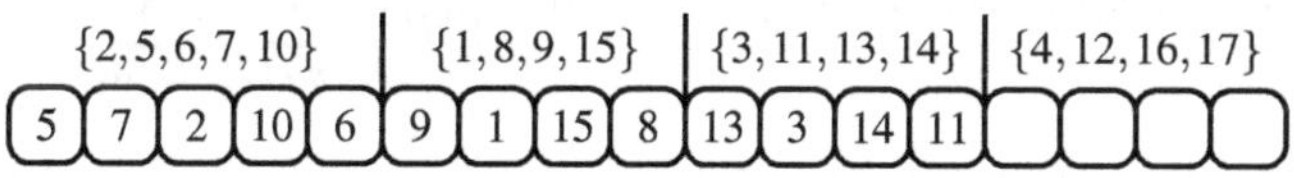

Repeating this logic with 11 being the second largest element of $\{11\} \cup T_4$, we can construct a maximum packing for α using $T_1, \cdots, T_4$:

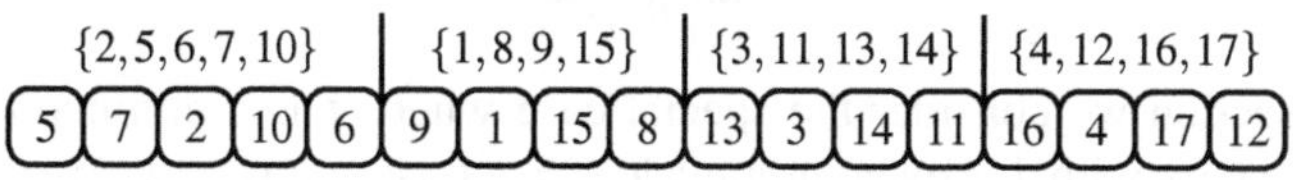

In general, if $\alpha_1 = \beta_1 = s$ and $\alpha_j = \beta_j = t$, then to be able to use T_2 to continue the construction of a maximum packing for either α or β, we must have that the tth largest element a_1 of T_1 is the sth largest element of $\{a_1\} \cup T_2$. If not, we cannot use $T_1, \ldots, T_n$ to construct a maximum packing for either α or β. If so, then the tth largest element a_2 of $\{a_1\} \cup T_2$ must be the s-largest element of $\{a_2\} \cup T_3$. If not, we cannot use $T_1, \cdots, T_n$ to construct a maximum packing for either α or β. If so, then the tth largest element a_3 of $\{a_2\} \cup T_3$ must be the s-largest element of $\{a_3\} \cup T_4$, and so on. Thus we can use $T_1, \ldots, T_n$ to construct a maximum packing σ for α if and only if we can use $T_1, \ldots, T_n$ to construct a maximum packing σ^* for β.

Moreover, if $q^{\mathrm{inv}(\alpha)} = q^{\mathrm{inv}(\beta)}$, then it will be the case that $q^{\mathrm{inv}(\sigma)} = q^{\mathrm{inv}(\sigma^*)}$, as needed to prove the theorem. □

7.4 Minimal Overlapping Patterns in Cycles

We can adapt definitions in the previous section to the cycle structure of permutations. Given a cycle $C = (c_0, \ldots, c_{p-1})$, we assume that the cycle has been rearranged so that c_0 is the smallest element of C. The reduction of C, $\text{red}(C)$, is the p-cycle in S_P where the ith smallest element in C is replaced by i. For example, $\text{red}(2,4,7,3) = (1,3,4,2)$.

Let $\tau = \tau_1 \cdots \tau_j \in S_j$ and $C = (c_0, \ldots, c_n)$ be an n-cycle in the cycle structure of some permutation σ. We say that τ consecutively occurs in σ if $j \leq n$ and there is an i such that $\text{red}(c_i c_{i+1} \cdots c_{i+j-1}) = \tau$ where we take the indices modulo p. Thus τ consecutively occurs in C if, as we traverse C, we see a consecutive sequence which reduces to τ. Let $\tau\text{-mch}(C)$ denote the number of τ-matches in C.

For example, if $\tau = 231$ and $C = (2,5,6,4,9)$ then there are two τ-matches in C, namely 564 and 492, so $\tau\text{-mch}(C) = 2$.

Our definitions ensure that τ occurs in C if and only if τ consecutively occurs in $\text{red}(C)$. Thus if σ is a permutation of S_n consisting of cycles $C_1, \ldots, C_k$, we define $\tau\text{-cmatch}(\sigma) = \sum_{i=1}^{k} \tau\text{-mch}(\text{red}(C_i))$. Let $\mathscr{C}_n$ be the set of all n-cycles in S_n. The exponential formula gives that

$$\sum_{n=0}^{\infty} \frac{z^n}{n!} \sum_{\sigma \in S_n} y^{\text{the number of cycles in } \sigma} x^{\tau\text{-cmatch}(\sigma)} = exp\left(y \sum_{n=1}^{\infty} \frac{z^n}{n!} \sum_{C \in \mathscr{C}_n} x^{\tau\text{-mch}(C)} \right). \tag{7.13}$$

Looking at the right-hand side of this equation, we are motivated to study the generating function

$$C_\tau(z,x) = \sum_{n \geq 1} \frac{z^n}{n!} \sum_{C \in \mathscr{C}_n} x^{\tau\text{-mch}(C)}.$$

In this subsection we consider the special case where τ is a minimal overlapping pattern that starts with 1. Given an n-cycle $C = (1, \sigma_2, \ldots, \sigma_n)$, we associate a permutation $\sigma(C) = 1\sigma_2 \cdots \sigma_n$. We begin with the following lemma:

Lemma 7.1. *If $\tau = \tau_1 \cdots \tau_j \in S_j$ and $\tau_1 = 1$, then for all n-cycles C, τ-cmatch$(C) =$ τ-mch$(\sigma(C))$.*

Proof. If $C = (c_0, \ldots, c_{n-1}) = (1, \sigma_2, \ldots, \sigma_n)$ and $c_i c_{i+1} \cdots c_{i+j-1}$ is a cycle τ-match in C where we take the indices modulo m, then c_0 is an element of the cycle τ-match if and only if $i = 0$. That is, since τ starts with 1, the only role that $c_0 = 1$ can play in a cycle τ-match is 1.

This means that if $c_i c_{i+1} \cdots c_{i+j-1}$ is a cycle τ-match in C, then $1 \leq i < i+j-1 \leq n$, thereby implying that $c_i c_{i+1} \cdots c_{i+j-1}$ is also τ-match in $\sigma(C)$. Thus every τ-match in $\sigma(C)$ gives rise to a τ-match in C. □

Another useful property of minimal overlapping patterns $\tau \in S_j$ which start with 1 is the following lemma whose proof we leave to the reader.

Lemma 7.2. *If $\tau = \tau_1 \cdots \tau_j \in S_j$ is a minimal overlapping permutation with $\tau_1 = 1$, then for all $n \geq 1$, every $\sigma \in MP_{\tau,n(j-1)+1}$ starts with 1.*

Lemma 7.1 gives that if $\tau \in S_j$ is a permutation that starts with 1, then

$$C_\tau(z,x) = \sum_{n=1}^{\infty} \frac{z^n}{n!} \sum_{\sigma \in S1_n} x^{\tau\text{-mch}(\sigma)}$$

where $S1_n$ is set of all permutations $\sigma \in S_n$ which start with 1.

Theorem 7.10. *Suppose that $\tau = \tau_1 \cdots \tau_j \in S_j$ is a minimal overlapping pattern that starts with 1 where $j \geq 3$. Then*

$$\sum_{n=1}^{\infty} \frac{z^n}{(n-1)!} \sum_{\sigma \in S1_n} x^{\tau-mch(\sigma)} = \frac{z + \sum_{n\geq 1} \frac{z^{n(j-1)+1}}{(n(j-1))!}(x-1)^n mp_{\tau,n(j-1)+1}}{1-(z+\sum_{n=1}^{\infty} \frac{z^{n(j-1)+1}}{(n(j-1)+1)!}(x-1)^n mp_{\tau,n(j-1)+1}}.$$

Proof. Let φ be the ring homomorphism in Theorem 7.8 after taking $q = 1$. The generating function in the statement of the theorem will follow from applying φ to the modified basis $p_{\nu,n}$ where

$$\nu(n) = \begin{cases} n & \text{if } n = s(j-1)+1 \text{ for some } s \geq 0, \\ 0 & \text{if otherwise.} \end{cases}$$

Expanding $p_{\nu,n}$ in terms of the elementary symmetric functions using Theorem 4.1, we have

$$\begin{aligned}(n-1)!\varphi(p_{\nu,n}) &= \frac{1}{n} \sum_{\lambda \vdash n} \binom{n}{\lambda} w_\nu(B_{\lambda,(n)}) f(\lambda_1) f(\lambda_2) \cdots \\ &= \sum_{\substack{T \in B_{\lambda,(n)} \text{ for some } \lambda \vdash n \\ \text{has bricks } b_1,\ldots,b_\ell}} \binom{n-1}{b_1-1,\ldots,b_{\ell(\lambda)}-1} f(b_1)f(b_2)\cdots.\end{aligned}$$

We have consciously chosen to use ν to change the weight on the first brick instead of the last brick.

Next we give a combinatorial interpretation to this last equation. Suppose we have a brick tabloid $B = (b_1,\ldots,b_\ell)$ of size n such that $b_i \in \{1\} \cup \{s(j-1)+1 : s \geq 1\}$ for all i. Interpret the $\binom{n-1}{b_1-1,b_2,\ldots,b_{\ell(\lambda)}}$ term as filling the cells of B with a permutation $\alpha = \alpha_1 \cdots \alpha_n$ such that $\alpha_1 = 1$ and α is increasing in each brick of B. That is, we choose $b_1 - 1$ elements for the first brick which we use to fill cells $2,\ldots,b_1$ in B and we choose b_i elements to fill b_i for each $i \geq 2$.

If $b_i = s(j-1)+1 > 1$, then let $1 \leq a_1^i < \cdots < a_{s(j-1)+1}^i \leq n$ be the elements of α which are in the cells of b_i reading from left to right. Interpret the term

$$f(b_i) = (x-1)^{(b_i-1)/(j-1)} mp_{\tau,b_i}$$

as the number of ways of filling b_i with a permutation $\beta_i \in MP^k_{\tau,s(j-1)+1}$ and then labeling each cell in b_i which is the start of a τ-match in β_i with either x or -1. For brick b_1, clearly 1 is in cell 1 if b_1 equals 1 and if $b_1 > 1$, then 1 is in cell 1 since by Lemma 7.2, every maximum packing in MP_{τ,b_1} starts with 1. Finally, we replace the numbers by $1,\ldots,s(j-1)+1$ that occur in β_i by $a_1,\ldots,a_{s(j-1)+1}$, respectively. Doing this for each brick will result in a filling of the cells of B with a permutation $\sigma \in S1_n$.

Let $\mathcal{Q}1_{\tau,n}$ denote the set of all labeled brick tabloids that can be constructed in this way. An element in $Q \in \mathcal{Q}1_{\tau,n}$ can be considered a triple $T = (B,\sigma,L)$ where

1. $B = (b_1,\ldots,b_\ell)$ is a brick tabloid of shape (n) such that $b_i \in \{1\} \cup \{s(j-1)+1 : s \geq 1\}$ for all i,
2. $\sigma \in S_n$ is a permutation which starts with 1, and
3. if $b_i = s(j-1)+1 > 1$, then cells of b_i are filled with a sequence γ_i such that $\mathrm{red}(\gamma_i)$ is a maximum packing for τ of size $s(j-1)+1$ and each cell of b_i which corresponds to the start of τ-match in γ_i is labeled with either -1 or x.

The weight of T is defined to be the product of the x labels in T and the sign of T, $sign(T)$, is the product of the -1 labels in T.

At the point we have shown

$$(n-1)!\varphi(p_{\nu,n}) = \sum_{T=(B,\sigma,L)\in\mathcal{Q}1_{\tau,n}} sgn(T)w(T).$$

We have arrived at the same situation that had in the proof of Theorem 7.8 with the exception that all the permutations σ start with 1. We can then apply exactly the same involution I that we applied in the proof of Theorem 7.8 and then apply φ to (4.1) in order to find the generating function in the statement of the theorem. □

By dividing both sides of the generating function in the statement of the last theorem by z and then integrating with respect to z, we see that

$$\begin{aligned} C_\tau(z,x) &= \sum_{n\geq 1} \frac{z^n}{n!} \sum_{\sigma\in S1_n} x^{\tau\text{-mch}(\sigma)} \\ &= \int_0^z \frac{1+\sum_{n\geq 1}\frac{y^{n(j-1)}}{(n(j-1))!}(x-1)^n mp_{\tau,n(j-1)+1}}{1-(y+\sum_{n\geq 1}\frac{y^{n(j-1)+1}}{(n(j-1)+1)!}(x-1)^n mp_{\tau,n(j-1)+1})} dy. \end{aligned}$$

Using the substitution that $u = (y+\sum_{n\geq 1}\frac{y^{n(j-1)+1}}{(n(j-1)+1)!}(x-1)^n mp_{\tau,n(j-1)+1})$, we find

$$C_\tau(z,x) = -\ln\left(1-(z+\sum_{n\geq 1}\frac{z^{n(j-1)+1}}{(n(j-1)+1)!}(x-1)^n mp_{\tau,n(j-1)+1})\right)$$

which, in turn, can be used in (7.13).

7.5 Minimal Overlapping Patterns in Words

There are two natural ways to define pattern matchings in words, depending on whether or not a word is reduced before looking for a pattern match.

Making these ideas precise, we say that the word $u = u_1 \cdots u_j$ exactly consecutively occurs in the word $w = w_1 \cdots w_n$ if there is an i such that $w_i w_{i+1} \cdots w_{i+j-1} = u$. In such a situation, we say that w has an exact u-match starting at position i. We let u-ematch(w) denote the number of exact u-matches in w. This was our concept of pattern matches given when discussing Theorem 7.4.

A second way to define a pattern match in a word is to reduce the word first, just as we did for permutations. If u and w have letters in $\{1,2,\ldots\}$ such that $\mathrm{red}(u) = u$, then we say that u consecutively occurs in w if there is an i such that $\mathrm{red}(w_i w_{i+1} \cdots w_{i+j-1}) = u$. In such a situation, w has a u-match starting at position i. Let u-match(w) denote the number of u-matches in w.

The two corresponding analogues of minimal overlapping permutations for words are as follows. We say that u has the exact match minimal overlapping property if the smallest i such that there exists a word w of length i with u-ematch$(w) = 2$ is $2j-1$. If $\mathrm{red}(u) = u$, then we say that u has the minimal overlapping property (for words) if the smallest i such that there exists a word w of length i with $u-\mathrm{match}(w) = 2$ is $2j-1$. As before, this means that for a word w, two u-matches in w can share at most one letter occurring at the end of the first u-match and at the start of the second u-match.

For example, 121 has both the minimal overlapping property and the exact match minimal overlapping property whereas 1122 does not have the minimal overlapping property.

A new phenomenon happens in the case of pattern matching in words. Let us say that a word u of length j with letters in $\{0,\ldots,k-1\} = [k]$ has the $[k]$-nonoverlapping property ($[k]$-exact match nonoverlapping property) if the smallest i such that there exists a word of length i with u-match$(w) = 2$ (u-ematch$(w) = 2$) is $2j$. Thus u has the $[k]$-nonoverlapping ($[k]$-exact match nonoverlapping) property if no two u-matches (exact u-matches) can share a letter.

For example, if $k = 3$ and $u = 00112$, then u has both the $[k]$-nonoverlapping property and the $[k]$-exact match nonoverlapping property. However, if $k = 3$ and $v = 011$, then v has the $[k]$-exact match nonoverlapping property but does not have the nonoverlapping property since $w = 01122$ has two v-matches. We note that whether a word u has the $[k]$-minimal overlapping property, the $[k]$-exact minimal overlapping property, or the $[k]$-exact nonoverlapping property can depend on k. For example, 1122 is not $[k]$-minimal overlapping if $k \geq 3$ but it has the nonoverlapping property in the alphabet $[2] = \{1,2\}$.

The goal of this section is to find generating functions for the number of matches of minimal overlapping patterns in words. To this end, suppose that $u \in A^j$ where $A = \{1,2,\ldots\}$ or $A = [k]$ for some $k \geq 2$. If $\mathrm{red}(u) = u$ and u has the A-minimal overlapping property, then the shortest words $w \in A^*$ such that u-match$(w) = n$ have length $n(j-1)+1$ so we let $MP^A_{u,n(j-1)+1}$ denote the set of words $w \in A^{n(j-1)+1}$ with u-match$(w) = n$. We will refer to elements of $MP^A_{u,n(j-1)+1}$ as A-maximum packings for u. Then we let $mp^A_{u,n(j-1)+1} = |MP^A_{u,n(j-1)+1}|$ if A is finite,

$$mp^A_{u,n(j-1)+1}(r) = \sum_{w \in MP^A_{u,n(j-1)+1}} r^{\Sigma w},$$

and

$$mp^A_{u,n(j-1)+1}(x_1,x_2,\dots) = \sum_{w \in MP^A_{u,n(j-1)+1}} x(w)$$

where $x(w) = x_1^{\text{the number of } w_1\text{'s in } w} x_2^{\text{the number of } w_2\text{'s in } w} \cdots$.

If u has the A-exact match minimal overlapping property, then the shortest words $w \in A^*$ such that u-ematch$(w) = n$ have length $n(j-1)+1$ so we let $EMP^A_{u,n(j-1)+1}$ denote the set of words $w \in A^{n(j-1)+1}$ such that u-ematch$(w) = n$. We will refer to elements of $EMP^A_{u,n(j-1)+1}$ as A-exact match maximum packings for u. Furthermore, we define $emp^A_{u,n(j-1)+1} = |EMP^A_{u,n(j-1)+1}|$ if A is finite,

$$emp^A_{u,n(j-1)+1}(r) = \sum_{w \in EMP^A_{u,n(j-1)+1}} r^{\Sigma w},$$

and

$$emp^A_{u,n(j-1)+1}(x_1,x_2,\dots) = \sum_{w \in EMP^A_{u,n(j-1)+1}} x(w).$$

For example, if $u = 121$ and $k \geq 2$, then the only $w \in [k]^{2n+1}$ such that u-ematch $(w) = n$ is $1(21)^n$. This means $emp^{[k]}_{121,2n+1} = 1$, $emp_{121,2n+1}(r) = r^{3n+1}$, and $emp^{[k]}_{121,n(j-1)+1}(x_1,x_2) = x_1^{n+1}x_2^n$ for all $n \geq 1$ and $k \geq 2$.

If we are just considering u-matches instead of exact u-matches in $[k]^{2n+1}$, then the only words $w \in [k]^{2n+1}$ such that u-match$(w) = n$ are of the form $si_1si_2j\cdots si_ns$ where $s \in \{1,\dots,k-1\}$ and $i_1,\dots,i_n \in \{s+1,\dots,k\}$. Thus

$$mp^{[k]}_{121,2n+1} = \sum_{s=1}^{k-1}(k-s)^n = \sum_{i=1}^{k-1} i^n,$$

$$mp^{[k]}_{121,2n+1}(r) = \sum_{s=1}^{k-1} r^{(n+1)s}(r^{s+1}[k-s]_r)^n = \sum_{i=1}^{k-1} r^{(2n+1)(k-i)}[i]_r^n, \qquad \text{and}$$

$$mp^{[k]}_{121,2n+1}(x_1,\dots,x_k) = \sum_{s=1}^{k-1} x_s^{n+1}(\sum_{t=s+1}^{k} x_t)^n.$$

Similarly, if we are just considering u-matches instead of exact u-matches for words with letters in $\{1,2,\dots\}$, then the only words w such that u-match$(w) = n$ are of the form $si_1si_2j\cdots si_ns$ where $i_1,\dots,i_n \in \{s+1,s+2,\dots,\}$. Thus $MP_{121,2n+1}$ is infinite and

$$mp^{\{1,2,\dots\}}_{121,2n+1}(r) = \sum_{s=1}^{\infty} r^{(n+1)s}(\frac{r^{s+1}}{1-r})^n = \sum_{s\geq 1} \frac{r^{s(2n+1)+n}}{(1-r)^n} \text{ and}$$

$$mp^{\{1,2,\dots\}}_{121,2n+1}(x_1,x_2,\dots) = \sum_{s=1}^{\infty} x_s^{n+1}(\sum_{t=s+1}^{\infty} x_t)^n.$$

Theorem 7.11. *Let u be a word of length j with letters in $\{1,2,\dots\}$ where $j \geq 3$. If u has the minimal overlapping property and $red(u) = u$, then*

$$\sum_{n\geq 0} z^n \sum_{w\in\{1,2,\dots\}^n} x^{u\text{-}match(w)} x(w)$$
$$= \frac{1}{1-((\sum_{i=1}^{\infty} x_i)\,z + \sum_{n\geq 1} z^{n(j-1)+1}(x-1)^n mp^{\{1,2,\dots\}}_{u,n(j-1)+1}(x_1,x_2,\dots))}.$$

If u has the exact match minimal overlapping property, then

$$\sum_{n\geq 0} z^n \sum_{w\in\{1,2,\dots\}^n} x^{u\text{-}ematch(w)} x(w) =$$
$$\frac{1}{1-((\sum_{i=1}^{\infty} x_i)\,z + \sum_{n\geq 1} z^{n(j-1)+1}(x-1)^n emp^{\{1,2,\dots\}}_{u,n(j-1)+1}(x_1,x_2,\dots))}.$$

Proof. This proof is a relatively straightforward modification of the proof given in Theorem 7.8.

Suppose u is a word of length j with letters in $\{1,2,\dots\}$ such that u has the minimal overlapping property and $\text{red}(u) = u$. Define a ring homomorphism φ by $\varphi(e_n) = (-1)^{n-1} f(n)$ where

$$f(n) = (x-1)^{(n-1)/(j-1)} mp^{\{1,2,\dots\}}_{u,n}(x_1,x_2,\dots)$$

if $n = s(j-1)+1$ for some $s \geq 0$ and $f(n) = 0$ otherwise. Applying φ to h_n gives

$$\varphi(h_n) = \sum_{\substack{T\in B_{\lambda,(n)} \text{ for some } \lambda\vdash n\\ \text{has bricks } b_1,\dots,b_\ell}} f(b_1)f(b_2)\cdots. \tag{7.14}$$

As usual, our next task is to give a combinatorial interpretation to the right-hand side of this last equation. Start with a brick tabloid $B = (b_1,\dots,b_\ell)$ of size n such that each brick b_i has a length which is of the form $s(j-1)+1$ for some $s \geq 0$. Interpret the term $f(b_i)$ as filling b_i with a word $v_i \in MP^{\{1,2,\dots\}}_{u,s(j-1)+1}$ and then labeling each cell in b_i which is the start of a u-match in v_i with either x or -1.

For example, one different possible combinatorial objects created in this way when $u = 121$ is

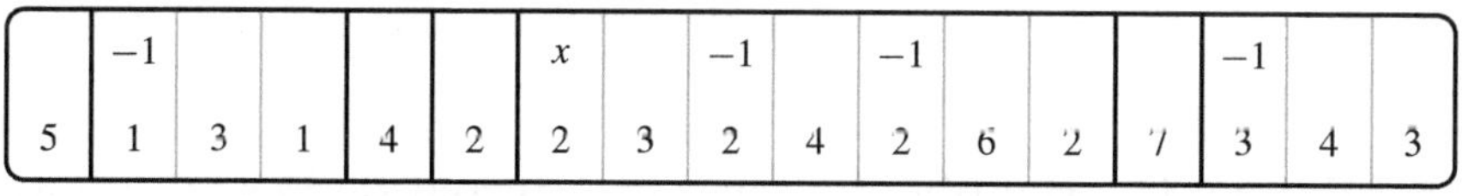

The weight of this object is $+x$. The signed weighted sum over all combinatorial objects is equal to (7.14).

These are the exact same combinatorial objects as found in proof of 7.8 with the exception that our brick tabloids contain words instead of permutations. From here on out the proof follows in exactly the same way as the proof of theorem 7.8:

apply the exact same sign reversing weight preserving involution I on the collection of such combinatorial objects T as found in the proof of theorem 7.8 (except with every instance of "τ" found in the description of the involution I replaced with the word "u"). For example, the image of the combinatorial object shown above is

						x		-1		-1				-1		
5	1	3	1	4	2	2	3	2	4	2	6	2	7	3	4	3

Fixed points under this involution have a weight of $x^{u\text{-match}(w)}x(w)$. The sum of all fixed points, and therefore $\varphi(h_n)$, is equal to $\sum_{w\in\{1,2,\dots\}^n} x^{u\text{-match}(w)}x(w)$. Applying φ to Theorem 2.5 gives the first generating function in the statement of the theorem, and repeating the argument yet again for exact matches gives the second generating function in the statement of the theorem. □

The proof of the last theorem still works if we restrict all the entries in the filled brick tabloids to be from the alphabet $[k]=\{1,\dots,k\}$ for any $k\geq 2$, and so we can replace $\{1,2,\dots\}$ in the statement of the last theorem with $[k]$ if desired.

Another variant on Theorem 7.11 can be stated for colored permutations. A colored permutation is a pair (σ,u) where σ is a permutation in S_n and u is a word of length n with letters in $\{0,\dots,k-1\}$. We say that a k-colored permutation (σ,u) has (τ,w)-match starting at position i if $\mathrm{red}(\sigma_i\cdots\sigma_{i+j-1})=\tau$ and $\mathrm{red}(u_iu_{i+1}\cdots u_{i+j-1})=w$. This is an exact (τ,w)-match if $u_1u_{i+1}\cdots u_{i+j-1}=w$. A number of papers in the literature have addressed matches in colored permutations, such as [34, 85, 84, 86].

The definitions for maximal packings and the minimal overlapping property for colored permutations follow by analogy to the situation for regular permutations or words. We record these results here, with the details carefully spelled out in [32].

Theorem 7.12. *If (τ,u) is a colored permutation with $red(u)=u$ and (τ,u) has the minimal overlapping property for colored permutations, then the generating function*

$$\sum_{n=0}^{\infty}\frac{z^n}{n!}\sum_{(\sigma,w)\text{ is a colored permutation}} x^{(t,u)-match(\sigma,u)}q^{inv(\sigma)}x(w)$$

is equal to

$$\frac{1}{1-(x_0+\cdots+x_{k-1})z-\sum_{n=1}^{\infty}\frac{z^{n(j-1)+1}}{[n(j-1)+1]_q!}(x-1)^n mp^{[k]}_{(\tau,u),n(j-1)+1}(q,x_0,\dots,x_{k-1})}.$$

7.6 Minimal Overlapping Patterns in Alternating Permutations

This section is devoted to finding an analogue of Theorem 7.8 for alternating permutations. In contrast to our earlier definition of an alternating permutation, in this section we will consider $\sigma=\sigma_1\cdots\sigma_n\in S_n$ an alternating permutation if

$$\sigma_1<\sigma_2>\sigma_3<\sigma_4\cdots.$$

That is, in this section only, we will consider an alternating permutation as beginning with an increase rather than a decrease. Let A_n be the set of all alternating permutations in S_n.

An alternating permutation $\tau \in A_{2j}$ has the alternating minimal overlapping property if the smallest i such that there is a permutation $\sigma \in A_{2i}$ with $\tau\text{-mch}(\sigma) = 2$ is $2(j-1)$. This means that in any permutation $\sigma = \sigma_1 \cdots \sigma_{2n} \in A_{2n}$, any two τ-matches in σ can share at most two letters which must be at the end of the first τ-match and the start of the second τ-match.

For example, $\tau = 142536$ does not have the alternating minimal overlapping property since $\tau\text{-match}(15263748) = 2$ and the τ-match starting at position 1 and the τ-match starting at position 3 share four letters, namely, 2637. However, the permutation $\tau = 143526$ does have the alternating minimal overlapping property.

If $\tau \in A_{2j}$ has the alternating minimal overlapping property, then the shortest permutations $\sigma \in A_{2m}$ such that $\tau\text{-match}(\sigma) = n$ have length $2n(j-1)+2$. Thus we let $UDMP_{\tau,2n(j-1)+2}$ equal the set of permutations $\sigma \in A_{2n(j-1)+2}$ such that $\tau\text{-match}(\sigma) = n$. We shall refer to the permutations in $UDMP_{\tau,2n(j-1)+2}$ as alternating maximum packings for τ and we let $udmp_{\tau,2n(j-1)+2} = |UDMP_{\tau,2n(j-1)+2}|$.

It will turn out that the generating functions for the number of τ-matches in alternating permutations can be expressed in terms of what we call generalized maximum packings for τ. We say that $\sigma \in S_{2n}$ is a generalized maximum packing for τ if we can break σ into consecutive blocks $\sigma = B_1 \cdots B_k$ such that

1. for all $1 \leq j \leq k$, B_j is either an increasing sequence of length 2 or $\text{red}(B_j)$ is an alternating maximum packing for τ of length $2s$ for some s and
2. for all $1 \leq j \leq k-1$, the last element of B_j is less than the first element of B_{j+1}.

If σ is a generalized maximum packing for τ, there is only one possible block structure. That is, if $\sigma = \sigma_1 \cdots \sigma_{2n} \in S_{2n}$ is a generalized maximum packing for τ, our conditions force that $\sigma_{2j-1} < \sigma_{2j}$ for $i = j, \ldots, n$. Therefore $\sigma_{2j-1}\sigma_{2j}$ and $\sigma_{2j+1}\sigma_{2j+2}$ are in the same block if and only if $\sigma_{2j} > \sigma_{2j+1}$.

If σ is a generalized maximum packing for τ of length $2n$ with block structure $B_1 \cdots B_k$, then we define the weight $w(B_j)$ of block B_j to be 1 if B_j has size 2 and to be $(x-1)^s$ if B_j has length $2s(j-1)+2$ where $s \geq 1$. Then we define the weight $w(\sigma)$ of σ to be $(-1)^{k-1} w(B_1) w(B_2) \cdots W(B_k)$.

For example, if $\tau = 143526$, then

$$\sigma = 1\ 2\ 3\ 8\ 7\ 9\ 4\ 10\ 6\ 11\ 5\ 12\ 13\ 14\ 15\ 18\ 17\ 19\ 16\ 20$$

is a generalized maximum packing for τ where $B_1 = 1\ 2$, $B_2 = 3\ 8\ 7\ 9\ 4\ 10\ 6\ 11\ 5\ 12$, $B_3 = 13\ 14$, $B_4 = 15\ 18\ 17\ 19\ 16\ 20$. Thus $w(B_1) = w(B_3) = 1$, $w(B_2) = (x-1)^2$, and $w(B_4) = (x-1)$ so that $w(\sigma) = (-1)^3 (x-1)^3 = -(x-1)^3$.

Let $GMP_{\tau,2n}$ denote the set of $\sigma \in S_{2n}$ which is a generalized maximum packings for τ and let

$$GMP_{\tau,2n}(x) = \sum_{\sigma \in GMP_{\tau,2n}} w(\sigma).$$

Similar definitions for generalized maximum packing can be given for permutations of an odd number. Specifically, $\sigma \in S_{2n+1}$ is a generalized maximum packing for τ if we can break σ into consecutive blocks $\sigma = B_1 \cdots B_k$ such that

1. for all $1 \le j < k$, B_j is either an increasing sequence of length 2 or $\mathrm{red}(B_j)$ is alternating maximum packing of τ of length $2s$ for some s,
2. B_k is a block of size 1, and
3. for all $1 \le j \le k-1$, the last element of B_j is less than the first element of B_{j+1}.

From here the sets $GMP_{\tau,2n+1}$ and $GMP_{\tau,2n+1}(x)$ are defined analogously as in the case of even length permutations.

The main theorem in this section is as follows.

Theorem 7.13. *For $\tau \in A_{2j}$, an alternating minimal overlapping permutation with $j \ge 3$, we have*

$$\sum_{n=0}^{\infty} \frac{z^{2n}}{(2n)!} \sum_{\sigma \in A_{2n}} x^{\tau\text{-}mch(\sigma)} = \frac{1}{1 - \sum_{n=1}^{\infty} GMP_{\tau,2n}(x) \frac{z^{2n}}{(2n)!}}.$$

and

$$\sum_{n=1}^{\infty} \frac{z^{2n-1}}{(2n-1)!} \sum_{\sigma \in A_{2n-1}} x^{\tau\text{-}mch(\sigma)} = \frac{\sum_{n=1}^{\infty} GMP_{\tau,2n-1}(x) \frac{z^{2n-1}}{(2n-1)!}}{1 - \sum_{n=1}^{\infty} GMP_{\tau,2n}(x) \frac{z^{2n}}{(2n)!}}.$$

Proof. We begin by finding the first of the two generating functions in the statement of the theorem. Define a ring homomorphism φ by setting $\varphi(e_{2n+1}) = 0$ and

$$\varphi(e_{2n}) = \frac{(-1)^{n-1}}{n!} GMP_{\tau,2n}(x)$$

for all n.

Since an integer partition $\lambda \vdash (2n+1)$ must have at least one odd part, $\varphi(e_\lambda)$, and therefore $\varphi(h_{2n+1})$, must both equal 0. Expanding $(2n)!\varphi(h_{2n})$ in the usual way, we have

$$(2n)!\varphi(h_{2n}) = \sum_{\substack{T \in B_{\lambda,(n)} \text{ for some } \lambda \vdash n \\ \text{has bricks } b_1,\ldots,b_\ell}} \binom{2n}{2b_1,\ldots,2b_\ell} GMP_{\tau,2b_1}(x) GMP_{\tau,2b_2}(x) \cdots.$$

Create combinatorial objects from this equation by selecting a brick tabloid of shape $(2n)$ with only even length bricks and associate with each brick a disjoint subset of $\{1,\ldots,2n\}$. Use the factor of the form $GMP_{\tau,2b_1}(x)GMP_{\tau,2b_2}(x)\cdots$ to select a sequence of permutations $\sigma^{(1)},\ldots,\sigma^{(\ell(\lambda))}$ such that $\sigma^{(j)} \in A_{2b_j}$ is a generalized maximum packing for τ for all j. Then for each j, we let $\alpha^{(j)}$ the sequence that arises by replacing the rth largest element of $\sigma^{(j)}$ by the rth largest element of the subset associated with each brick. Place these sequences $\alpha^{(j)}$ into the cells of the bricks.

We still must account for the x weights coming from $GMP_{\tau,b_i}(x)$. To do so, place a weight of 1 on top of each block of size 2 that ends a brick. Place a weight of -1 on each block of size 2 which does not end a brick. For blocks of size $\geq 2j$, we place an $(x-1)$ at start of each τ-match in the block and, in addition, we add a factor of -1 to the first match in the block if the block is not the last block in a brick.

For example, we have pictured one combinatorial object created in this way when $\tau^{(1)} = 231546$ where the underlying brick tableau has bricks of length 2, 8, and 6. We have also indicated the block structure in each brick by underlying those elements in a common block.

1		-1		$x-1$						$x-1$					
1	3	4	5	9	10	7	12	11	13	6	8	2	15	14	16

If $T_{\tau,2n}$ denotes the set of all combinatorial objects constructed in this way, then

$$(2n)!\varphi(h_{2n}) = \sum_{T \in T_{\tau,2n}} w(T)$$

where the weight of T is the product of the weights above the integers in T.

We now define two involutions, I and J, on the set of combinatorial objects we have created. The first involution I is as follows. Read the bricks from left to right until finding the first brick d_j such that either

1. the generalized maximum packing corresponding to the elements in d_j consists of more than one block or
2. the generalized maximum packing corresponding to the elements in d_j consists of a single block and the last element of d_j is less than the first element of the following brick d_{j+1}.

In the first situation, split d_j into two bricks d^* and d^{**} where d^* contains the cells of the first block in the generalized maximum packing corresponding to the elements in d_j and d^{**} contains the remaining cells of d_j. Keep all the labels the same except change the label on the first cell of d^* from -1 to 1 if the first block of d_j is of size 2 and from $-(x-1)$ to $(x-1)$ if the first block of d_j has size ≥ 4.

In the second situation, combine bricks d_j and d_{j+1} into a single brick d. Since the last element of d_j is less than the first element of d_{j+1}, the elements in the new brick d will still reduce to a generalized maximum packing. Keep all the labels the same except change the label on the first cell of d_j from 1 to -1 if d_j is of size 2 and from $(x-1)$ to $-(x-1)$ if d_j has size ≥ 4.

In either situation, do not change the underlying permutation α. If neither situation applies, define T to be a fixed point under I.

For example, if T is the combinatorial object depicted earlier in the proof, then the second situation applies. The effect of I on T is that we combine the first and second bricks in T to create the combinatorial object shown below:

-1		-1		$x-1$						$x-1$					
1	3	4	5	9	10	7	12	11	13	6	8	2	15	14	16

This function I is a sign reversing involution. Therefore, in order to understand $n!\varphi(h_{2n})$, we only need to understand the fixed points T under I. Such a fixed point T must have elements in each brick d which reduce to a generalized maximum packing of τ consisting of a single block. Furthermore, we cannot combine any consecutive bricks d_j and d_{j+1} such that the last element in d_j is greater than or equal to the first element of d_{j+1} for any j. This means that the underlying permutation α must be an alternating permutation.

To recap, fixed points T under I must satisfy these conditions:

1. The permutation α is an alternating permutation of length $2n$,
2. T has bricks $d_1,\ldots,d_k$ where each d_j has even length and the elements in d_j reduce to a generalized maximum packing of τ within a single block, and
3. the label in the jth cell of T is $(x-1)$ if j is the start of τ-match in α and is equal to 1 otherwise.

Instead of these exact fixed points, take any combinatorial object T with a cell with a label $(x-1)$ and create two combinatorial objects T_1 and T_2 such that the label $(x-1)$ is changed to x in T_1 and -1 in T_2. Therefore instead of considering fixed points which satisfy the three above conditions, change the third condition to

3. the label in the jth cell of T is x or -1 if j is the start of τ-match in α and is equal to 1 otherwise.

For example, one such fixed point T under I is when $\tau = 231546$ is

1		1		-1				x						1	
1	4	3	9	6	7	2	10	8	11	5	13	12	14	15	16

On these fixed points we will apply the involution J, which is fundamentally the same as the involution given in the proof of Theorem 7.8, with modifications only made to account for the fact that we are dealing with alternating permutations.

Define a sign reversing weight preserving involution J on the collection of such combinatorial objects T by scanning the cells from left to right looking for the first time we are in one of the following cases:

Case 1. There is a brick b_i of size $2j$ whose first cell is labeled with -1. In this case, let $J(T)$ be the object T with this brick of size j replaced with j bricks of size 2 and the -1 sign removed.

Case 2. There are j consecutive bricks of size 2 in T, namely $b_i, b_{i+1}, \ldots, b_{i+j-1}$, such that the letters in these cells form a τ-match. In this case, $J(T)$ is T with the bricks $b_i, b_{i+1}, \ldots, b_{i+j-1}$ combined into a single brick b of size $2j$ and the first cell of b is labeled with -1.

Case 3. There is a brick b_i of size $2(s+1)(j-1)+2$ where $s \geq 1$ such that all the labels on b_i are xs except for the cell that is $2j$ cells from the right which is labeled with -1. In this case, $J(T)$ is found by replacing the brick b_i by a brick of size $2s(j-1)+2$ followed by $j-1$ bricks of size 2 and removing the -1 label that was in b_i.

Case 4. There are $2j$ consecutive bricks in T, namely $b_i, b_{i+1}, \ldots, b_{i+j-1}$ such that $b_i = 2s(j-1)+2 > 2$ and $b_{i+1}, \ldots, b_{i+j-1}$ are of size 1, all the labels on b_i are xs, and the letters in these bricks form an alternating maximum packing for τ of size $2(s+1)(j-1)+2$. In this case, $J(T)$ is found by replacing the bricks $b_i, b_{i+1}, \ldots, b_{i+j-1}$ by a single brick b of size $2(s+1)(j-1)+2$ and by labeling the last cell of b_i with a -1.

Case 5. There is a brick b_i of size $2(s+1)(j-1)+2$ where $s \geq 1$ such that the first cell of b_i is labeled with -1. In this case, $J(T)$ is found by replacing the brick b_i by $j-1$ bricks of size 2 followed by a brick of size $2s(j-1)+2$ and by removing the -1 label that was on the first cell of b_i.

Case 6. There are j consecutive bricks in T, namely $b_i, b_{i+1}, \ldots, b_{i+j-1}$, such that $b_i, \ldots, b_{i+j-2}$ are of size 2, $b_{i+j-1} = 2s(j-1)+2 > 2$, and the letters in these bricks form an alternating maximum packing for τ of size $2(s+1)(j-1)+2$. In this case, $J(T)$ is found by replacing the bricks $b_i, b_{i+1}, \ldots, b_{i+j-1}$ by a single brick b of size $2(s+1)(j-1)+2$ and by labeling the first cell of b with a -1.

Case 7. There is a brick b_i of size $2s(j-1)+2$ where $s \geq 3$ such that the first cell is labeled with an x and there is a cell which has a label -1 which is not the jth cell from the right. Let t be the left-most cell of b_i which is labeled with -1. In this case, $J(T)$ is found by replacing the brick b_i with j consecutive bricks $c_1, c_2, \ldots, c_{j-1}, c_j$ where c_1 contains all the cells of b_i up to and including cell t, $c_2, \ldots, c_{j-1}$ are bricks of size 2, and c_j contains the remaining cells of b_i. Remove the -1 label from cell t.

Case 8. There are j consecutive bricks in T, namely $b_i, b_{i+1}, \ldots, b_{i+j-1}$, such that $b_i = 2c(s-1)+2 > 2$ and has no -1 labels, $b_{i+1}, \ldots, b_{i+j-2}$ are bricks of size 2, $b_{i+j-1} = d(j-1)+1 > 1$, and the letters in these three bricks form an alternating maximum packing for τ of size $2(c+d+1)(j-1)+2$. In this case, $J(T)$ is found by replacing the j bricks $b_i, b_{i+1}, \ldots, b_{i+j-1}$ by a single brick b and adding a label -1 on the last cell of b_i.

Case 9. If none of the previous 8 cases apply, set $J(T) = T$.

The remainder of the proof in the case of alternating permutations in A_{2n} follows in a completely analogous way as the proof of Theorem 7.8, that is, it can be shown that J is an involution and that the fixed points under J correspond to alternating permutations σ with a power of x corresponding to the number of τ matches in σ. In turn, this means that

$$(2n)!\varphi(h_{2n}) = \sum_{\sigma \in A_{2n}} x^{\tau\text{-match}(\sigma)}.$$

The first generating function in the statement of the theorem follows from applying φ to the relationship between e_n and h_n found in Theorem 2.5.

As for the alternating permutations of an odd length in A_{2n+1}, we use the new basis $p_{\nu,n}$ for an appropriate choice of the function ν. We would like to select ν so that we find the exact same combinatorial objects for alternating permutations in A_{2n} except that the entries in the final cell of the brick tabloid are erased. This is the idea we used in the proofs of Theorems 3.5 and 4.6 when we first found generating functions for alternating permutations and j-alternating permutations.

In this situation, the appropriate choice for ν is to take $\nu(2n-1)=0$ and

$$\nu(2n) = \frac{(-1)^{2n-1} GMP_{\tau,2n-1}(x)/(2n-1)!}{\varphi(e_{2n})}$$

for all n. After applying φ to $(2n-1)!p_{\nu,2n}$, this choice of ν will change the last brick in one of our combinatorial objects T described earlier in the proof in the following ways:

1. Since we are dividing by $\varphi(e_{2n})$ in the definition of ν, we erase all labels in a last brick of length $2n$ in T.
2. Since we are dividing by a $1/(2n-1)!$ instead of the usual $1/(2n!)$ in the numerator of the function ν, $(2n-1)!\varphi(p_{2n,\nu})$ will produce multinomial coefficients of the form $\binom{2n-1}{2b_1,\ldots,2b_e-1}$ in our usual expansion of $(2n-1)!\varphi(p_{2n,\nu})$. This will allow us to leave the last cell blank.
3. Use the remaining $(-1)^{2n-1} GMP_{\tau,2n-1}(x)$ term in ν to fill the first $2n-1$ cells in the last brick with blocks in the usual way.

For example, one such combinatorial object created in this way when $\tau = 231546$, and which is counted by $(2n)!\varphi(p_{\nu,2n})$, is shown below:

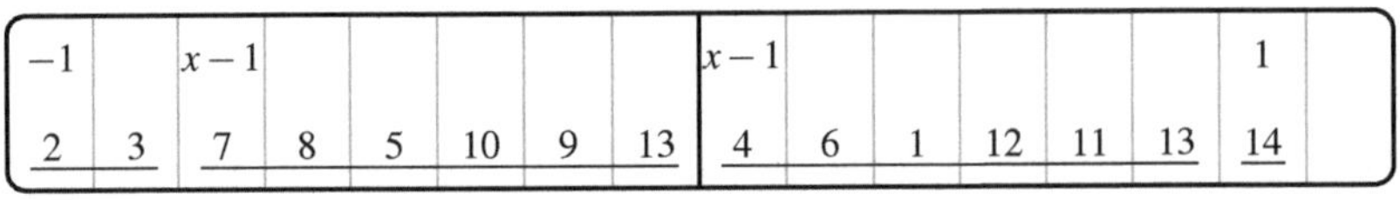

The only difference between the fillings of even length and our current fillings is that our current fillings must have a last brick which ends in a block of size 1 and the last cell of that brick is blank.

It is not difficult to check that the involutions I and J are not affected by these changes to our combinatorial objects T. Therefore we can proceed exactly as in the case of even length alternating permutations: first apply I to T, then apply J to the fixed points under I. The fixed points under J will correspond to permutations in A_{2n-1}, which in turn will show that

$$(2n-1)!\varphi(p_{2n,\nu}) = \sum_{\alpha \in A_{2n-1}} x^{\tau\text{-match}(\alpha)}.$$

From here, the second generating function in the statement of the theorem follows immediately from applying φ to (4.1). $\square$

We end this section with a discussion on the problem of actually finding $udmp_{\tau,2n}$, $GMP_{\tau,2n}(x)$, and $GMP_{\tau,2n+1}(x)$. In general, finding $GMP_{\tau,2n}(x)$ and $GMP_{\tau,2n+1}(x)$ is more difficult than finding $mp_{\tau,2n}$ and $mp_{\tau,2n+1}$. Even if we cannot easily provide a closed expression for $GMP_{\tau,2n}(x)$ or $GMP_{\tau,2n+1}(x)$ as a function of n for any given alternating minimal overlapping permutation τ, we can still compute $GMP_{n,\tau}(x)$ using recursions.

Let $F_{n,k}$ denote the set of all fillings of a $k \times n$ rectangular array with the integers $1,\ldots,kn$ such that the elements increase from bottom to top in each column. We let (i,j) denote the cell in the ith row from the bottom and the jth column from the left of the $k \times n$ rectangle and we let $F(i,j)$ denote the element in cell (i,j) of $F \in F_{n,k}$.

If F is any filling of a $k \times n$ rectangle with distinct positive integers such that elements in each column increase, reading from bottom to top, then we let $\mathrm{red}(F)$ denote the element of $F_{n,k}$ which results from F by replacing the ith smallest element of F by i. For example, below we show a filling F alongside $\mathrm{red}(F)$.

$F =$

12	16	22
8	15	17
6	10	13
1	7	5

$\mathrm{red}(F) =$

7	10	12
5	9	11
3	6	8
1	4	2

For $F \in F_{n,k}$ and $1 \leq c_1 < \cdots < c_j \leq n$, we let $F[c_1,\ldots,c_j]$ the filling of the $k \times j$ rectangle where the elements in column a of $F[c_1,\ldots,c_j]$ equal the elements in column c_a in F for $a = 1,\ldots,j$. We can then extend the usual pattern matching definitions from permutations to elements of $F_{n,k}$ as follows.

Let P be an element of $F_{j,k}$ and $F \in F_{n,k}$ where $j \leq n$. Then P occurs in F if there are $1 \leq i_1 < i_2 < \cdots < i_j \leq n$ such that $\mathrm{red}(F[i_1,\ldots,i_j]) = P$, F avoids P if there is no occurrence of P in F, and there is a P-match in F starting at position i if $\mathrm{red}(F[i,i+1,\ldots,i+j-1]) = P$. When $k = 1$, then $F_{n,1} = S_n$, and our definitions reduce to the standard definitions that have appeared in the pattern matching literature.

We let P-match(F) denote the number of P-matches in F. For example, below we have displayed a P, F, and G for which there are no P-matches in F but there is an occurrence of P in F, since $\mathrm{red}(F[1,2,5]) = P$. Also, there are 2 P-matches in G starting at positions 1 and 2, respectively, so P-match$(G) = 2$:

$P =$

3	6	9
2	5	8
1	4	7

$F =$

4	11	12	16	18	14
2	10	8	13	17	9
1	5	6	3	15	7

$G =$

4	7	11	16	18	14
2	6	10	13	17	9
1	5	8	12	15	3

If $P \in F_{j,k}$, then let $MP^P_{n(j-1)+1}$ to be the set of $F \in F_{n(j-1)+1,k}$ with P-match$(F) = n$, i.e., the set of $F \in F_{n,k}$ with the property that there are P-matches in F starting at positions $1, j, 2j-1, \ldots, nj-(j-1)$. We let $mp^P_{n(j-1)+1} = |MP^P_n|$, and by convention, define $mp^P_1 = 1$.

Given a permutation $\sigma = \sigma_1 \cdots \sigma_{2j} \in A_{2n}$, one can construct a column strict array $P_\sigma \in F_j$ by letting the ith column of P_σ consist of $\sigma_{2(i-1)+1}\sigma_{2(i-1)+2}$, reading from bottom to top, for $i = 1, \ldots, j$. Then if $\sigma \in A_{2j}$ is an alternating minimal overlapping permutation and $\tau \in A_{2n(j-1)+2}$ is an alternating maximum packing for σ, then P_τ will be a maximum packing for P_σ. For example if $\sigma = 1\ 3\ 2\ 5\ 4\ 6$ and $\tau = 2\ 3\ 1\ 7\ 5\ 8\ 4\ 10\ 9\ 11\ 6\ 13\ 12\ 14$, then P_σ and P_τ are pictured below:

$P_\sigma =$

3	5	6
2	1	4

$P_\tau =$

3	7	8	10	11	13	14
2	1	5	4	9	6	12

Suppose that we are given an alternating minimal overlapping permutation $\tau \in A_{2j}$. To help us visualize the order relationships within P_τ, we form a directed graph G_{P_τ} on the cells of the $2 \times j$ rectangle by drawing a directed edge from the position of the number s to the position of the number $s+1$ in P_τ for $j = 1, \ldots, 2j-1$. For example, here is the graph G_{P_τ} pictured immediately to the right of P_τ for $\tau = 231546$:

$P_\tau =$

3	5	6
2	1	4

$G_P =$ 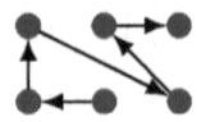

The graph G_{P_τ} determines the order relationships between all the cells in P_τ since $P_\tau(r,s) < P_\tau(u,v)$ if there is a directed path from cell (r,s) to cell (u,v) in G_{P_τ}.

Now suppose that $F \in MP^{P_\tau}_{2n(j-1)+2}$ where $n \geq 2$. Because there is a P_τ-match starting in column j, we can superimpose G_{P_τ} on the cells in columns $j, j+1, \ldots, 2j-1$ to determine the order relations between the elements in those j columns. If we do this for j-tuple of columns, $a(j-1)-(a-1)$ and $(a+1)j-a$ for $a = 1, \ldots, n-1$, we end up with a directed graph on the cells of the $2 \times (n(j-1)+1)$ rectangle which we will call $G_{n(j-1)+1,P_\tau}$. For example, here is $G_{9,P_{231546}}$ for the graph G_P shown above:

$G_{9,P} =$ 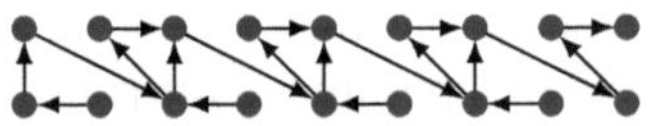

If $F \in MP^{P_\tau}_{n(j-1)+1}$ and there is a directed path from cell (r,s) to cell (u,v) in $G_{n(j-1)+1,P_\tau}$, then it must be the case that $F(r,s) < F(u,v)$. Note that $G_{n(j-1)+1,P_\tau}$ will always be a directed acyclic graph without multiple edges.

The graph $G_{n(j-1)+1,P}$ induces a poset

$$W_{n(j-1)+1,P} = (\{(i,j) : 1 \leq i \leq 2 \ \&\ 1 \leq j \leq n(j-1)+1\}, <_W)$$

on the cells of the $2 \times (n(j-1)+1)$ rectangle by defining $(i,j) <_W (s,t)$ if and only if there is a directed path from (i,j) to (s,t) in $G_{n(j-1)+1,P}$.

We define the graph $G^+_{n(j-1)+1,P_\tau}$ to be the Hasse diagram for the covering relation given by the partially ordered set indicated by $G_{n(j-1)+1,P_\tau}$. In our running example, this is

$G^+_{9,P} =$

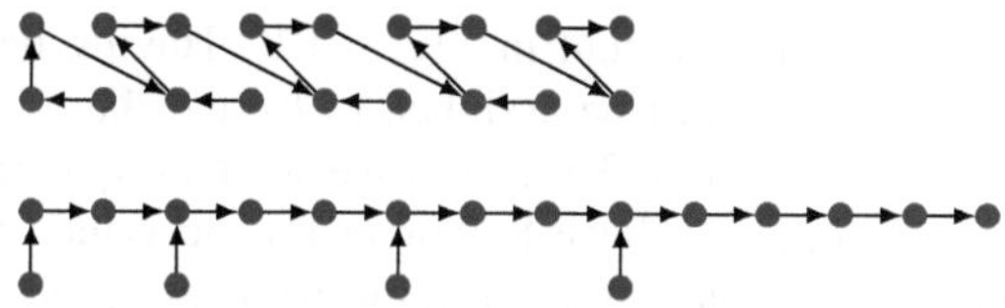

The following lemma gives information about directed acyclic graphs with no multiple edges, which will allow us to replace $G_{n(j-1)+1,P}$ with $G^+_{n(j-1)+1,P}$. Given a directed acyclic graph $G = (V,E)$ with no multiple edges, let $Con(G)$ equal the set of all pairs $(i,j) \in V \times V$ such that there is a directed path in G from vertex i to vertex j. The proof of the lemma and the next theorem can be found in [59].

Lemma 7.3. *Let $G = (V,E)$ be a directed acyclic graph with no multiple edges. Let H be the subgraph of G that results by removing all edges $e = (i,j) \in E$ such that there is a directed path from i to j in G that does not involve e. Then $Con(G) = Con(H)$.*

Using Lemma 7.3, one can prove the following characterization of when the maximum packings for P are unique. That is, we say that a pattern $P \in F_{j,k}$ with the minimal overlapping property is degenerate if it satisfies

1. $P(1,1) = 1$ or $P(1,j) = 1$,
2. $P(k,1) = jk$ or $P(k,j) = jk$, and
3. at least one of $P(i,1)+1 = P(i+1,1)$ or $P(i,j)+1 = P(i+1,j)$ holds, for each $1 \le i < k$.

Theorem 7.14. *Suppose $k \ge 2$ and $P \in F_{j,k}$ is a pattern with the minimal overlapping property. Then $mp_{P,n(j-1)+1} = 1$ for all $n \ge 1$ if and only if P is degenerate.*

Unfortunately there are no alternating minimal overlappings $\sigma \in A_{2j}$ that correspond to degenerate patterns in $F_{j,2}$ when $j = 3$: If $P \in F_{j,2}$ where $j \ge 3$ and is degenerate, either the first column of P is equal to 12 or the last column of P is $(2n-1)2n$, both of which violate the alternating condition.

For nondegenerate patterns, Lemma 7.3 allows us to replace the graph $G_{n(j-1)+1,P}$ by the simpler graph $G^+_{n(j-1)+1,P}$ without losing an information about the possible maximum packings of P_{231546}.

The number of linear extensions of the poset $W_{G^+_{9,P_{231546}}}$ is not hard to find. A linear extension of $W_{G^+_{9,P_{231546}}}$ is just a labeling L of vertices of $G^+_{9,P_{231546}}$ with the numbers $1,\ldots,18$ in such a way such that if there is a directed path from vertex v to vertex w in $G^+_{9,P_{231546}}$, then $L(v) < L(w)$.

Define the branch points of $G^+_{9,P_{231546}}$ to be those vertices whose in degree is ≥ 2. Suppose that b_1, b_2, b_3 are the branch points of $G^+_{9,P_{231546}}$, reading from left to right, and a_i is the vertex connected to b_i from below for $i = 1,2,3$. The rightmost branch point b_3 and the 5 vertices to its right must be assigned the numbers $13,\ldots,18$ in a linear extension L, reading from left to right. Taking away these 6 vertices, we see that we have 12 vertices left and that the vertex a_3 is independent of the rest of the remaining graph.

Thus $L(a_3)$ can be any element in $\{1,\ldots,12\}$. Having fixed $L(a_3)$, the largest three remaining numbers must be assigned to b_2 and the two vertices on the path from b_2 to b_3. Removing these 3 vertices, we see that a_2 is independent of the remaining 8 vertices so that we have 8 choices for the value of $L(a_2)$. Having fixed $L(a_2)$, the largest three remaining numbers must be assigned to b_1 and the two vertices on the path from b_1 to b_2. Removing these 3 vertices, we see that a_1 is independent of the remaining 4 vertices so that we have 4 choices for the value of $L(a_2)$. In conclusion, the number of linear extensions of $W_{G^+_{9,P_{231546}}}$ is $12\cdot 8\cdot 4 = 4^3 3!$.

This argument can be applied in general to $W_{G^+_{2n+1,P_{231546}}}$. That is, there will $n-1$ branch points $b_1,\ldots,b_{n-1}$ in $G^+_{2n+1,P_{231546}}$, reading from left to right. We then let a_i be the vertex connected to b_i from below in $G^+_{2n+1,P_{231546}}$. Then there will be $4n+2$ vertices in $G^+_{2n+1,P_{231546}}$ and there are 6 vertices on the path from b_{n-1} moving to the right and these elements must be assigned the largest 6 numbers in any linear extension of $W_{G^+_{2n+1,P_{231546}}}$.

Taking away these 6 vertices, we see that we have $4(n-1)$ vertices left and that the vertex a_{n-1} is independent of the rest of the remaining graph. Thus $L(a_{n-1})$ can be any element in $\{1,\ldots,4(n-1)\}$. Having fixed $L(a_{n-1})$, the largest three remaining numbers must be assigned to b_{n-2} and the two vertices on the path from b_{n-2} to b_{n-3}. Removing these 3 vertices, we see that a_{n-2} is independent of the remaining $4(n-2)$ vertices so that we have $4(n-2)$ choices for the value of $L(a_2)$. Having fixed $L(a_{n-2})$, the largest three remaining numbers must be assigned to b_{n-3} and the two vertices on the path from b_{n-3} to b_{n-2}. Removing these 3 vertices, we see that a_{n-2} is independent of the remaining $4(n-3)$ vertices so that we have $(n-3)$ choices for the value of $L(a_{n-2})$.

Continuing on in this way, we see that the number of linear extensions of $W_{G^+_{2n+1,P_{231546}}}$ is $\prod_{i=1}^{n-1} 4i = 4^{n-1}(n-1)!$. Therefore

$$udmp_{4n+2,231546} = mp_{2n+1,P_{231546}} = 4^{n-1}(n-1)!.$$

More examples of explicit formulas for $udmp_{2n(j-1)+2,\tau}$ when $\tau \in A_{2j}$ has the alternating minimal overlapping property which can be found in the thesis of Adrian Duane [31].

In general, it is a difficult problem to find explicit formulas for $GMP_{\tau,2n}(x)$ and $GMP_{\tau,2n+1}(x)$. However if $\tau = \tau_1\cdots\tau_{2j} \in A_{2j}$ has the alternating minimal overlapping property where $\tau_1 = 1$ and $\tau_{2j} = 2j$, then $GMP_{\tau,2n}(x)$ satisfies a simple recursion, as shown in the following theorem due to Adrian Duane and Jeffrey Remmel [32].

Theorem 7.15. *Suppose that $\tau = \tau_1\cdots\tau_{2j} \in A_{2j}$ with $j \geq 3$ has the alternating minimal overlapping property where $\tau_1 = 1$ and $\tau_{2j} = 2j$. Then for $n < j$, $G_{\tau,2n}(x) = (-1)^{n-1}$. Additionally, $G_{\tau,2n}(x) = (x-1)+(-1)^{j-1}$, and $G_{\tau,2n}(x)$ is equal to*

$$(x-1)^s udmp_{\tau,2n} - GMP_{\tau,2n-2}(x) - \sum_{i=1}^{s-1}(x-1)^i udmp_{\tau,2i(j-1)+2} GMP_{\tau,2(s-i)(j-1)}(x)$$

if $2n = 2s(j-1)+2$ *for some* $s \geq 1$, *and* $G_{\tau,2n}(x)$ *is equal to*

$$-GMP_{\tau,2n-2}(x) - \sum_{i=1}^{\lfloor \frac{n-1}{j-1} \rfloor} (x-1)^i udmp_{\tau,2i(j-1)+2} GMP_{\tau,2n-(2i(j-1)+2)}(x)$$

if $j-1$ *does not divide* $n-1$.

Proof. If $n < j$, then any generalized maximum packing for τ of length $2n$ consists of n blocks $B_1 \cdots B_n$ of size 2. Thus the only generalized maximum packing for τ of length $2n$ is the identity permutation $12 \cdots 2n$ which has weight $(-1)^{n-1}$.

If $n = j$, any generalized maximum packing for τ of length $2n$ consists of either (i) n blocks $B_1 \cdots B_n$ of size 2 or (ii) one block B_1 of size $2n$ which must be equal to τ. The only permutation for (i) is the identity which contributes $(-1)^{j-1}$ to $G_{\tau,2j}(x)$ and the only for (ii) is τ which contributes $(x-1)$ to $G_{\tau,2j}(x)$.

Suppose that $n > j$. Since τ starts with 1 and ends with $2j$, any maximum packing $\sigma \in MP_{\tau,2s(j-1)+2}$ must start with 1 and end with $2s(j-1)+2$. Now suppose that σ is a generalized maximum packing for τ of length $2n$ which has block structure $B_1B_2 \cdots B_k$. Our observation about maximum packings for τ ensures that each block B_i must start with the smallest element in block B_i and end with the largest element in B_i. Since for any $i < k$ the last element of B_i is less than the smallest element of B_{i+1}, it follows that if $|B_i| = b_i$ for $i = 1,\ldots,k$, then the elements of B_1 are just $1,\ldots b_1$ and the elements of B_{i+1} are just $1+\sum_{a=1}^{i} b_i, \ldots, b_{i+1}+\sum_{a=1}^{i} b_i$. Thus we can classify such generalized maximum packings σ by the size of the first block.

Case 1. Suppose $k \geq 2$ and $b_1 = 2$. In this case, $B_1 = 12$ and the reduced permutation $\text{red}(B_2 \cdots B_k)$ is a generalized maximum packing for τ of length $2n-2$. Moreover, $w(B_1 \cdots B_k) = -w(\text{red}(B_2 \cdots B_k))$. Thus the σ in this case contributes $-GMP_{\tau,2n-2}(x)$ to $GMP_{\tau,2n}(x)$.

Case 2. Suppose $k \geq 2$ and $b_1 = 2s(j-1)+2$ for some $s < \frac{n-1}{j_1}$. In this case, B_1 is an element of $MP_{2s(j-1)+2}$ and $\text{red}(B_2 \cdots B_k)$ is a generalized maximum packing for τ of length $2n-(2s(j-1)+2)$. Moreover $w(B_1 \cdots B_k) = -(x-1)^s w(\text{red}(B_2 \cdots B_k))$. Hence, such permutations contribute $-\sum_{s=1}^{n-1}(x-1)^s udmp_{\tau,2s(j-1)+2} GMP_{\tau,2n-2}(x)$ to $GMP_{\tau,2n}(x)$.

Case 3. Suppose $k = 1$. In this case, $2n$ must be of the form $2s(j-1)+2$ for some s and $\sigma \in MP_{\tau,2n}$. Such permutations contribute $(x-1)^s udmp_{\tau,2s(j-1)+2}$ to $GMP_{\tau,2n}(x)$.

This completes the proof. □

In the case where $\tau \in A_{2j}$ has the minimal overlapping property and τ starts with 1 and ends with $2j$, it is possible to compute $G_{\tau,2n+1}(x)$. We do so in the following theorem.

Theorem 7.16. *Suppose that* $\tau = \tau_1 \cdots \tau_{2j} \in A_{2j}$ *with* $j \geq 3$ *has the alternating minimal overlapping property where* $\tau_1 = 1$ *and* $\tau_{2j} = 2j$. *Then* $G_{\tau,1}(x) = 1$ *and* $G_{\tau,2n+1}(x) = -G_{\tau,2n}(x)$ *for* $n \geq 1$.

Proof. The only generalized maximum packing for τ of length 1 is 1 (weight 1).

Our previous observations show that if $B_1 \cdots B_k$ is the block structure of a generalized maximum packing for τ of length $2n+1$, then B_k is of size 1 and its element must be the largest element $2n+1$. Thus $B_1 \cdots B_{k-1}$ is a generalized maximum packing for τ of length $2n$ and $w(B_1 \cdots B_k) = -w(B_1 \cdots B_{k-1})$. □

There are more cases where we can derive similar recursions for $GMP_{\tau,2n}(x)$ and $GMP_{\tau,2n+1}(x)$ when τ either starts with one or ends with the largest elements. For example, consider our previous example of $\tau = 231546$. Then using the graphical reasoning above, one can show that a typical generalized maximum packing σ with block structure $B_1, \ldots, B_k$ is just a linear extension of a poset like

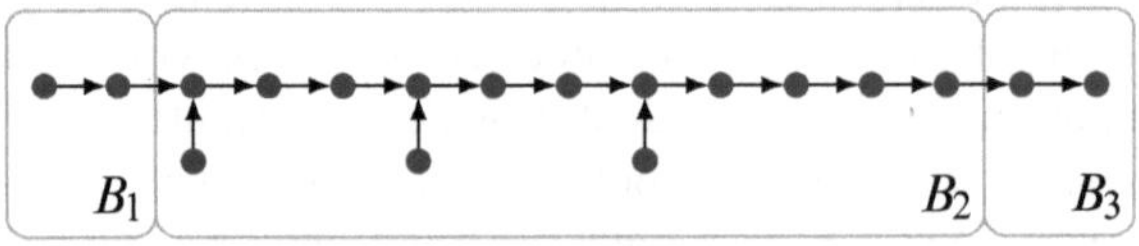

or possibly

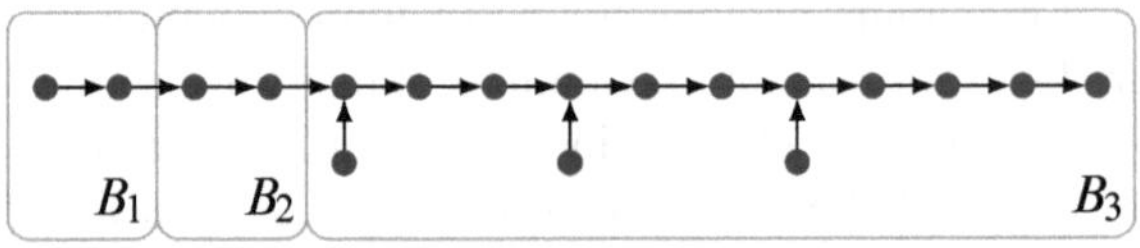

That is, if the last block B_k is of size 2, then the two elements in B_k are two largest elements in σ. If the last block B_k is of size $2(2k+1)$, then all the elements in the previous blocks are smaller than all the elements in B_k except those elements connected to the main chain of $G^+_{P(231546),2k+1}$ by vertical arrows.

Theorem 7.17. *Define* $f(i,n)$ *by* $f(1,n) = 2n-5$, $f(2,n) = (2n-9)(2n-6)$, *and for* $s \geq 3$, $f(s,n) = (2n-4s-1)(2n-4s+2)\prod_{j=1}^{s-2}(2n-4s+2+4j)$. *If* $\tau = 231546$, *then* $G_{\tau,2n}(x) = (-1)^{n-1}$ *for* $n < j$, $G_{\tau,2n}(x) = (x-1)+(-1)^{j-1}$, $G_{\tau,2n}(x)$ *is equal to*

$$4^{s-1}(s-1)!(x-1)^s udmp_{\tau,2n} - GMP_{\tau,2n-2}(x) - \sum_{i=1}^{s-1}(x-1)^i f(i,n) udmp_{\tau,4i+2} GMP_{\tau,2n-4i-2}(x)$$

if $n = 2s+1$ *for some* $s \geq 1$, *and* $G_{\tau,2n}(x)$ *is equal to*

$$-GMP_{\tau,2n-2}(x) - \sum_{i=1}^{s-1}(x-1)^i f(i,s) udmp_{\tau,4i+2} GMP_{\tau,2n-4i-2}(x)$$

if $n = 2s$.

Proof. If $n < j$, then any generalized maximum packing for τ of length $2n$ consists of n blocks $B_1 \cdots B_n$ of size 2. Thus the only generalized maximum packing for τ of length $2n$ is the identity permutation $12 \cdots 2n$ which has weight $(-1)^{n-1}$.

If $n = j$, any generalized maximum packing for τ of length $2n$ consists of either (i) of n blocks $B_1, \ldots, B_n$ of size 2 or (ii) one block B_1 of size $2n$ which must be equal to τ. The only permutation for (i) is the identity which contributes $(-1)^{j-1}$ to $G_{\tau,2j}(x)$ and the only for (ii) is τ which contributes $(x-1)$ to $G_{\tau,2j}(x)$.

Suppose that $n > j$. We classify the generalized maximum packings of τ by the size of the last block of its block structure $B_1, B_2, \ldots, B_k$. We have three cases.

Case 1. Suppose $k \geq 2$ and $b_k = 2$. In this case, $B_k = (2n-1)(2n)$ and the reduced permutation $\text{red}(B_2, \ldots, B_{k-1})$ is a generalized maximum packing for τ of length $2n-2$. Moreover $w(B_1 \cdots B_k) = -w(\text{red}(B_2 \cdots B_k)$. Thus the σ in this case contributes $-GMP_{\tau,2n-2}(x)$ to $GMP_{\tau,2n}(x)$.

Case 2. Suppose $k \geq 2$ and $b_k = 4i+2$ for some $i < s$. By our remarks above, all the elements of $B_1, \ldots, B_{k-1}$ are less than elements along central chain of the poset $G^+_{p(231546),2i+1}$ except for the elements connected to the central chain by vertical arrows. We have pictured this situation in the case where $|B_k|$ is of size 6, 10, 14, and 18 below:

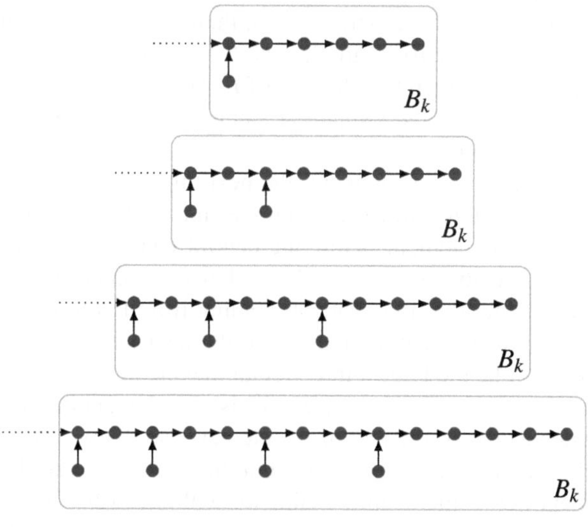

The dotted line represents that we are only displaying the final connection between the elements of $B_1, \ldots, B_{k-1}$ and B_k. In such a situation, each linear extension of the poset corresponds to a generalized maximum packing of τ with block structure $B_1, \ldots, B_k$ with weight $(-1)^{k-1} \prod_{i=1} w(B_i)$. We have a few sub-cases to consider:

1. If B_k is of size 6, then the last 5 elements of the central chain of B_k must be the largest 5 elements of the poset. If we remove these elements, then the element of B_k below the central chain is independent of the remaining elements so that we have $2n-5$ choices for this element. Once we choose this element, the rest of σ just corresponds to a linear extension of the poset associated with the block structure $B_1 \cdots B_{k-1}$. Such permutations contribute

$$-(x-1)(2n-5)GMP_{\tau,2n-6}(x) = -(x-1)f(1,n)GMP_{\tau,2n-6}(x)$$

to $GMP_{\tau,2n}(x)$.

2. If B_k is of size 10, then the last 6 elements of the central chain of B_k must be the largest 6 elements of the poset. If we remove these elements, then the element of B_k below the central chain on the right is independent of the remaining elements so that we have $2n-6$ choices for this element.
Once this choice is made, the first two elements of the central chain must be the largest two remaining elements. After removing these two elements, the element below the central chain is independent of the rest of the elements so we have $(2n-9)$ ways to choose this element. Once that is chosen, the rest of σ just corresponds to a linear extension of the poset associated with the block structure $B_1 \cdots B_{k-1}$. Such permutations contribute
$$-(x-1)^2(2n-6)(2n-9)GMP_{\tau,2n-10}(x) = -(x-1)^2 f(2,n)GMP_{\tau,2n-10}(x)$$
to $GMP_{\tau,2n}(x)$.
3. If B_k is of size $4s+2$ where $s \geq 3$, then the last 6 elements of the central chain of B_k are must be the largest 6 elements of the poset. If we remove these elements, then the element of B_k below the central chain on the right is independent of the remaining elements so that we have $2n-6$ choices for this element.
Once we choose this element, the next three elements of the central chain must be the largest three remaining elements. Once we remove those three elements, the element below the central chain is independent of the rest of the elements so there are $(2n-10)$ ways choose this element. If this element is not the first element, reading from left to right, of the elements of B_k below the central chain, then the next three elements of the central chain must be the largest three remaining elements.
Once we remove those three elements, the element below the central chain is independent of the rest of the elements so that we have $(2n-14)$ ways to choose this element. We can continue in this way until we reach the first element below the central chain; we will have $(2n-6)(2n-10)\cdots(2n-6+4(s-2))$ ways to choose the values of the elements below the central chain.
Having chosen all these, then the first two elements of the central chain will be the largest of the remaining elements. Once we have chosen those two elements, we will have $(2n-6+4(s-2))-3$ ways to the first element below the chain and then the rest of σ just corresponds to a linear extension of the poset associated with the block structure $B_1 \cdots B_{k-1}$. Such permutations contribute
$$-(x-1)^s f(s,n)GMP_{\tau,2n-2s-2}(x)$$
to $GMP_{\tau,2n}(x)$.

Case 3. Suppose $k = 1$. In this case, $2n$ must be of the form $2s+2$ for some s and $\sigma \in MP_{\tau,2n}$. Such permutations contribute $(x-1)^s udmp_{\tau,2s+2} = 4^{s-1}(s-1)!(x-1)^s$ to $GMP_{\tau,2n}(x)$.

This completes the proof. □

We can also compute $GMP_{231546,2n+1}(x)$ in terms of $GMP_{231546,2n+1}(x)$, as we show in our final result in this chapter.

Theorem 7.18. *Suppose that* $\tau = 231546$ *has the alternating minimal overlapping property. Then* $G_{\tau,1}(x) = 1$ *and* $G_{\tau,2n+1}(x) = -G_{\tau,2n}(x)$ *for* $n \geq 1$.

Proof. Clearly 1 is the only generalized maximum packing for τ of length 1 and it has weight 1.

Our previous observations show that if $B_1, \ldots, B_k$ is the block structure of a generalized maximum packing for τ of length $2n+1$, then B_k is of size 1 and its element must be the largest element $2n+1$. Thus $B_1, \ldots, B_{k-1}$ is a generalized maximum packing for τ of length $2n$ and $w(B_1 \cdots B_k) = -w(B_1 \cdots B_{k-1})$. □

Exercises

7.1. Given $A(z) = \sum_{n=0}^{\infty} |\{\sigma \in S_n \text{ does not have a } \tau\text{-match}\}| z^n/n!$ for some $\tau \in S_j$, approximate the expected number of nonoverlapping τ-matches in a randomly selected permutation in S_n.

7.2. Find a normal distribution (see Exercise 3.4) which gives a reasonable approximation for the distribution of nonoverlapping descents in S_n.

7.3. Let $\mathcal{T}$ be the set of permutations in $\tau = \tau_1 \cdots \tau_{j+1} \in S_{j+1}$ which satisfy $\tau_1 > \tau_2 > \cdots > \tau_j$ and $\tau_j < \tau_{j+1}$. Find generating functions for

$$\sum_{n=0}^{\infty} \frac{z^n}{n!} \sum_{\sigma \in S_n} x^{\mathcal{T}\text{-nlap}(\sigma)} \quad \text{and} \quad \sum_{n=0}^{\infty} \frac{z^n}{n!} \sum_{\sigma \in S_n} x^{\mathcal{T}\text{-match}(\sigma)}.$$

Solutions

7.1 The probability that j consecutive integers is not a τ-match is $1 - 1/j!$. Grouping consecutive integers in a permutation into sets of size j, the probability that a permutation in S_n will not have any τ-matches is at most $(1-1/j!)^{\lfloor n/j \rfloor}$. Since the coefficient of z^n in $A(z)$ is the probability that a permutation in S_n will not have a τ-match, this means that the sum $A(1)$ converges since it is not greater than the convergent series $\sum_{n=0}^{\infty}(1-1/j!)^{\lfloor n/j \rfloor}$. The number $A'(1)$ is also well defined.

Using Theorem 7.1 and the ideas in Exercise 3.3, the expected number of τ-matches is

$$\frac{\partial}{\partial x}\frac{A(z)}{(1-x)+x(1-z)A(z)}\bigg|_{x=1} = \frac{1-A(z)(1-z)}{A(z)(z-1)^2}$$
$$= \frac{1/A(1)}{(z-1)^2}+\frac{1-A'(1)/A(1)^2}{(z-1)}+g(z),$$

where $g(z)$ is a function with radius of convergence larger than 1. Using Exercise 1.20, the expected number of nonoverlapping τ-matches in a randomly selected permutation in S_n is approximately $(n+1)/A(1)+A'(1)/A(1)^2-1$.

7.2 The function $A(z)$ in this example is e^z. Using the result in the solution to exercise 7.1, the expected number of nonoverlapping descents in S_n is approximately $(n+2)/e-1$.

As for the variance,

$$A_x^2 f(z,x)\big|_{x=1} = -\frac{2e^{-2z}}{(z-1)^3}-\frac{3e^{-z}}{(z-1)^2}-\frac{1}{z-1}$$
$$= -\frac{2e^{-2}}{(z-1)^3}-\frac{(3e^{-1}-4e^{-2})}{(z-1)^2}-\frac{4e^{-2}+3e^{-1}-1}{(z-1)}+g(z)$$

for some function $g(z)$ with no singularities. Using the result in Exercise 1.20, we can use the singular part of the above equation to find a very good estimate for the second moment. Subtracting the square of $(n+2)/e-1$ from this second moment gives that the variance is very close to $(3-e)(n+2)/e^2$.

The normal distribution which best approximates the number of descents in S_n is therefore $e^{1-\frac{(ex+e-n-2)^2}{2(3-e)(n+2)}}/\sqrt{2\pi(3-e)(n+2)}$. Below we plot this normal distribution when $n=25$ along with bars showing the exact probabilities that a permutation in S_{25} has x nonoverlapping descents:

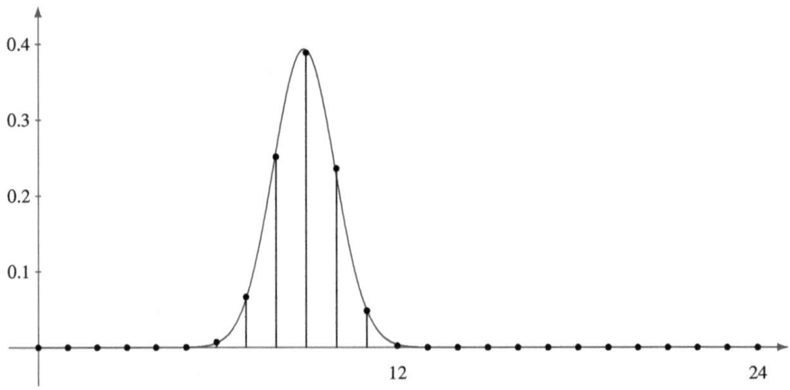

7.3 Every permutation in S_n without a $\mathcal{T}$-match must be the concatenation of integers which reduce to a permutation of $n-\ell$ without a $j\ \cdots\ 2\ 1$-match with a decreasing string of length ℓ. For example, when $j=3$, the permutation

$$\sigma = 9\ 1\ 7\ 8\ 4\ 5\ 2\ 11\ 12\ 10\ 6\ 3$$

does not have a $\mathcal{T} = \{3214, 4213, 4312\}$ match and is the concatenation of the integers 9 1 7 8 4 5 2 11 12 and the decreasing sequence 10 6 3. This concatenation is unique if we require ℓ to be a multiple of j, so we have

$$\begin{aligned} A(z) &= \sum_{n=0}^{\infty} \frac{z^n}{n!} |\{\sigma \in S_n \text{ has no } \mathcal{T} \text{ matches}\}| \\ &= \sum_{n=0}^{\infty} \frac{z^n}{n!} \sum_{\ell=0}^{n} \binom{n}{\ell} \left((n-\ell)!B(z))\right|_{z^{n-\ell}} \left(\ell! \sum_{i=0}^{\infty} \frac{z^{ij}}{(ij)!} \right) \Bigg|_{z^\ell}, \end{aligned}$$

where $B(z)$ is the exponential generating function for the permutations in S_n without a $j\ \cdots\ 2\ 1$ match found in the statement of Theorem 3.4. Therefore

$$A(z) = B(z) \sum_{i=0}^{\infty} \frac{z^{ij}}{(ij)!} = \frac{e^{\zeta^0 z} + \cdots + e^{\zeta^{j-1} z}}{(1-\zeta^{j-1})e^{\zeta z} + \cdots + (1-\zeta)e^{\zeta^{j-1} z}},$$

where $\zeta = e^{2\pi i/j}$ is a primitive j^{th} root of unity. The desired generating functions follow by using this particular $A(z)$ in Theorems 7.2 and 7.6.

Notes

Theorem 7.1 was proved by Kitaev [70]. The proof that we have presented, which also gives a method to calculate the function $A(z)$ in the statement of the theorem, is found in [88]. This method of finding $A(z)$ was used to prove more explicit formulas for the generating functions that avoid certain consecutive patterns in [77].

As stated in section 7.2, the cluster method is due to Goulden and Jackson and it has many applications. Our proof of the cluster method is new.

The results in sections 7.3 and 7.5 are based on [32], the results in section 7.4 are new, and the results in section 7.5 are based on [31, 100].

Many of the results on consecutive patterns in permutations were first proved by Sergey Kitaev and Sergey Kitaev with Toufik Mansour. The generating function in (7.3) is due to Elizalde and Noy.

Equation (7.12) generalizes a result due to Kitaev [69], which was proved using an inclusion–exclusion argument, when proving the results due to Elizalde and Noy.

The problem of computing $udmp_{\tau,2n}$ has been studied in [59] in a different context. There maximum packings in column strict arrays were considered. Here we studied the problem of computing $mp^P_{n(j-1)+1}$ for $PF_{2,j}$ is shown to reduced to finding the number of linear extensions of a certain poset associated with P.

Chapter 8
The Reciprocity Method

In Chapters 3, 4, and 7, we showed that the generating function for the distribution of many generating functions for permutations and words can be proved by applying ring homomorphisms φ to symmetric function identities. Many different generating functions came from applying φ to the identity in Theorem 2.5, namely

$$H(z) = \sum_{n\geq 0} h_n z^n = \frac{1}{1+\sum_{n=1}^{\infty}(-z)^n e_n} = \frac{1}{E(-z)}.$$

Suppose that we are given some statistic $s(\sigma)$ on permutations and wish to study the generating function

$$\sum_{n=0}^{\infty} \frac{z^n}{n!} \sum_{\sigma\in S_n} x^{s(\sigma)}.$$

In order to use our methods, we would like to come up with a ring homomorphism φ which is defined on the elementary symmetric functions such that

$$n!\varphi(h_n) = \sum_{\sigma\in S_n} x^{s(\sigma)},$$

which in turn would make $\varphi(H(t))$ equal to the desired generating function.

But how do we find an appropriate homomorphism φ? The reciprocity method, the subject of this chapter, provides one possible answer to this question.

The reciprocity method begins by assuming that we can write the desired generating function as

$$\sum_{n=0}^{\infty} \frac{z^n}{n!} \sum_{\sigma\in S_n} x^{s(\sigma)} = \frac{1}{1+\sum_{n=1}^{\infty} U_n(x)\frac{z^n}{n!}}$$

for some function $U_n(x)$ where we assume that the generating function

$$U(z,x) = \sum_{n=0}^{\infty} U_n(x)\frac{z^n}{n!} = \frac{1}{1+\sum_{n=1}^{\infty}\frac{z^n}{n!}\sum_{\sigma\in S_n} x^{s(\sigma)}}$$

A. Mendes, J. Remmel, *Counting with Symmetric Functions*, Developments in Mathematics 43, DOI 10.1007/978-3-319-23618-6_8

is the result of applying a second homomorphism Γ to the identity $H(z) = 1/E(-z)$. This assumption implies that $\Gamma(e_0) = 1$ and

$$\Gamma(e_n) = \frac{(-1)^n}{n!} \sum_{\sigma \in S_n} x^{s(\sigma)}$$

for $n \geq 1$. Expanding $n!h_n$ in terms of brick tabloids,

$$\begin{aligned} U_n(x) &= n!\Gamma(h_n) \\ &= n! \sum_{\lambda \vdash n} (-1)^{n-\ell(\lambda)} |B_{\lambda,(n)}| \Gamma(e_\lambda) \\ &= \sum_{\substack{T \in B_{\lambda,(n)} \text{ for some } \lambda \vdash n \\ \text{has bricks } b_1,\dots,b_\ell}} (-1)^{\ell(\lambda)} \binom{n}{b_1,\dots,b_\ell} \prod_{i=1} \sum_{\sigma \in S_{b_i}} x^{s(\sigma)}. \end{aligned} \tag{8.1}$$

The right-hand side of this expression has a natural combinatorial interpretation, giving us information about $U_n(x)$ and information about how to define the appropriate ring homomorphism φ.

Let us illustrate this idea with example. Suppose that we want to find the generating function for

$$\sum_{n=0}^{\infty} \frac{z^n}{n!} \sum_{\sigma \in S_n} x^{\text{des}(\sigma)+1} = \frac{1}{1 + \sum_{n=1}^{\infty} U_n(x) \frac{z^n}{n!}}.$$

Looking at (8.1), we can say that

$$U_n(x) = \sum_{\substack{T \in B_{\lambda,(n)} \text{ for some } \lambda \vdash n \\ \text{has bricks } b_1,\dots,b_\ell}} (-1)^{\ell(\lambda)} \binom{n}{b_1,\dots,b_\ell} \prod_{i=1} \sum_{\sigma \in S_{b_i}} x^{\text{des}(\sigma)+1}.$$

A combinatorial interpretation to the right-hand side of (8.1) is as follows.

First select a brick tabloid with bricks $b_1,\dots,b_k$. Interpret the multinomial coefficient as picking an ordered set partition $T_1,\dots,T_k$ of $\{1,\dots,n\}$ such that the size of T_i equals b_i for $i = 1,\dots,n$. We interpret $\prod_{i=1} \sum_{\sigma \in S_{b_i}} x^{\text{des}(\sigma)+1}$ as the number of ways of placing a permutation $\sigma^{(i)}$ of T_i in the cells of brick b_i and labeling each cell which starts a descent of $\sigma^{(i)}$ with an x and labeling the last cell of b_i with an x for $i = 1,\dots,k$. Finally, we interpret the term $(-1)^{\ell(\lambda)}$ as changing the label of the last cell of each brick from x to $-x$.

One such combinatorial object is shown below, where we have chosen $T_1 = \{1,4,7,9\}$, $T_2 = \{2,12,13\}$, $T_3 = \{2,5,10\}$, $T_4 = \{6,8,11\}$, $\sigma^{(1)} = 3124$, $\sigma^{(2)} = 132$, $\sigma^{(3)} = 213$, and $\sigma^{(4)} = 123$:

x			$-x$		x	$-x$	x		$-x$			$-x$
7	1	4	9	3	13	12	5	2	10	6	8	11

.

Let $\mathcal{O}_n$ denote the set of all such combinatorial objects. Define the weight of such an object to be the product of all x or $-x$ labels. Then $U_n(x)$ is equal to the weighted sum over $\mathcal{O}_n$.

As usual, we can define an involution I on $\mathcal{O}_n$ by scanning the cells from right to left looking for the first cell c such that either

i c is labeled with an x or
ii cell c is labeled with $-x$ so that c is the last cell in some brick b_i and the element in cell c is larger the first cell of brick b_{i+1}.

In the first case, split the brick b containing c into two bricks b^* and b^{**} such that b^* contains the cells of b up to and including cell c and b^{**} contains the remaining cells of b. Change the label on cell c from x to $-x$. In the second case, replace b_i and b_{i+1} by a single brick and change the label on cell c from $-x$ to x. If neither case applies, do nothing.

It follows that I is sign reversing weight preserving involution and $U_n(x)$ is therefore a weighted sum over the fixed points of I. In such a fixed point, there can be no cells labeled x in the fixed point, so σ is increasing within each brick b of B and there can be no decreases between bricks.

This means that the permutation σ in a fixed point must be the identity permutation. We can have any brick structure on $\sigma = 12\cdots n$. Therefore, for each $i < n$, we can decide to have brick end at cell i, in which case we label cell i as $-x$, or we cannot have a brick end at cell i, in which case the label is 1. Since the last cell is labeled with $-x$, we have

$$U_n(x) = -x(1-x)^{n-1}.$$

Therefore

$$\begin{aligned}\sum_{n=0}^{\infty} \frac{z^n}{n!} \sum_{\sigma \in S_n} x^{\mathrm{des}(\sigma)+1} &= \frac{1}{1+\sum_{n=1}^{\infty} U_n(x)\frac{z^n}{n!}} \\ &= \frac{1}{1-\sum_{n=1}^{\infty} x(1-x)^{n-1}\frac{z^n}{n!}} \\ &= \frac{1-x}{1-xe^{(1-x)z}}.\end{aligned}$$

We found the desired generating function without knowing the appropriate ring homomorphism φ from the outset.

8.1 The Reciprocity Method for Pattern Avoiding Permutations

Let $NM_n(\tau)$ denote the set of permutations in S_n with no consecutive τ-matches where τ is a permutation which begins with 1. We begin by assuming that

$$\sum_{n=0} \frac{z^n}{n!} \sum_{\sigma \in NM_n(\tau)} y^{\mathrm{des}(\sigma)+1} = \frac{1}{1+\sum_{n=1}^{\infty} U_{\tau,n}(y)\frac{z^n}{n!}}$$

for some function $U_{\tau,n}(y)$.

In this section we discuss how the reciprocity method can be used to find recursions for $U_\tau(y)$ in the case when τ is of the form $1324\cdots p$ for $p \geq 4$, that is, we are considering the case where τ is created by from applying the transposition (23) to the identity permutation. This will enable us to find the generating function for the number of descents in permutations in $NM_n(\tau)$.

Theorem 8.1. *Let $\tau = 1324\cdots p$ where $p \geq 5$. Then $U_{\tau,1}(y) = -y$ and*

$$U_{\tau,n}(y) = (1-y)U_{\tau,n-1}(y) + \sum_{k=2}^{\lfloor \frac{n-2}{p-2} \rfloor +1} (-y)^{k-1} U_{\tau,n-((k-1)(p-2)+1)}(y)$$

for $n \geq 2$.

Proof. Beginning with equation 8.1, we create combinatorial objects which have a weighted sum equal to $U_{\tau,n}(y)$. Looking at (8.1), we select a brick tabloid with bricks $b_1,\ldots,b_\ell$ and then interpret the multinomial coefficient as selecting an ordered set partition $T_1,\ldots,T_\ell$ of $\{1,\ldots,n\}$ such that $|T_i| = b_i$ for each i. Use the factor of the form $\prod_{i=1}^{\ell} NM_{\tau,b_i}(y)$ to select permutations $\sigma^{(1)},\ldots,\sigma^{\ell}$ such that $\sigma^{(i)} \in NM_{b_i}(\tau)$ for each i. Use these permutations to rearrange the ordered set partition associated with each brick.

Place a power of y in each cell which contains a descent of $\sigma^{(i)}$ and label the last cell of each brick with a y in order to account for the factor $y^{\mathrm{des}(\sigma^{(i)})+1}$ in equation (8.1). Lastly, change the final y to $-y$, accounting for the $(-1)^{\ell(\lambda)}$ in (8.1). Below we display one possible combinatorial object created in this manner when $\tau = 13245$:

	y		y		y			$-y$		y		y	$-y$	y	$-y$
2	12	5	16	13	14	6	7	9	15	1	3	11	8	10	4

Define the weight of such an object T, $w(T)$, to be the product of the y and $-y$ labels and let $\mathcal{O}_{\tau,n}$ be the set of all possible combinatorial objects created in this way. It follows that

$$U_{\tau,n}(y) = \sum_{T \in \mathcal{O}_{\tau,n}} w(T).$$

Time for an involution! Define I on $T \in \mathcal{O}_{\tau,n}$ in the following way. Scan the cells of T from left to right looking for the first cell c such that either

i c is labeled with a y or
ii c is a cell at the end of a brick b_i, $\sigma_c > \sigma_{c+1}$, and there is no τ-match of σ that lies entirely in the cells of bricks b_i and b_{i+1}.

In the case of (i), if c is a cell in brick b_j, split b_j into two bricks b'_j and b''_j where b'_j contains all the cells of b_j up to an including cell c and b''_j consists of the remaining cells of b_j. Change the label on cell c from y to $-y$.

In the case of (ii), combine the two bricks b_i and b_{i+1} into a single brick b and change the label on cell c from $-y$ to y.

If neither case (i) or case (ii) applies, define $I(T) = T$.

The fact that τ has only one descent is the key to ensuring that I is an weight preserving sign reversing involution. That is, suppose that we are in case (i) and we split brick b_j into two bricks b'_j and b''_j at cell c. Then if $j > 1$, we have to consider the brick b_{j-1} that proceeds b_j. Let d be the last cell of brick b_{j-1} and suppose that $\sigma_d > \sigma_{d+1}$. Then the reason that we did not apply case (ii) of the definition of I at cell d to define $I(T)$ is that there must be a τ-match that straddles the cells of b_{j-1} and b_j. Since τ has one descent this τ-match can not extend beyond cell c since then there would be two descents in the τ-match, namely $\sigma_d > \sigma_{d+1}$ and $\sigma_c > \sigma_{c+1}$. Thus our τ-match that straddles the bricks b_{j-1} and b_j, is contained in the bricks b_{j-1} and b'_j. It then easily follows that the first cell where we can apply our involution to $I(T)$ is again at cell c which implies that $I^2(T) = T$. Verifying that $I^2(T) = T$ when either $\sigma_d < \sigma_{d+1}$ or $j = 1$ is straightforward. Similarly, it is easy to check that $I^2(T) = T$ in the case where we apply case (ii) to define I.

A fixed point under I cannot have any cells labeled with y, and so elements within each brick must be increasing. Similarly, if b_i and b_{i+1} are two consecutive bricks in a fixed point T, then either there is an increase between bricks b_i and b_{i+1} or there is a τ-match contained in the elements of the cells of b_i and b_{i+1} which must necessarily involve both the last element in b_i and the first element of b_{i+1}.

We claim that, in addition, the numbers in the first cells of the bricks must form an increasing sequence reading from left to right. That is, suppose that b_i and b_{i+1} are two consecutive bricks in a fixed point T of I and that $a > a'$ where a is the number in the first cell of b_i and a' is the number in the first cell of b_{i+1}. Then the number in the last cell of b_i must be greater than a' so there is a τ-match in the cells of b_i and b_{i+1}.

However, a' is the least number that resides in the cells of b_i and b_{i+1}, meaning that the only way that a' could be part of a τ-match that occurs in the cells of b_i and b_{i+1} is to have a' play the role of 1. But since we are assuming that τ starts with 1, this means a' is part of a τ-match and that τ-match must be entirely contained in b_{i+1}—an impossibility. Thus a' cannot be part of any τ-match that occurs in the cells of b_i and b_{i+1}.

However, this would mean that the τ-match that occurs in the cells of b_i and b_{i+1} must be contained entirely either in the cells of b_i or in the cells of b_{i+1} which again is impossible. We can now conclude that $a < a'$.

To recap, the fixed points T under I must satisfy

i there are no cells labeled with y in T,
ii the integers in each brick of T form an increasing sequence, and

iii if b_i and b_{i+1} are two consecutive bricks in T, then either there is increase between b_i and b_{i+1} or there is a decrease between b_i and b_{i+1} but there is a τ-match contained in the elements of the cells of b_i and b_{i+1} which straddle these two bricks.

The results so far in the proof are valid for $\tau = 1324\ldots p$ where $p \geq 4$, but from this point on we specialize to the case of $\tau = 1324\cdots p$ where $p \geq 5$.

Let T be a fixed point of I. Number 1 must be in the first cell of T. We claim that 2 must be in the second or third cell of T. Suppose that 2 is in cell c where $c > 3$. Since there are no descents within any brick, 2 must be the first cell in a brick. Moreover, since the minimal numbers in the bricks of T form an increasing sequence, 2 must be in the first cell of the second brick. Thus if b_1 and b_2 are the first two bricks in T, then 1 is in the first cell of b_1 and 2 is in the first cell of b_2. But then there is no τ-match in the cells of b_1 and b_2. We now have two cases, depending on whether 2 is in the second cell of T or if 2 is in the third cell of T.

Case 1. Suppose 2 is in the second cell of T. There are two possibilities, namely either

i 1 and 2 are both in the first brick b_1 of T or
ii brick b_1 is a single cell filled with 1 and 2 is in the first cell of the second brick b_2 of T.

In either case, 1 is not part of a τ-match in T and if we remove cell 1 from T and subtract 1 from the numbers in the remaining cells, we would end up with a fixed point T' of I in $\mathcal{O}_{\tau,n-1}$.

In case (i), we have the weight of T and T' the same. In case (ii), since b_1 will have a label $-y$ on the first cell, so $w(T) = (-y)w(T')$. It follows that fixed points in Case 1 will contribute $(1-y)U_{\tau,n-1}(y)$ to $U_{\tau,n}(y)$.

Case 2. Suppose 2 is in the third cell of T. Let $T(i)$ denote the number in cell i of T and $b_1, b_2, \ldots$ be the bricks of T. Since there are no descents within bricks in T and the first numbers of each brick are increasing, it must that 2 is in the first cell of b_2. Thus b_1 has two cells because there must be a τ-match straddling the cells of b_1 and b_2 where the 1 in the first cell of b_1 plays the roll of 1 and the 2 in the first cell of b_2 plays the roll of 2.

Then b_2 must have at least $p-2$ cells because if not, there would not be a τ-match straddling the cells of b_1 and b_2. But then the only reason that we cannot combine bricks b_1 and b_2 is that there is a τ-match in the cells of b_1 and b_2 which could only start at position 1.

Next we claim that $T(p-1) = p-1$. Since there is a τ-match starting at position 1 and $p \geq 5$, we know that all the numbers in the first $p-2$ cells of T are strictly less than $T(p-1)$. Thus $T(p-1) \geq p-1$. Now if $T(p-1) > p-1$, then let i be least number in the set $\{1,\ldots,p-1\}$ that is not contained in bricks b_1 and b_2. Since the numbers in each brick are increasing and the minimal numbers of the bricks are increasing, the only possible position for i is the first cell of brick b_3. From here it follows that there is a decrease between bricks b_2 and b_3.

Since T is a fixed point of I_τ, there is a τ-match in the cells of b_2 and b_3. But since τ has only one descent, this τ-match can only start at the cell c which is the second to the last cell of b_2. Thus c could be $p-1$ if b_2 has $p-2$ cells or $c > p-1$ if b_2 has more than $p-2$ cells. In either case, $p-1 < T(p-1) \leq T(c) < T(c+1) > T(c+2) = i$. But this is impossible since to have a τ-match starting at cell c, we must have $T(c) < T(c+2)$. Thus it must be the case that $T(p-1) = p-1$ and $\{T(1),\ldots,T(p-1)\} = \{1,\ldots,p-1\}$.

We now have two sub-cases.

Case 2a. Suppose there is no τ-match in T starting at cell $p-1$. We claim that $T(p) = p$. That is, if $T(p) \neq p$, then p cannot be in b_2, so p must be in the first cell of the brick b_3. But then we claim that we could combine bricks b_2 and b_3. That is, there will be a decrease between bricks b_2 and b_3 since $p < T(p)$ and $T(p)$ is in b_2. Since there is no τ-match in T starting at cell $p-1$, the only possible τ-match among the cells of b_2 and b_3 would have to start at a cell $c \neq p-1$. But it cannot be the case that $c < p-1$ since then $T(c) < T(c+1) < T(c+2)$.

Similarly, it cannot be that $c > p-1$ since then $T(c) > p$ and p has to be part of the τ-match—this is impossible since $T(c)$ must play the role of 1 in the τ-match. Thus $T(p) = p$. Thus if T' is the result of removing the first $p-1$ cells from T and subtracting $p-1$ from the remaining numbers, then T' will be a fixed point of I in $\mathcal{O}_{\tau,n-(p-1)}$.

If b_2 has $p-2$ cells, then T' will start with a brick with one cell and if b_2 has more than $p-2$ cells, then T' will start with a brick with at least two cells. Since there is $-y$ coming from the brick b_1, the fixed points in Case 2a. will contribute $-yU_{\tau,n-(p-1)}(y)$ to $U_{\tau,n}(y)$.

Case 2b. Suppose there is a τ-match starting at cell $p-1$ in T. In this case, $T(p-1) < T(p) > T(p+1)$ and so b_2 must have $p-2$ cells and brick b_3 starts at cell $p+1$. We claim that b_3 must have at least $p-2$ cells. That is, if b_3 has less than $p-2$ cells, then there could be no τ-match among the cells of b_2 and b_3 so then we could combine b_2 and b_3 violating the fact that T is a fixed point of I.

In the general case, assume that the bricks $b_2,\ldots,b_{k-1}$ in T all have $(p-2)$ cells. Then let $r_1 = 1$ and for $j = 2,\ldots,k-1$, let $r_j = 1+(j-1)(p-2)$. Thus r_j is the position of the second to last cell of brick b_j for $1 \leq j \leq k-1$. Furthermore, assume that there is a τ-match starting at cell r_j for $1 \leq j \leq k-1$. It follows that $T(r_{k-1}) < T(r_{k-1}+1) > T(r_{k-1}+2)$ and so brick b_k must start at cell $r_{k-1}+2$ and there is a decrease between bricks b_{k-1} and b_k.

But then b_k has at least $p-2$ cells since otherwise we could combine bricks b_{k-1} and b_k violating the fact that T is a fixed point of I. Let $r_k = 1+(k-1)(p-2)$ and assume that T does not have a τ-match starting at position r_k. This is the situation pictured below:

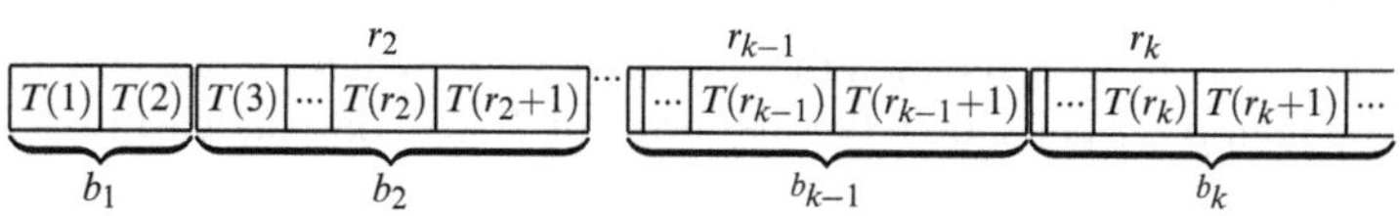

First we claim $T(r_j) = r_j$ and $\{1,\ldots,r_j\} = \{T(1),\ldots,T(r_j)\}$ for $j = 1,\ldots,k$. We have shown that $T(1) = 1$ and that $T(r_2) = T(p-1) = p-1$ and $\{T(1),\ldots,T(p-1)\} = \{1,\ldots,p-1\}$. Assume by induction that $T(r_{j-1}) = r_{j-1}$ and $\{1,\ldots,r_{j-1}\} = \{T(1),\ldots,T(r_{j-1})\}$. Since there is a τ-match that starts at cell r_{j-1} and $p \geq 5$, we know that all the numbers

$$T(r_{j-1}),T(r_{j-1}+1),\ldots,T(r_{j-1}+p-3)$$

are $\leq T(r_j) = T(r_{j-1}+p-2)$. The sets $\{1,\ldots,r_{j-1}\}$ and $\{T(1),\ldots,T(r_{j-1})\}$ must be equal, and so we know $T(r_j) \geq r_j$.

Next suppose that $T(r_j) > r_j$ and let i be the least number that does not lie in the bricks $b_1,\ldots,b_j$. The numbers in each brick increase and the minimal numbers in the bricks are increasing, so i is in the first cell of the next brick b_{j+1}. Now if $j < k$, then $i = T(r_j+2) \leq r_j < T(r_j) < T(r_{j+1})$ which would violate the fact that there is a τ-match in T starting at cell r_j.

If $j = k$, then it follows that there is a decrease between bricks b_k and b_{k+1} since b_{k+1} starts with $i \leq r_k < T(r_k)$. Since T is a fixed point of I, this must mean that there is a τ-match in the cells of b_k and b_{k+1}. But since τ has only one descent, this τ-match can only start at the cell c which is the second to the last cell of b_k. Thus c must be greater than r_k because there cannot be a τ-match starting at cell r_k.

So b_{k+1} must have more than $p-2$ cells. In this case, we have that $i \leq r_k < T(r_k) \leq T(c) < T(c+1) > T(c+2) = i$. But this cannot be since to have a τ-match starting at cell c, we must have $T(c) < T(c+2)$. Thus it must be the case that $T(r_j) = r_j$. But then $r_{j-1} = T(r_{j-1}) < T(d) < T(r_j) = r_j$ for $r_{j-1} < d < r_j$ so that $\{T(1),\ldots,T(r_j)\} = \{1,\ldots,r_j\}$, as desired. Thus we have proved by induction that $T(r_j) = r_j$ and $\{1,\ldots,r_j\} = \{T(1),\ldots,T(r_j)\}$ for $j = 1,\ldots,k$.

This means that the sequence $T(1),\ldots,T(r_k)$ is completely determined. Next we claim that since there is no τ-match starting at position r_k, it must be the case that $T(r_k+1) = r_k+1$. That is, if $T(r_k+1) \neq r_k+1$, then r_k+1 cannot be in brick b_k so then r_k+1 must be in the first cell of the brick b_{k+1}.

But then we could combine bricks b_k and b_{k+1}. That is, there will be a decrease between bricks b_k and b_{k+1} since $r_k+1 < T(r_k+1)$ and $T(r_k+1)$ is in b_k. Since there is no τ-match starting in T at cell r_k, the only possible τ-match among the cells of b_k and b_{k+1} would have to start at a cell $c \neq r_k$.

Now it cannot be that $c < r_k$ since then $T(c) < T(c+1) < T(c+2)$. If $c > r_k$, then $T(c) > r_k+1$ and r_k+1 would have to be part of the τ-match, meaning $T(c)$ could not play the role of 1 in the τ-match. Thus it must be the case that $T(r_k+1) = r_k+1$.

It now follows that if we let T' the result of removing the first r_k cells from T and subtracting r_k from each number in the remaining cells, then T' will be a fixed point I in $\mathcal{O}_{\tau,n-r_k}$.

If b_k has $p-2$ cells, then the first brick of T' will have one cell and if b_k has more than $p-2$ cells, then the first brick of T' will have at least two cells. Since there is a factor $-y$ coming from each of the bricks $b_1, \ldots, b_{k-1}$, the fixed points in Case 2b. will contribute $\sum_{k\geq 3}(-y)^{k-1}U_{\tau,n-((k-1)(p-2)+1)}(y)$ to $U_{\tau,n}(y)$.

We have now proved the theorem. □

The statement of Theorem 8.1 changes slightly in the case of taking $\tau = 1324$, as we record in the following theorem.

Theorem 8.2. *Let $\tau = 1324$. Then $U_{\tau,1}(y) = -y$ and*

$$U_{\tau,n}(y) = (1-y)U_{1234,n-1}(y) + \sum_{k=2}^{\lfloor \frac{n}{2} \rfloor} (-y)^{k-1} C_{k-1} U_{1324,n-2k+1}(y)$$

for $n \geq 2$ where $C_k = \binom{2k}{k}/(k+1)$ is the kth Catalan number.

Proof. The beginning of this proof is the same as the proof of Theorem 8.1. The difference is in the analysis of the fixed points under the involution I.

Let T be a fixed point under I. We know that 1 is in the first cell of T. In the same way as in the proof of theorem 8.1, either 2 is in the second or third cell in T. We break the situation into the two different cases.

Case 1. Suppose 2 is in the second cell of T. In this case there are two possibilities, namely either

i 1 and 2 lie in the first brick b_1 of T or
ii brick b_1 has one cell and 2 is the first cell of the second brick b_2 of T.

In either case, 1 is not part of a 1324-match and if we remove cell 1 from T and subtract 1 from the elements in the remaining cells, we would end up with a fixed point T' of I in $\mathcal{T}_{1324,n-1}$.

In case (i), the weights of T and T' are the same. In case (ii), since b_1 will have a label $-y$ on the first cell, $w(T) = (-y)w(T')$. Therefore fixed points in Case 1 will contribute $(1-y)U_{1324,n-1}(y)$ to $U_{1324,n}(y)$.

Case 2. Suppose 2 is in cell 3 of T. Let $T(i)$ denote the element in i cell of T and let $b_1, b_2, \ldots$ be the bricks of T. There are no descents within bricks in T and the minimal elements in the bricks are increasing, so 2 is in the first cell of a brick b_2. As in the proof of Theorem 8.1, it follows that b_1 must have two cells. But then b_2 must have at least two cells since if b_2 has one cell, there could be no 1324-match contained in the cells of b_1 and b_2 and we could combine bricks b_1 and b_2 which would mean that T is not a fixed point of I.

Thus b_1 has two cells and b_2 has at least two cells. But then the only reason that we could not combine bricks b_1 and b_2 is that there is a 1324-match in the cells of b_1 and b_2 which could only start at the first cell.

We now have two sub-cases.

Case 2a. There is no 1324-match in T starting at cell 3. Suppose for the moment that $\{T(1),T(2),T(3),T(4)\}$ is not equal to $\{1,2,3,4\}$ and let

$$i = min(\{1,2,3,4\} \setminus \{T(1),T(2),T(3),T(4)\}).$$

Since there is a 1324-match starting at position 1, it follows that $T(4) > 4$ since $T(4)$ is the fourth largest element in $\{T(1), T(2), T(3), T(4)\}$. Since the minimal elements of the bricks of T are increasing, it must be that i is the first element in brick b_3. But then we could combine bricks b_2 and b_3 because there will be a decrease between bricks b_2 and b_3. Since there is no 1324-match in T starting at cell 3, the only possible 1324-match among the elements in b_2 and b_3 would have start at a cell $c > 3$. But then $T(c) > i$, which is impossible since it would have to play the role of 1 in the 1324-match and i would have to play the role of 2 in the 1324-match. Thus it must be that $T(1) = 1$, $T(2) = 3$, $T(3) = 2$, and $T(4) = 4$.

If we let T' be the result of removing the first 3 cells from T and subtract 3 from the remaining elements, then T' will be a fixed point in $\mathcal{O}_{1324,n-3}$. Since there is $-y$ coming from the brick b_1, the fixed points in Case 2a. contribute $-yU_{1324,n-3}(y)$ to $U_{1324,n}(y)$.

Case 2b. Suppose there is a 1324-match starting at 3 in T. In this case, $T(3) < T(4) > T(5)$, so b_2 must have two cells and brick b_3 starts at cell 5. We claim that b_3 must have at least two cells. Indeed, if b_3 has one cell, then there could be no 1324-match among the cells of b_2 and b_3 so that we could combine b_2 and b_3 violating the fact that T is a fixed point.

In general, assume that the bricks $b_2, \ldots, b_{k-1}$ in T all have two cells and there are 1324-matches starting at cells $1, 3, \ldots, 2k-3$ but there is no 1324-match starting at cell $2k-1$ in T. Then we know that b_k has at least two cells. Let $c_i < d_i$ be the numbers in the first two cells of brick b_i for $i = 1, \ldots, k$. Then we know $\mathrm{red}(c_i d_i c_{i+1} d_{i+1}) = 1324$ for $1 \leq i \leq k-1$ and so $c_i < c_{i+1} < d_i < d_{i+1}$.

It must be the case that $\{T(1), \ldots, T(2k)\} = \{1, \ldots, 2k\}$. If not, there is a number greater than $2k$ that occupies one of the first $2k$ cells. Let M be the greatest such number. If M occupies one of the first $2k$ cells then there must be a number less than $2k$ that occupies one of the last $n-2k$ cells. Let m be the least such number. Since numbers in bricks are increasing, M must occupy the last cell in one of the first $k-1$ bricks or occupy cell $2k$. If M occupies the last cell in one of the first $k-1$ bricks, then M is part of a τ-match

$$\cdots \boxed{c_i \,|\, M}\,\boxed{c_{i+1} \,|\, d_{i+1}} \cdots.$$

Then $\mathrm{red}(c_i\ M\ c_{i+1}\ d_{i+1}) = 1\ 3\ 2\ 4$ implies that $M < d_{i+1}$, contradicting our choice of M as the greatest number in the first $2k$ cells. Thus M cannot occupy the last cell in one of the first $k-1$ bricks. This means that M must occupy cell $2k$ in T.

Since numbers in bricks are increasing, m occupies the first cell of b_{k+1}. Thus there is a descent between bricks b_k and b_{k+1}, meaning m must be part of a 1324-match. But the only way this can happen is if in the 1324-match involving m, m plays the role of 2 and the numbers in the last two cells of brick b_k play the role of 1 3.

Since we are assuming that a 1324-match does not start at cell $2k-1$ which is the cell that the number c_k occupies, the numbers in the last two cells of brick b_k must be greater than or equal to $d_k = M$ which is impossible since $m < M$. Therefore $\{T(1),\ldots,T(2k)\} = \{1,\ldots,2k\}$ and $d_k = 2k$.

It now follows that if we remove the first $2k-1$ cells from T and replace each remaining number i in T by $i-(2k-1)$, then we end up with a fixed point in T' under I. Each such fixed point T will contribute $(-y)^{k-1}U_{n-2k+1}(y)$ to $U_n(y)$.

The only thing left to do is to count the number of such fixed points T. That is, we must count the number of sequences $c_1d_1c_2d_2\ldots c_kd_k$ such that

i $c_1 = 1$,
ii $c_2 = 2$,
iii $d_k = 2k$,
iv $\{c_1,d_1,\ldots,c_k,d_k\} = \{1,2,\ldots,2k\}$, and
v $\mathrm{red}(c_id_ic_{i+1}d_{i+1}) = 1324$ for each $1 \le i \le k-1$.

We aim to show there are C_{k-1} such sequences where C_{k-1} is the $k-1$-st Catalan number, which is also the number of Dyck paths of length $2k-2$. (See Exercise 3.9 for the definition of a Dyck path.) We will prove this bijectively.

Start with a Dyck path $P = (p_1,p_2,\ldots,p_{2k-2})$ of length $2k-2$. Label the segments $p_1,\ldots,p_{2k-2}$ with $2,\ldots,2k-1$, respectively. Define $\phi(P)$ to be the sequence $c_1d_1\ldots c_kd_k$ where $c_1 = 1$ and $c_2\ldots c_k$ are the labels of the up-steps of P, reading from left to right, $d_1\ldots d_{k-1}$ are the labels of the down steps, reading from left to right, and $d_{2k} = 2k$.

For example, bijection ϕ in the case $k = 7$ below when applied to the Dyck path shown here yields the permutation 1 4 2 7 3 8 5 9 6 12 10 13 11 14.

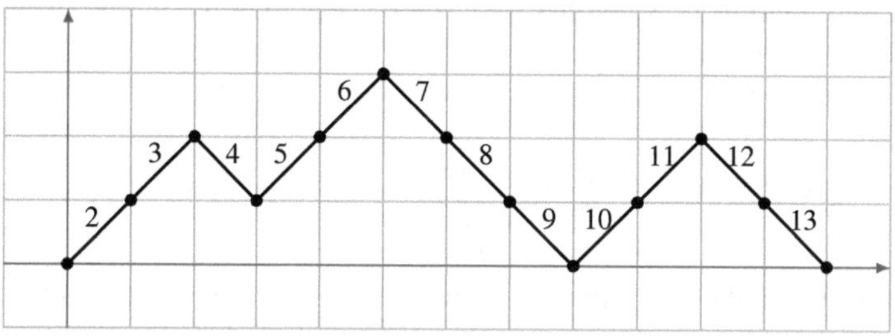

By construction, if P is a Dyck path of length $2k-2$ and $\phi(P) = c_1d_1\ldots c_kd_k$, then $c_1 < c_2 < \cdots < c_k$ and $d_1 < d_2 < \cdots < d_k$. Moreover, since each Dyck path must start with an up-step, we have that $c_2 = 2$. Clearly $c_1 = 1$, $d_k = 2k$, and $\{c_1,d_1,\ldots,c_k,d_k\} = \{1,\ldots,2k\}$ by construction. Thus $c_1d_1\ldots c_kd_k$ satisfies conditions (i)-(iv).

As for condition (v), note that $c_1 = 1 < d_1 > 2 = c_2 < d_2$ so $\mathrm{red}(c_1d_1c_2d_2) = 1\,3\,2\,4$. If $2 \le i \le k-1$, then c_i equals the label of the $(i-1)$th up-step, c_{i+1} equals the label of the ith up-step, and d_i is the label of ith down-step. Since in a Dyck path, the ith down-step must occur after the ith up-step, it follows that $c_i < c_{i+1} < d_i < d_{i+1}$ so that $\mathrm{red}(c_id_ic_{i+1}d_{i+1}) = 1\,3\,2\,4$.

Vice versa, if we start with a sequence $c_1d_1 \ldots c_kd_k$ satisfying conditions (i)-(v) and create a path $P = (p_1, \ldots, p_{2k-2})$ with labels $2, \ldots, 2k-1$ such that p_j is a up-step if $j+1 \in \{c_2, \ldots, c_k\}$ and p_j is an down-step if $j+1 \in \{d_1, \ldots, d_{k-1}\}$, then condition (iii) ensures P starts with an up-step and condition (v) ensures that the ith up-step occurs before the ith down step so that P will be a Dyck path. Thus ϕ is a bijection between the set of Dyck paths of length $2k-2$ and the set of sequence $c_1, d_1, \ldots, c_k, d_k$ satisfying conditions (i)-(v).

It follows that the fixed points T of I where the bricks $b_1, b_2, \ldots, b_{k-1}$ are of size 2 and there are 1324-matches starting at positions $1, 3, \ldots, 2k-3$ in T, but there is no 1324-match starting at position $2k-1$ in T contribute $C_{k-1}(-y)^{k-1}U_{n-2k+1}(y)$ to $U_n(y)$. This proves the theorem. □

The recursions for $U_{\tau,n}$ given in Theorems 8.1 and 8.2 allow us to find generating functions for

$$\sum_{n=0} \frac{z^n}{n!} \sum_{\sigma \in NM_n(\tau)} y^{\mathrm{des}(\sigma)+1}.$$

The exponential formula, our Theorem 4.8, can be used to provide a refinement of this generating function.

Let $\sigma = \sigma_1 \cdots \sigma_n$ be a permutation in S_n. We say that σ_j is a left-to-right minimum of σ if $\sigma_j < \sigma_i$ for all $i < j$. For example, the left-to-right minima of $\sigma = 938471625$ are 9, 3, and 1. Let lr-min(σ) denote the number of left-to-right minima of σ.

Given a cycle $C = (c_0, \ldots, c_{p-1})$ with smallest element c_0 written first, we let cyc-des$(C) = 1 + \mathrm{des}(c_0, \ldots, c_{p-1})$. The statistic cyc-des$(C)$ counts the number of descent pairs as we traverse once around the cycle (the extra 1 counts the descent pair $c_{p-1} > c_0$). By convention, if $C = (c_0)$ is a one cycle, we let cyc-des$C = 1$. If σ is a permutation in S_n with k cycles $C_1 \ldots C_k$, then we define cyc-des$(\sigma) =$ cyc-des$(C_1) + \cdots +$ cyc-des(C_k).

In [64], Miles Jones and Jeffrey Remmel studied generating functions of the form

$$NCM_\tau(z,x,y) = \sum_{n=0}^{\infty} \frac{z^n}{n!} \sum_{\sigma \in NCM_n(\tau)} x^{\#\text{ of cycles in}(\sigma)} y^{\text{cyc-des}(\sigma)},$$

where $NCM_n(\tau)$ is the set of permutations $\sigma \in S_n$ which have no cycle-τ-matches. The basic idea was to use the exponential formula to reduce the problem of computing $NCM_\tau(t,x,y)$ to the problem of computing similar generating functions for n cycles.

That is, let $NCM_{n,k}(\tau)$ be the set of permutations σ of S_n with k cycles such that σ has no cycle-τ-matches and $L_m^{ncm}(\tau)$ denote the set of m cycles γ in S_m such γ has no cycle-τ-matches. Then it can be shown with the exponential formula that

$$NCM_\tau(z,x,y) = \sum_{n=0}^{\infty} \frac{z^n}{n!} \sum_{k=1}^{n} x^k \sum_{\sigma \in NCM_{n,k}(\tau)} y^{\text{cyc-des}(\sigma)} = e^{x \sum_{m \geq 1} \frac{z^m}{m!} \sum_{C \in L_m^{ncm}(\tau)} y^{\text{cyc-des}C}}.$$

It turns out that if $\tau \in S_j$ is a permutation that starts with 1, then we can reduce the problem of finding $NCM_\tau(t,x,y)$ to the usual problem of finding the generating function of permutations that have no τ-matches. Let $\bar{\sigma}$ be the permutation that arises from $C_1 \cdots C_k$ by erasing all the parentheses and commas where $C_1,\ldots,C_k$ are ordered by decreasing minimal elements and each cycle starts with its minimal element. Then the minimal elements of the cycles correspond to left-to-right minima in $\bar{\sigma}$. Under the bijection $\sigma \to \bar{\sigma}$, cyc-des$(\sigma) = $ des$(\bar{\sigma}) + 1$ since every left-to-right minima other than the first element of $\bar{\sigma}$ is part of a descent pair in $\bar{\sigma}$.

In [64], it was proved that if $\tau \in S_j$ and τ starts with 1, then

i σ has k cycles if and only if $\bar{\sigma}$ has k left-to-right minima,
ii cyc-des$(\sigma) = 1 + des(\bar{\sigma})$, and
iii σ has no cycle-τ-matches if and only if $\bar{\sigma}$ has no τ-matches

for any $\sigma \in S_n$. These facts imply that $NCM_\tau(z,x,y) = NM_\tau(z,x,y)$. (However, if τ does not start with 1, then $|NM_n(\tau)|$ might not equal $|NCM_n(\tau)|$. For example, it can be checked that $|NCM_7(3142)| = 4236$ whereas $|NM_7(3142)| = 4237$.) Furthermore, these fact also imply that $NM_\tau(t,x,y) = F(t,y)^x$ for some function $F(t,y)$.

If we let $U_\tau(z,y)$ the generating function $\sum_{n=0}^{\infty} U_{\tau,n}(y)z^n/n!$, then

$$U_\tau(z,y) = \frac{1}{1+\sum_{n=1}^{\infty} \frac{z^n}{n!} \sum_{\sigma \in NM_{\tau,n}} y^{\text{des}(\sigma)+1}},$$

and so for any permutation τ which starts with a 1,

$$\sum_{n=0}^{\infty} \frac{z^n}{n!} NM_{\tau,n}(x,y) = \left(\frac{1}{U_\tau(z,y)}\right)^x.$$

Mathematica can be used to unwind the recursions in Theorems 8.1 and 8.2 in order to find the values of $U_{132\ldots p,n}(y)$. Once these polynomials are found, the polynomials $NM_{1324\ldots p,n}(x,y)$ can be computed. For example, the table below gives the coefficient of y^i in $U_{13245,n}(y)$ for $n = 1,\ldots,10$:

$n \backslash i$	1	2	3	4	5	6	7	8	9	10
1	−1									
2	−1	1								
3	−1	2	−1							
4	−1	3	−3	1						
5	−1	5	−6	4	−1					
6	−1	7	−12	10	−5	1				
7	1	9	−21	23	−15	6	−1			
8	−1	11	−34	47	−39	21	−7	1		
9	−1	13	−51	88	−90	61	−28	8	−1	
10	−1	15	−72	153	−189	156	−90	36	−9	1

Using this table, the polynomials $NM_{13245,n}(x,y)$ can also be found:

n	The polynomial $NM_{13245,n}(x,y)$
1	xy
2	$xy+x^2y^2$
3	$xy+xy^2+3x^2y^2+x^3y^3$
4	$xy+4xy^2+7x^2y^2+xy^3+4x^2y^3+6x^3y^3+x^4y^4$
5	$xy+10xy^2+15x^2y^2+11xy^3+30x^2y^3+25x^3y^3+xy^4+5x^2y^4+10x^3y^4+$ $10x^4y^4+x^5y^5$
6	$xy+24xy^2+31x^2y^2+62xy^3+140x^2y^3+90x^3y^3+26xy^4+91x^2y^4+$ $120x^3y^4+65x^4y^4+xy^5+6x^2y^5+15x^3y^5+20x^4y^5+15x^5y^5+x^6y^6$
7	$xy+54xy^2+63x^2y^2+273xy^3+553x^2y^3+301x^3y^3+292xy^4+840x^2y^4+$ $875x^3y^4+350x^4y^4+57xy^5+238x^2y^5+406x^3y^5+350x^4y^5+140x^5y^5+$ $xy^6+7x^2y^6+21x^3y^6+35x^4y^6+35x^5y^6+21x^6y^6+x^7y^7$

In every case that we have tested, the polynomials $U_{1324,n}(-y)$ and $U_{1324\ldots p,n}(-y)$ are unimodal, leading us to conjecture they are unimodal for all $p \geq 4$ and $n \geq 1$.

Exercises

8.1. Prove that the coefficient of x^ky^k in $NM_{1324\ldots p,n}(x,y)_{x^ky^k}$ is equal to $S(n,k)$ where $S(n,k)$ denotes the number of set partitions of $\{1,\ldots,n\}$ into k disjoint nonempty sets.

8.2. Prove that the coefficient of xy^2 in $NM_{1324\ldots p,n}(x,y)$ is equal to $2^{n-1}-n$ if $n<p$ and $2^{n-1}-2n+p-1$ if $n \geq p$.

8.3. How many permutations $\sigma \in S_n$ have exactly one descent and no consecutive 1324 matches for $n \geq 4$? How many permutations $\sigma \in S_n$ have exactly one descent and no consecutive $132\ldots p$ matches for $n \geq p$?

Solutions

8.1 A permutation $\sigma \in S_n$ that contributes to the coefficient x^ky^k in $NM_{1324\ldots p,n}(x,y)$ must have k left-to-right minima and $k-1$ descents. Since each left-to-right minima of σ which is not the first element is always the second element of descent pair, it follows that if $1=i_1<i_2<i_3<\cdots<i_k$ are the positions of the left-to-right minima, then σ must be increasing in each of the intervals $[1,i_2),[i_2,i_3),\ldots,[i_{k-1},i_k),[i_k,n]$. Therefore

$$\{\sigma_1,\ldots,\sigma_{i_2-1}\},\{\sigma_{i_2},\ldots,\sigma_{i_3-1}\},\ldots,\{\sigma_{i_{k-1}},\ldots,\sigma_{i_k-1}\},\{\sigma_{i_k},\ldots,\sigma_n\}$$

is just a set partition of $\{1,\ldots,n\}$ ordered by decreasing minimal elements. Moreover, no such permutation can have a $1324\ldots p$ match for any $p \geq 4$.

If we are given a set partition of $\{1,\ldots,n\}$ into the sets $A_1,\ldots,A_k$ such that $min(A_1) > \cdots > min(A_k)$, then the permutation σ created by listing the elements in each of the sets $A_k, A_{k-1},\ldots,A_1$ in increasing order is a permutation with k left-to-right minima and $k-1$ descents where for any set $A \subseteq \{1,\ldots,n\}$.

8.2 Suppose that $\sigma \in S_n$ contributes to the coefficient of xy^2 in $NM_{1324\ldots p,n}(x,y)$, meaning that σ has one left-to-right minima and one descent. Thus, σ must start with 1 and have one descent.

Now if A is any subset of $\{2,\ldots,n\}$ and $B = \{2,\ldots,n\} \setminus A$, then we let σ_A the permutation created by writing down 1, the elements in A in increasing order, and then the elements in B in increasing order. The only choices of A that do not give rise to a permutation with one descent are $\emptyset$ and $\{2,\ldots,i\}$ for $i = 2,\ldots,n$. It follows that there $2^{n-1} - n$ permutations that start with 1 and have 1 descent.

Next consider when such a σ_A could have a $1234\ldots p$ match. If the $1234\ldots p$-match starts at position i, then $\text{red}(\sigma_i\sigma_{i+1}\sigma_{i+2}\sigma_{i+3}) = 1324$. This means that the only descent is at position $i+1$ and all the elements σ_j for $j \geq i+3$ are greater than or equal to σ_{i+3}. But then all the elements between 1 and σ_{i+2} must appear in increasing order in $\sigma_2 \ldots \sigma_{i-1}$. It follows that σ_A is of the form $1\ldots(q-2)q(q+2)$ $(q+1)(q+2)\ldots n$. There are no such permutations if $n \leq p-1$ and there are $n-(p-1)$ such permutations if $n \geq p$ as q ranges between 1 and $n-(p-1)$.

8.3 The permutation σ with exactly one descent and no 1324 or $1324\ldots p$ matches must have either one or two left–right minima. Therefore in the case of 1324, the number of such permutations is equal to

$$\begin{aligned} NM_{1324,n}(x,y)|_{xy^2} - NM_{1324,n}(x,y)|_{x^2y^2} &= (2^{n-1} - 2n + 4 - 1) + (2^{n-1} - 1) \\ &= 2^n - 2n + 2, \end{aligned}$$

where we used exercises 8.1 and 8.2 together with the fact that there are $2^{n-1} - 1$ set partitions of $\{1,\ldots,n\}$ into two sets. Similarly, in the case of $132\ldots p$, we have

$$\begin{aligned} NM_{1324,n}(x,y)|_{xy^2} - NM_{1324,n}(x,y)|_{x^2y^2} &= (2^{n-1} - 2n + p - 1) + (2^{n-1} - 1) \\ &= 2^n - 2n + p - 2. \end{aligned}$$

Notes

The reciprocity method was first recorded in [65].

In [67], generating functions were explicitly calculated to find formulas for the permutations in S_n with exactly two descents and no consecutive $132\ldots p$ matches. This gave enumeration results similar, but more intricate, to our Exercise 8.3. For instance, for $n \geq 8$, the number of permutations in S_n with exactly 2 descents and no consecutive 13245 matches is equal to

$$3^n+(7-2n)2^n+25-\frac{53n}{3}+n^2-\frac{n^3}{3}.$$

The proof of this result relies on generating function manipulations (see [67]). A direct combinatorial proof is not known.

The methods in section 8.1 can be used to prove other results for collections of patterns that start with 1 and have one descent. In [66], permutations of the form $\tau = 1p2\ldots(p-1)$ were considered and it is shown that $NM_\tau(z,x,y) = \left(\frac{1}{U_\tau(z,y)}\right)^x$ where $U_\tau(z,y) = \sum_{n=0}^{\infty} U_{\tau,n}(y)\frac{z^n}{n!}$ with $U_{\tau,1}(y) = -y$, and for $n \geq 2$,

$$U_{\tau,n}(y) = (1-y)U_{\tau,n-1}(y) + \sum_{k=1}^{\lfloor \frac{n-2}{p-2} \rfloor} (-y)^k \binom{n-k(p-3)-2}{k-1} U_{\tau,n-(k(p-2)+1)}(y).$$

Similarly, if $p \geq 5$ and $\tau = 134\ldots(p-1)2p$, then it is shown that $U_{\tau,1}(y) = -y$, and for $n \geq 2$,

$$U_{\tau,n}(y) = (1-y)U_{\tau,n-1}(y) + \sum_{k=1}^{\lfloor \frac{n-2}{p-2} \rfloor} (-y)^k \frac{1}{(p-3)k+1}\binom{k(p-2)}{k} U_{\tau,n-k(p-2)-1}.$$

The reciprocity method can also be applied to find generating function of $x^{\text{des}(w)+1}$ over all words $w \in \{1,2,\ldots\}^*$ which have no u-matches where u is a word such that $\text{des}(u) = 1$. In this case explicit formulas for the generating function can be found instead of resorting to recursions. This work will appear in the forthcoming Ph.D. thesis of Sangha, who is a student of the second author.

The reciprocity method can also be extended to compute generating functions of the form

$$\sum_{n=0}^{\infty} \frac{z^n}{n!} \sum_{\sigma \in NM_u(\tau)} y^{\text{des}(\sigma)+1}.$$

where τ has more than one descent. This requires modifying the involution I in Theorem 8.1. This will be done in the forthcoming Ph.D. thesis of Bach, who is also a student of the second author.

Appendix A
Transition Matrices

The a to b edge in the following directed graph is labeled with the λ, μ entry of the a-to-b transition matrix $A_{\lambda,\mu}$. This means $a_\mu = \sum_{\lambda \vdash n} A_{\lambda,\mu} b_\lambda$.

A. Mendes, J. Remmel, *Counting with Symmetric Functions*, Developments in Mathematics 43, DOI 10.1007/978-3-319-23618-6

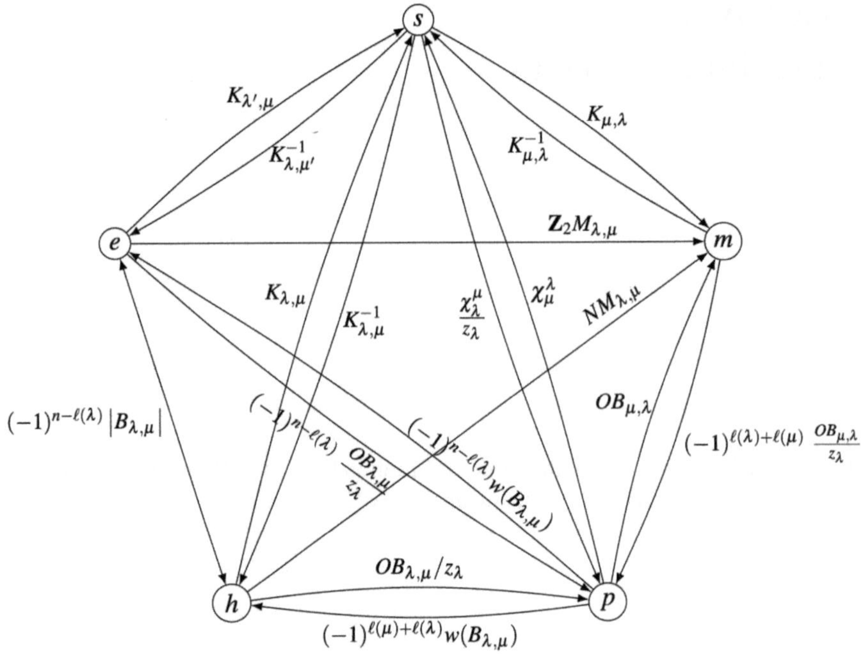

$\lvert B_{\lambda,\mu}\rvert$	Brick tabloids of content λ, shape μ (page 50)
χ^{λ}_{μ}	Signed sum of rim hook tableau of shape λ, content μ (Exercise 2.14)
$K_{\lambda,\mu}$	Column strict tableau of shape λ, content μ (page 48)
$K^{-1}_{\mu,\lambda}$	Signed sum of special rim hook tabloids, shape λ, content μ (Ex. 2.15)
$NM_{\lambda,\mu}$	Non-neg. integer matrices, row sum λ, column sum μ (Exercise 2.16)
$OB_{\mu,\lambda}$	Ordered brick tabloids of shape λ, content μ (page 55)
$w(B_{\lambda,\mu})$	Weighted brick tabloids of content λ, shape μ (page 53)
$\mathbf{Z}_2M_{\lambda,\mu}$	0,1 matrices with row sum λ, column sum μ (page 49)
z_λ	If λ has m_i parts of size i, $z_\lambda = 1^{m_1}2^{m_2}\cdots m_1!m_2!\cdots$

References

1. Aigner, M., Ziegler, G.M.: Proofs from the Book, 5th edn. Springer, Berlin (2014) [Including illustrations by Karl H. Hofmann]
2. André, D.: Developpements de sec x et de tang x. C. R. Acad. Sci. Paris **88**, 965–967 (1879)
3. André, D.: Mémoire sur les permutations alternées. J. Math. Pure Appl. **7**, 167–184 (1881)
4. Andrews, G.E.: The Theory of Partitions. Cambridge Mathematical Library. Cambridge University Press, Cambridge (1998) [Reprint of the 1976 original]
5. Bach, Q., Remmel, J.: Generating functions for descents over permutations which avoid sets of consecutive patterns. Aust. J. Comb. (to appear)
6. Beck, D.: Permutation enumeration of the symmetric and hyperoctahedral groups and the combinatorics of symmetric functions. Ph.D. thesis, University of California, San Diego (1993)
7. Beck, D.: The combinatorics of symmetric functions and permutation enumeration of the hyperoctahedral group. Discret. Math. **163**(1–3), 13–45 (1997)
8. Beck, D., Remmel, J.: Permutation enumeration of the symmetric group and the combinatorics of symmetric functions. J. Comb. Theory Ser. A **72**(1), 1–49 (1995)
9. Beck, D.A., Remmel, J.B., Whitehead, T.: The combinatorics of transition matrices between the bases of the symmetric functions and the B_n analogues. In: Proceedings of the 5th Conference on Formal Power Series and Algebraic Combinatorics, Florence, 1993, vol. 153, pp. 3–27 (1996)
10. Beissinger, J.S.: Factorization and enumeration of labeled combinatorial objects. Ph.D. thesis, University of Pennsylvania (1981)
11. Bender, E.A., Goldman, J.R.: Enumerative uses of generating functions. Indiana Univ. Math. J. **20**, 753–765 (1970/1971)
12. Benjamin, A.T., Quinn, J.J.: Proofs That Really Count: The Art of Combinatorial Proof. The Dolciani Mathematical Expositions, vol. 27. Mathematical Association of America, Washington, DC (2003)

A. Mendes, J. Remmel, *Counting with Symmetric Functions*, Developments in Mathematics 43, DOI 10.1007/978-3-319-23618-6

13. Bergeron, F., Labelle, G., Leroux, P.: Combinatorial Species and Tree-Like Structures. Encyclopedia of Mathematics and Its Applications, vol. 67. Cambridge University Press, Cambridge (1998) [Translated from the 1994 French original by Margaret Readdy, With a foreword by Gian-Carlo Rota]
14. Brenti, F.: Unimodal polynomials arising from symmetric functions. Proc. Am. Math. Soc. **108**(4), 1133–1141 (1990)
15. Brenti, F.: Permutation enumeration symmetric functions, and unimodality. Pac. J. Math. **157**(1), 1–28 (1993)
16. Carlitz, L.: Enumeration of up-down permutations by number of rises. Pac. J. Math. **45**, 49–58 (1973)
17. Carlitz, L.: Permutations and sequences. Adv. Math. **14**, 92–120 (1974)
18. Carlitz, L.: A combinatorial property of q-Eulerian numbers. Am. Math. Mon. **82**, 51–54 (1975)
19. Carlitz, L.: Permutations, sequences and special functions. SIAM Rev. **17**, 298–322 (1975)
20. Carlitz, L., Scoville, R.: Enumeration of permutations by rises, falls, rising maxima and falling maxima. Acta Math. Acad. Sci. Hungar. **25**, 269–277 (1974)
21. Carlitz, L., Scoville, R.: Generalized Eulerian numbers: combinatorial applications. J. Reine Angew. Math. **265**, 110–137 (1974)
22. Carlitz, L., Scoville, R.: Enumeration of up-down permutations by upper records. Monatsh. Math. **79**, 3–12 (1975)
23. Carlitz, L., Scoville, R., Vaughan, T.: Enumeration of pairs of permutations. Discret. Math. **14**(3), 215–239 (1976)
24. Chen, Y.M.: Combinatorial Algorithms for Plethysm. ProQuest LLC, Ann Arbor (1982). Thesis (Ph.D.), University of California, San Diego
25. Comtet, L.: Advanced Combinatorics: The Art of Finite and Infinite Expansions, Enlarged edn. D. Reidel Publishing Co., Dordrecht (1974)
26. Désarménien, J., Foata, D.: Fonctions symétriques et séries hypergéométriques basiques multivariées. Bull. Soc. Math. Fr. **113**(1), 3–22 (1985)
27. Désarménien, J., Foata, D.: Fonctions symétriques et séries hypergéométriques basiques multivariées, II. In: Combinatoire énumérative, Montreal, 1985/Quebec, 1985. Lecture Notes in Mathematics, vol. 1234, pp. 68–90. Springer, Berlin (1986)
28. Désarménien, J., Foata, D.: Statistiques d'ordre sur les permutations colorées. Discret. Math. **87**(2), 133–148 (1991)
29. Dotsenko, V., Khoroshkin, A.: Anick-type resolutions and consecutive pattern avoidance (2010) [ArXiv:1002.2761v1]
30. Doubilet, P., Rota, G.C., Stanley, R.: On the foundations of combinatorial theory, VI. The idea of generating function. In: Proceedings of the 6th Berkeley Symposium on Mathematical Statistics and Probability, University of California, Berkeley, 1970/1971. Probability Theory, vol. II, pp. 267–318. University of California Press, Berkeley (1972)

31. Duane, A.: Pattern matching statistics in the permutations s_n and the alternating permutations a_n for minimally overlapping patterns. Ph.D. thesis, University of California, San Diego (2013)
32. Duane, A., Remmel, J.: Minimal overlapping patterns in colored permutations. Electron. J. Comb. **18**(2), Paper 25, 38 (2011)
33. Eğecioğlu, Ö., Remmel, J.B.: Brick tabloids and the connection matrices between bases of symmetric functions. Discret. Appl. Math. **34**(1–3), 107–120 (1991) [Combinatorics and Theoretical Computer Science (Washington, DC, 1989)]
34. Elizalde, S.: Consecutive patterns and statistics on restricted permutations. Ph.D. thesis, Universitat Politecnica de Catalunya (2004)
35. Elizalde, S., Noy, M.: Consecutive patterns in permutations. Adv. Appl. Math. **30**(1–2), 110–125 (2003) [Formal Power Series and Algebraic Combinatorics (Scottsdale, AZ, 2001)]
36. Elizalde, S., Noy, M.: Clusters, generating functions and asymptotics for consecutive patterns in permutations. Adv. Appl. Math. **49**(3–5), 351–374 (2012)
37. Entringer, R.: A combinatorial interpretation of the Euler and Bernoulli numbers. Nieuw Arch. Wisk. (3) **14**, 241–246 (1966)
38. Entringer, R.: Enumeration of permutations of $(1,\cdots,n)$ by number of maxima. Duke Math. J. **36**, 575–579 (1969)
39. Euler, L.: Foundations of Differential Calculus. Springer, New York (2000) [Translated from the Latin by John D. Blanton]
40. Fédou, J.M., Rawlings, D.: More statistics on permutation pairs. Electron. J. Comb. **1**, Research Paper 11, approx. 17 pp. (electronic) (1994)
41. Fédou, J.M., Rawlings, D.: Statistics on pairs of permutations. Discret. Math. **143**(1–3), 31–45 (1995)
42. Flajolet, P.: Combinatorial aspects of continued fractions. Ann. Discret. Math. **9**, 217–222 (1980) [Combinatorics 79 (Proc. Colloq., Université de Montréal, Montreal, 1979), Part II]
43. Foata, D.: Sur un énoncé de MacMahon. C. R. Acad. Sci. Paris **258**, 1672–1675 (1964)
44. Foata, D.: On the Netto inversion number of a sequence. Proc. Am. Math. Soc. **19**, 236–240 (1968)
45. Foata, D., Schützenberger, M.P.: Théorie Géométrique des Polynômes Eulériens. Lecture Notes in Mathematics, vol. 138. Springer, Berlin (1970)
46. Frame, J.S., Robinson, G.d.B., Thrall, R.M.: The hook graphs of the symmetric groups. Can. J. Math. **6**, 316–324 (1954)
47. Fuller, E., Remmel, J.: Symmetric functions and generating functions for descents and major indices in compositions. Ann. Comb. **14**(1), 103–121 (2010)
48. Garsia, A., Gessel, I.: Permutation statistics and partitions. Adv. Math. **31**(3), 288–305 (1979)
49. Garsia, A., Joni, S.N.: Composition sequences. Commun. Algebra **8**(13), 1195–1266 (1980)
50. Gessel, I.: Counting permutations by descents, greater index, and cycle structure. Preprint, Department of Mathematics M.I.T., Cambridge (1977)

51. Gessel, I.: Generating functions and enumeration of sequences. Ph.D. thesis, Massachusetts Institute of Technology (1977)
52. Goulden, I.P., Jackson, D.M.: An inversion theorem for cluster decompositions of sequences with distinguished subsequences. J. Lond. Math. Soc. (2) **20**(3), 567–576 (1979)
53. Goulden, I.P., Jackson, D.M.: Combinatorial Enumeration. A Wiley-Interscience Publication. Wiley, New York (1983) [With a foreword by Gian-Carlo Rota, Wiley-Interscience Series in Discrete Mathematics]
54. Greene, C.: An extension of Schensted's theorem. Adv. Math. **14**, 254–265 (1974)
55. Haglund, J.: The q,t-Catalan Numbers and the Space of Diagonal Harmonics. University Lecture Series, vol. 41. American Mathematical Society, Providence (2008) [With an appendix on the combinatorics of Macdonald polynomials]
56. Hall, J.I., Palmer, E.M., Robinson, R.W.: Redfield's lost paper in a modern context. J. Graph Theory **8**(2), 225–240 (1984)
57. Harary, F., Palmer, E.: The enumeration methods of Redfield. Am. J. Math. **89**, 373–384 (1967)
58. Harary, F., Robinson, R.W.: The rediscovery of Redfield's papers. J. Graph Theory **8**(2), 191–193 (1984)
59. Harmse, J., Remmel, J.: Patterns in column strict fillings of rectangular arrays. Pure Math. Appl. **22**(2), 141–171 (2011)
60. Henle, M.: Dissection of generating functions. Stud. Appl. Math. **51**, 397–410 (1972)
61. Hillman, A.P., Grassl, R.M.: Reverse plane partitions and tableau hook numbers. J. Comb. Theory Ser. A **21**(2), 216–221 (1976)
62. Jacobi, C.G.: Zur combinatorischen analysis. J. Reine Angew. Math. **22**, 372–374 (1841)
63. James, G., Liebeck, M.: Representations and Characters of Groups, 2nd edn. Cambridge University Press, New York (2001)
64. Jones, M.E., Remmel, J.: Pattern matching in the cycle structures of permutations. Pure Math. Appl. **22**(2), 173–208 (2011)
65. Jones, M.E., Remmel, J.: A reciprocity approach to computing generating functions for permutations with no pattern matches. In: 23rd International Conference on Formal Power Series and Algebraic Combinatorics (FPSAC 2011). Discrete Mathematics and Theoretical Computer Science Proceedings, AO, pp. 551–562. Association of Discrete Mathematics and Theoretical Computer Science, Nancy (2011)
66. Jones, M.E., Remmel, J.B.: Applying a reciprocity method to count permutations avoiding two families of consecutive patterns. Pure Math. Appl. **24**(2), 151–178 (2013)
67. Jones, M.E., Remmel, J.B.: A reciprocity method for computing generating functions over the set of permutations with no consecutive occurrence of a permutation pattern. Discret. Math. **313**(23), 2712–2729 (2013)

68. Joyal, A.: Une théorie combinatoire des séries formelles. Adv. Math. **42**(1), 1–82 (1981)
69. Kitaev, S.: Generalized patterns in words and permutations. Ph.D. thesis, Chalmers University of Technology and Göteborg University (2003)
70. Kitaev, S.: Introduction to partially ordered patterns. Discret. Appl. Math. **155**(8), 929–944 (2007)
71. Kitaev, S.: Patterns in Permutations and Words. Monographs in Theoretical Computer Science. An EATCS Series. Springer, Heidelberg (2011) [With a foreword by Jeffrey B. Remmel]
72. Knuth, D.E.: Permutations, matrices, and generalized Young tableaux. Pac. J. Math. **34**, 709–727 (1970)
73. Kulikauskas, A., Remmel, J.: Lyndon words and transition matrices between elementary, homogeneous and monomial symmetric functions. Electron. J. Comb. **13**(1), Research Paper 18, 30 (2006)
74. Langley, T.: The plethysm of two schur functions at hook, near-hook, and two row shapes. Ph.D. thesis, University of California, San Diego (2001)
75. Langley, T.: Alternate transition matrices for Brenti's q-symmetric functions and a class of (q,t)-symmetric functions on the hyperoctahedral group. In: Proceedings of Formal Power Series and Algebraic Combinatorics (FPSAC), Melbourne (2002)
76. Langley, T.M., Remmel, J.B.: Enumeration of m-tuples of permutations and a new class of power bases for the space of symmetric functions. Adv. Appl. Math. **36**(1), 30–66 (2006)
77. Liese, J., Remmel, J.: Generating functions for permutations avoiding a consecutive pattern. Ann. Comb. **14**(1), 123–141 (2010)
78. Liese, J., Remmel, J.: Q-analogues of the number of permutations with k-excedances. Pure Math. Appl. **21**(2), 285–320 (2010)
79. LoBue Tiefenbruck, J., Remmel, J.B.: A Murnaghan-Nakayama rule for generalized Demazure atoms. In: 25th International Conference on Formal Power Series and Algebraic Combinatorics (FPSAC 2013). Discrete Mathematics and Theoretical Computer Science Proceedings, AS, pp. 969–980. Association of Discrete Mathematics and Theoretical Computer Science, Nancy (2013)
80. Loehr, N.A.: Abacus proofs of Schur function identities. SIAM J. Discret. Math. **24**(4), 1356–1370 (2010)
81. Loehr, N.A.: Bijective Combinatorics. Discrete Mathematics and Its Applications (Boca Raton). CRC Press, Boca Raton (2011)
82. Macdonald, I.G.: Symmetric Functions and Hall Polynomials, 2nd edn. Oxford Mathematical Monographs. The Clarendon Press/Oxford University Press, New York (1995) [With contributions by A. Zelevinsky, Oxford Science Publications]
83. MacMahon, P.A.: Combinatory Analysis. Two Volumes (Bound as One). Chelsea, New York (1960)
84. Mansour, T.: Pattern avoidance in coloured permutations. Sém. Lothar. Comb. **46**, Art. B46g, 12 pp. (electronic) (2001/2002)

85. Mansour, T.: Coloured permutations containing and avoiding certain patterns. Ann. Comb. **7**(3), 349–355 (2003)
86. Mansour, T., West, J.: Avoiding 2-letter signed patterns. Sém. Lothar. Comb. **49**, Art. B49a, 11 (2002/2004)
87. Mendes, A., Remmel, J.: Generating functions for statistics on $C_k \wr S_n$. Sém. Lothar. Comb. **54A**, Art. B54At, 40 (2005/2007)
88. Mendes, A., Remmel, J.: Permutations and words counted by consecutive patterns. Adv. Appl. Math. **37**(4), 443–480 (2006)
89. Mendes, A., Remmel, J.: Descents, inversions, and major indices in permutation groups. Discret. Math. **308**(12), 2509–2524 (2008)
90. Mendes, A., Remmel, J.B., Riehl, A.: Permutations with k-regular descent patterns. In: Permutation Patterns. London Mathematical Society Lecture Note Series, vol. 376, pp. 259–285. Cambridge University Press, Cambridge (2010)
91. Murnaghan, F.D.: The characters of the symmetric group. Am. J. Math. **59**(4), 739–753 (1937)
92. Nakayama, T.: On some modular properties of irreducible representations of a symmetric group, I. Jpn. J. Math. **18**, 89–108 (1941)
93. Nakayama, T.: On some modular properties of irreducible representations of symmetric groups, II. Jpn. J. Math. **17**, 411–423 (1941)
94. Niedermaier, A., Remmel, J.: Analogues of up-down permutations for colored permutations. J. Integer Seq. **13**(5), Article 10.5.6, 32 (2010)
95. Pólya, G.: Kombinatorische anzahlbestimmungen für gruppen, graphen und chemische verbindungen. Acta Math. **68**(1), 145–254 (1937)
96. Ram, A., Remmel, J.B., Whitehead, T.: Combinatorics of the q-basis of symmetric functions. J. Comb. Theory Ser. A **76**(2), 231–271 (1996)
97. Redfield, J.H.: The theory of group-reduced distributions. Am. J. Math. **49**(3), 433–455 (1927)
98. Remmel, J.: Permutation statistics and (k,ℓ)-hook schur functions. Discret. Math. **67**, 271–298 (1987)
99. Remmel, J., Riehl, M.: Generating functions for permutations which contain a given descent set. Electron. J. Comb. **17**(1), Research Paper 27, 33 (2010)
100. Remmel, J.B.: Consecutive up-down patterns in up-down permutations. Electron. J. Comb. **21**(3), Paper 3.2, 35 (2014)
101. Riddell, R., Uhlenbeck, G.: On the theory of the virial development of the equation of state of monoatomic gases. J. Chem. Phys. **21**, 2056–2064 (1953)
102. Riordan, J.: An Introduction to Combinatorial Analysis. Dover, Mineola (2002) (Reprint of the 1958 original [Wiley, New York; MR0096594 (20 #3077)])
103. Robinson, G.d.B.: On the representations of the symmetric group. Am. J. Math. **60**(3), 745–760 (1938). doi:10.2307/2371609. http://www.dx.doi.org/10.2307/2371609
104. Sagan, B.E.: The Symmetric Group: Representations, Combinatorial Algorithms, and Symmetric Functions, 2nd edn. Graduate Texts in Mathematics, vol. 203. Springer, New York (2001)

105. Schensted, C.: Longest increasing and decreasing subsequences. Can. J. Math. **13**, 179–191 (1961)
106. Stanley, R.P.: Theory and application of plane partitions, I, II. Stud. Appl. Math. **50**, 167–188 (1971); Stud. Appl. Math. **50**, 259–279 (1971)
107. Stanley, R.P.: Binomial posets, Möbius inversion, and permutation enumeration. J. Comb. Theory Ser. A **20**(3), 336–356 (1976)
108. Stanley, R.P.: Enumerative Combinatorics, vol. 2. Cambridge Studies in Advanced Mathematics, vol. 62. Cambridge University Press, Cambridge (1999)
109. Stanley, R.P.: Enumerative Combinatorics, vol. 1, 2nd edn. Cambridge Studies in Advanced Mathematics, vol. 49. Cambridge University Press, Cambridge (2012)
110. Sylvester, J.J.: Note on the numbers of Bernoulli and Euler and a new theorem concerning prime numbers. In: The Collected Mathematical Papers of James Joseph Sylvester, vol. 2. Chelsea, New York (1973)
111. Touchard, J.: Sur les cycles des substitutions. Acat. Math. **70**, 243–297 (1939)
112. Viennot, G.: Une forme géométrique de la correspondence de Robinson-Schensted. In: Combinatoire et Représentation du Groupe Symétrique (Actes Table Ronde CNRS, Université Louis-Pasteur Strasbourg, Strasbourg, 1976). Lecture Notes in Mathematics, vol. 579, pp. 29–58. Springer, Berlin (1977)
113. Wagner, J.: The combinatorics of the permutation enumeration of wreath products between cyclic and symmetric groups. Ph.D. thesis, University of California, San Diego (2000)
114. Wagner, J.: The permutation enumeration of wreath products $C_k \S S_n$ of cyclic and symmetric groups. Adv. Appl. Math. **30**(1–2), 343–368 (2003) [Formal power series and algebraic combinatorics (Scottsdale, AZ, 2001)]
115. White, D.: Orthogonality of the characters of S_n. J. Comb. Theory Ser. A **40**(2), 265–275 (1985)
116. White, D.E.: A bijection proving orthogonality of the characters of S_n. Adv. Math. **50**(2), 160–186 (1983)
117. Wilf, H.S.: Generatingfunctionology, 3rd edn. A K Peters Ltd., Wellesley (2006)

Index

A. Mendes, J. Remmel, *Counting with Symmetric Functions*, Developments in Mathematics 43, DOI 10.1007/978-3-319-23618-6

MIX
Papier aus verantwortungsvollen Quellen
Paper from responsible sources
FSC® C105338
FSC
www.fsc.org

If you have any concerns about our products,
you can contact us on
ProductSafety@springernature.com

In case Publisher is established outside the EU,
the EU authorized representative is:
Springer Nature Customer Service Center GmbH
Europaplatz 3, 69115 Heidelberg, Germany

Printed by Libri Plureos GmbH
in Hamburg, Germany